Intelligent Marine Robotics Modelling, Simulation and Applications

Intelligent Marine Robotics Modelling, Simulation and Applications

Special Issue Editors

Cheng Siong Chin
Rongxin Cui

MDPI • Basel • Beijing • Wuhan • Barcelona • Belgrade

Special Issue Editors

Cheng Siong Chin
Newcastle University
Singapore

Rongxin Cui
Northwestern Polytechnical University
China

Editorial Office
MDPI
St. Alban-Anlage 66
4052 Basel, Switzerland

This is a reprint of articles from the Special Issue published online in the open access journal *Journal of Marine Science and Engineering* (ISSN 2077-1312) from 2018 to 2019 (available at: https://www.mdpi.com/journal/jmse/special_issues/marine_robotics).

For citation purposes, cite each article independently as indicated on the article page online and as indicated below:

LastName, A.A.; LastName, B.B.; LastName, C.C. Article Title. *Journal Name* **Year**, *Article Number*, Page Range.

ISBN 978-3-03928-132-9 (Pbk)
ISBN 978-3-03928-133-6 (PDF)

Cover image courtesy of Cheng Siong Chin.

Contents

About the Special Issue Editors

Cheng Siong Chin received the B.Eng. degree (Hons.) in mechanical and aerospace engineering from Nanyang Technological University (NTU) in 2000, the M.Sc. degree (Hons.) in advanced control and systems engineering from The University of Manchester in 2001, and the Ph.D. degree in applied control engineering from the Research Robotics Centre, NTU, in 2008. He is currently an Adjunct Professor with the School of Automotive Engineering, Chongqing University, and the Director of Innovation and Engagement at Newcastle University, Singapore. He was a lecturer in mechatronics engineering with Temasek Polytechnic in 2008. He worked in the industry for a few years before moving into academia. He has published over 100 journal papers, books, book chapters, and conference papers. He currently holds three U.S. patents, two provisional U.S. patent applications, one Singapore provisional patent, and two trade secrets in electronics and measurement systems. He obtained two research grants from SMI and four EDB-Industrial Postgraduate Programme grants in the areas of intelligent systems design and predictive analytics. His current research interests include intelligent systems design and simulation of complex systems in uncertain environment. He is a fellow of the Higher-Education Academy and IMarEST, a Senior Member of the IET, and a Chartered Engineer. He received the Best Paper Award for the Virtual Reality of Autonomous Vehicle in the 2018 10th International Conference on Modelling, Identification and Control sponsored by IEEE.

Rongxin Cui received the B.Eng. degree in autonomic control and the Ph.D. degree in control science and engineering from Northwestern Polytechnical University, Xi'an, China, in 2003 and 2008, respectively. From August 2008 to August 2010, he was a Research Fellow with the Centre for Offshore Research and Engineering, National University of Singapore, Singapore. He is currently a Professor with the School of Marine Science and Technology, Northwestern Polytechnical University, Xi'an, China. His current research interests include control of nonlinear systems, cooperative path planning and control for multiple robots, control and navigation for underwater vehicles, and system development. Dr. Cui is an Editor for the *Journal of Intelligent and Robotic Systems*.

Preface to "Intelligent Marine Robotics Modelling, Simulation and Applications"

The biennial Congress of the Italian Society of Oral Pathology and Medicine (SIPMO) is an International meeting dedicated to the growing diagnostic challenges in the oral pathology and medicine field. The III International and XV National edition will be a chance to discuss clinical conditions which are unusual, rare, or difficult to define. Many consolidated national and international research groups will be involved in the debate and discussion through special guest lecturers, academic dissertations, single clinical case presentations, posters, and degree thesis discussions. The SIPMO Congress took place from the 17th to the 19th of October 2019 in Bari (Italy), and the enclosed copy of Proceedings is a non-exhaustive collection of abstracts from the SIPMO 2019 contributions.

Cheng Siong Chin, Rongxin Cui
Special Issue Editors

Journal of
Marine Science and Engineering

Article

Sliding Mode Control in Backstepping Framework for a Class of Nonlinear Systems

He Shen [1], Joseph Iorio [1] and Ni Li [2,*]

[1] Department of Mechanical Engineering, California State University, Los Angeles, CA 90032, USA;
    he.shen@calstatela.edu (H.S.); jiorio@calstatela.edu (J.I.)
[2] School of Aeronautics, Northwestern Polytechnical University, Xian 710072, China
* Correspondence: lini@nwpu.edu.cn; Tel.: +86-029-8849-1982

Received: 11 October 2019; Accepted: 13 November 2019; Published: 9 December 2019

Abstract: Both backstepping control (BC) and sliding mode control (SMC) have been studied extensively over the past few decades, and many variations of controller designs based on them can be found in the literature. In this paper, sliding mode control in a backstepping framework (SBC) for a class of nonlinear systems is proposed and its connections to SMC studied. SMC is shown to be a special case of SBC. Without losing generality, the regulation control problem is studied, while tracking control is achieved by replacing the states with the difference between the states and their desired values. The SBCs are designed for nonlinear single-input-single-output (SISO) and multiple-input-multiple-output (MIMO) systems with the presence of bounded uncertainties from unmodeled dynamics, parametric variations, disturbances, and measurement noise, and the closed loop systems are proven to be asymptotically stable using the Lyapunov stability theory. The comparison of SBC to SMC from the design process, chattering effects, and chatter reduction are also discussed. SBC inherits the merits of backstepping control in choosing gains independently, while leveraging useful nonlinear dynamics for controller design simplification. Hence, it provides more flexibility in controller design in the sense of controlling coverage speed and making use of useful nonlinearities in the dynamics. To demonstrate the effectiveness of SBC, an application on cruise tracking control of an autonomous underwater vehicle was studied.

Keywords: robust control; nonlinear systems; backstepping control; sliding mode control; Lyapunov stability; autonomous underwater vehicle

---

1. Introduction

The backstepping method breaks the problem of controlling complex higher order systems into a sequence of lower order control problems through a recursive procedure. By doing this, the flexibility in these lower order systems can be explored for controller design, which makes the control less restrictive in system complexity requirements, compared to other methods [1]. Over the past few decades, backstepping control has been studied extensively with applications for the control of spacecraft and aircraft [2–5], marine vessels [6–8], various motors [9–11], robotic manipulators and systems [12–14], and many others [15–18].

Numerous variations of backstepping control can be found in the literature. Specific formulations exist for nonlinear systems with time delays [19], systems with structural obstacles in obtaining continuous feedback laws [20], time-varying systems [21], nonlinear systems with block structures [22], plants with unknown backlash nonlinearities [23], nonholonomic systems with strong nonlinear drifts [24], systems with unknown dead-zone non-linearity [25], nonlinear systems with delay of neutral type [26], and so on. The backstepping method is also used to assist in design for fuzzy control [27–29], neural network control [25,30–34], and wavelet adaptive control [35,36]. Another related topic is robust backstepping control (RBC), which focuses on handling uncertainties from un-modeled dynamics (i.e., imperfection or simplification in modeling and variations in system parameters), noise in measurements, and disturbances [37–39].

Aside from RBC, sliding mode control (SMC) is another Lyapunov-based robust control method, which has attracted much attention from both the industry and academia [40]. SMC produces a switching control law to force the system to converge to the sliding surface while trapping the system within a boundary layer near the sliding surface under the guidance of the Lyapunov stability theory [41,42]. Due to the nature of switching control law, the chattering problem associated with the controller has also caught the attention of researchers. Many solutions can be found in [43–46]. There is also some research that combines backstepping and sliding mode methods to design controllers for special applications such as the adaptive backstepping sliding mode control for linear induction motor drives presented in [47].

In this paper, a systematic SBC design is formulated for both single-input-single-output (SISO) systems and multiple-input-multiple-output (MIMO) systems. The general nonlinear system considered here is in a recursive form. Systems not in the standard form can be converted to the standard form using input-output linearization or other methods. Starting from the recursive model, an SBC is designed to guarantee that the closed loop system is asymptotically stable. The proposed controller is also compared to the popular SMC on simplicity of design, uncertainty handling, and chatter reduction. Further, the design of integral SBC (ISBC) is presented. It is found that SBC and ISBC retain the advantages of both the robustness of SMC and the simplicity of backstepping control. The proposed SBC/ISBC methods provide more flexibility in controlling the convergence speed of the closed loop system, while retaining the ability to leverage useful nonlinear dynamics toward a simpler control law. The main contributions of this paper are as follows: (1) SBC and ISBC for a class of nonlinear systems are proposed; (2) the controller design formulations reveal both the equivalence and difference between SBC/ISBC and SMC/ISMC; and (3) it demonstrates that the proposed SBC/ISBC are more flexible design and can achieve better performance than the traditional SMC/ISMC.

The rest of the paper is organized as follows: in Section 2, the SBC for nonlinear SISO system is presented. The results for MIMO system are given in Section 3. In Section 4, the SBC is compared to SMC in terms of simplicity of design, chatter effects, and some additional performance metrics. Section 5 details a benchmark control example using the proposed SBC method to illustrative the effectiveness of the controller. Finally, conclusions are drawn in Section 6.

2. SBC of SISO Systems

2.1. Problem Formulation

The nonlinear SISO system considered in this paper is assumed to have the following form (or, the system is assumed to be input-output linearizable to have the following form):

$$x^{(n)} = f(x) + g(x)u \tag{1}$$

where x is the state variable, composed of the state and its derivatives up to the order of n, and both $f(x)$ and $g(x)$ are nonlinear functions of x. Without special notices, all scalar functions are in regular style, while vectors and matrices will be in bold.

Remark 1. For a general nonlinear system given in a more general form as:

$$\begin{cases} \dot{x}_1 = f_1(x) \\ \dot{x}_2 = f_2(x) \\ \quad\vdots \\ \dot{x}_n = f_n(x) + g_1(x)u \end{cases} \tag{2}$$

and assuming the output function is given as $y = h(x)$, then, the system can be linearized using input-output linearization, which results in

$$\begin{cases} \dot{y} = \ell_f h(x) + \ell_g h(x) u \\ \ddot{y} = \ell_{f^{(2)}} h(x) + \ell_g \ell_f h(x) u \\ \quad \vdots \\ y^{(r)} = \ell_{f^{(r)}} h(x) + \ell_g \ell_{f^{(r-1)}} h(x) u \end{cases} \tag{3}$$

where ℓ represents the Lie derivative operator and r is the relative degree of the system. We know that:

$$\ell_g h(x) = \ell_g \ell_f h(x) = \cdots \ell_g \ell_{f^{(r-2)}} h(x) = 0 \tag{4}$$

and

$$\ell_g \ell_{f^{(r-1)}} h(x) \neq 0 \tag{5}$$

By defining the new states $\xi_1 = y$, $\xi_2 = \dot{y}, \cdots$, $\xi_r = y^{(r-1)}$, the general nonlinear system in (2) can be written in the form of (1).

Remark 2. Without losing generality, the regulation problem (i.e., $\lim_{t\to\infty} x = 0$ for SISO system or $\lim_{t\to\infty} x = 0 \in \mathbb{R}^m$ for MIMO systems) will be considered in this paper. For tracking problems where $\lim_{t\to\infty} x = x_d$ or $\lim_{t\to\infty} x = x_d \in \mathbb{R}^m$, new state variables $\widetilde{x} = x - x_d$ or $\widetilde{x} = x - x_d$ can be defined to convert the tracking problem to a regulation problem.

Since the standard backstepping method breaks a complex system into a lower order system to simplify the controller design process, system (1) is firstly written into the following canonical form as:

$$\begin{cases} \dot{x}_1 = x_2 \\ \dot{x}_2 = x_3 \\ \quad \vdots \\ \dot{x}_n = f(x) + g(x) u \end{cases} \tag{6}$$

Following the standard Backstepping control design, the first new state z_1 is defined as $z_1 = x_1$. Then, we take the derivative of z_1 to get $\dot{z}_1 = -\gamma_1 z_1 + \underbrace{\left(\gamma_1 z_1 + \dot{x}_1 \right)}_{z_2}$. Hence, z_2 can be defined as $z_2 = \gamma_1 z_1 + \dot{x}_1 = \gamma_1 x_1 + x_2$. We repeat this process until z_{n-1}. Since the system in (6) is already in canonical form, this process can be described using the following transformation:

$$z_i = \left(\prod_{j=1}^{i-1} \left(\frac{d}{dt} + \gamma_j \right) \right) x, \quad 1 < i \leq n \tag{7}$$

It is easy to verify that under the transformation given in (7), the system in Equation (6) can be converted into the following recursive structure:

$$\begin{cases} \dot{z}_1 = -\gamma_1 z_1 + z_2 \\ \dot{z}_2 = -\gamma_2 z_2 + z_3 \\ \quad \vdots \\ \dot{z}_n = v \end{cases} \tag{8}$$

where $\gamma_i > 0$, $i = 1, 2, \cdots, n$ are positive real numbers. For simplicity, we can write $v = \dot{z}_n = \frac{d}{dt}\left(x_n + \sum_{j=1}^{n-1} \alpha_j x_j\right) = \dot{x}_n + \sum_{j=1}^{n-1} \alpha_j x_{j+1}$, and α_j are functions of an independent scalar set $\{\gamma_n, \gamma_{n-1}, \cdots, \gamma_1\}$, i.e., $\alpha_1 = \gamma_1 \gamma_2 \cdots \gamma_n$ and $\alpha_{n-1} = \sum_{j=1}^{n} \gamma_j$.

Without consideration for robustness, the traditional backstepping controller can be derived, such that $v = -\gamma_n z_n$, which results in a controller of the form:

$$u = \frac{1}{g(x)}\left(-f(x) - d(x) - \gamma_n z_n\right) \tag{9}$$

where $d(x) = \sum_{j=1}^{n-1} \alpha_j x_{j+1}$. This makes the system "theoretically" exponentially stable. Here, "theoretically" means that this is only guaranteed to be true when all sources of uncertainties do not exist. Alternatively, the controller that ensures $v = -\gamma_n sgn(z_n)$ is given by:

$$u = \frac{1}{g(x)}\left(-f(x) - d(x) - \gamma_n sgn(z_n)\right) \tag{10}$$

Since γ_n is a user-defined constant parameter, the control formulations in (9) and (10) are not robust.

2.2. SBC Design

Before moving on to the SBC design, the following assumption is made to quantify the uncertainties from un-modeled dynamics, parameter variation, noise, and disturbances.

Assumption 1. The matching conditions requirement assumes that in (1) the uncertainty of f and g satisfy the following:

$$\begin{cases} \left|f - \hat{f}\right| \leq F_f \\ \left|d - \hat{d}\right| \leq F_d, \ d = \sum_{j=1}^{n-1} \alpha_j x_{j+1} \\ \hat{g} = (1 + \Delta)g, \ |\Delta| \leq G < 1 \end{cases} \tag{11}$$

where F_f and F_d are the upper bound of the uncertainties of $f(x)$ and $d(x)$, and Δ is the difference between $g(x)$ and $\hat{g}(x)$, and G is the upper bound of $|\Delta|$.

Theorem 1. For a class of nonlinear system given in form of (1) or the equivalent canonical form of (8), which meets the uncertainty bounds or matching condition of (11), a nonlinear SBC control is given by:

$$u = \frac{1}{\hat{g}(x)}\left[-\hat{f}(x) - \hat{d}(x) - ksgn(s)\right] \tag{12}$$

where

$$s = z_n = \left(\prod_{j=1}^{n-1}\left(\frac{d}{dt} + \gamma_j\right)\right)x = x_n + \sum_{j=1}^{n-1} \alpha_j x_j, \tag{13}$$

and the control gain is chosen as

$$k \geq \frac{1}{1-G}\left(F_f + F_d + G\left|\hat{f}(x) + \hat{d}(x)\right| + \eta\right) \tag{14}$$

where η is an arbitrary positive scalar, and the sign function is given by

$$sgn(s) = \begin{cases} -1 & if \ s < 0 \\ 0 & if \ s = 0 \\ 1 & if \ s > 0 \end{cases} \tag{15}$$

Then, the system is asymptotically stable.

Remark 3. Since $G < 1$, the positive control gain in (14) always exists and can further be chosen as equality for simplicity, which gives:

$$k = \frac{1}{1-G}\left(F_f + F_d + G\left|\hat{f}(x) + \hat{d}(x)\right| + \eta\right) \tag{16}$$

Proof. To prove Theorem 1, we choose the positive definite Lyapunov function $V = \frac{1}{2}s^2$. Its derivative becomes:

$$
\begin{aligned}
\dot{V} \; &= s\dot{s} = s\left(\dot{z}_n\right) = s\left(d(x) + \dot{x}_n\right) \\
&= s(d(x) + f(x) + g(x)u) \\
&= s\left(\begin{array}{c} d(x) + f(x) + g(x)\frac{1}{\hat{g}(x)} \cdot \\ \left[-\hat{f}(x) - \hat{d}(x) - k sgn(s)\right] \end{array}\right) \\
&= s\left(\begin{array}{c} d(x) + f(x) + (1+\Delta)\cdot \\ \left[-\hat{f}(x) - \hat{d}(x) - k sgn(s)\right] \end{array}\right) \\
&= s\left(\begin{array}{c} \left[f(x) - \hat{f}(x)\right] + \left[d(x) - \hat{d}(x)\right] \\ -\Delta\left[\hat{f}(x) + \hat{d}(x)\right] - (1+\Delta)k sgn(s) \end{array}\right) \\
&\leq |s|\left(F_f + F_d + G\left|\hat{f}(x) + \hat{d}(x)\right| - (1-G)k\right)
\end{aligned} \tag{17}
$$

If k is chosen large enough such that Equation (14) is satisfied, then, $\dot{V} \leq -\eta|s|$, and thus the Lyapunov rate is strictly decreasing. According to the Lyapunov stability theorem, we can conclude that the system is asymptotically stable. $\square$

2.3. Integral SBC Control

To eliminate the steady-state bias or shorten the rising time, an integral term is typically considered in the controller design. The importance of integral terms in backstepping control design has been demonstrated in [48]. Considering the system in (6), the integral sliding mode control in backstepping framework control (ISBC) can be implemented by adding another state $x_0 = \int_{t_0}^{t} x(\tau)d\tau$ to the system, which results in:

$$
\begin{cases}
\dot{x}_0 = x_1 \\
\dot{x}_1 = x_2 \\
\quad \vdots \\
\dot{x}_n = f(x) + g(x)u
\end{cases} \tag{18}
$$

The results in Section 2 are valid for system (18), if we redefine the function $s(x)$ and $d(x)$ as follows:

$$s = z_n = \left(\prod_{j=0}^{n-1}\left(\frac{d}{dt} + \gamma_j\right)\right)x = x_n + \sum_{j=0}^{n-1}\alpha_j^* x_j \tag{19}$$

$$d = \sum_{j=0}^{n-1}\alpha_j^* x_{j+1} \tag{20}$$

Then, the control law and the control gain update law are the same as those in Equations (13) and (14). For systems that need higher order integral control, the same method can be used to add new states.

2.4. Chattering Effects

As the sliding surface function used in the SBC design, denoted by s in Equation (13), approaches zero, the sign function tends to switch at a high frequency. This chattering phenomenon and its reduction has been extensively studied for SMC, where there are two primary methods: the boundary layer method [42] and other surface approach dynamics (e.g., using saturation, arctangent function instead of sign function) [49], and higher order or nonlinear sliding surface functions [50,51]. Algorithms to eliminate chatter were also studied in [52] and the existence of a solution for nonlinear convergent chatter-free sliding mode control was studied in [53]. The same ideas can be used to reduce chatter in SBC design.

3. SBC of MIMO Systems

In this paper, a MIMO system of the following form is considered:

$$x_i^{(n_i)} = f_i(x) + \sum_{j=1}^{m} g_{i,j}(x)\, u_j \text{ and } i, j = 1, 2, \cdots, m \tag{21}$$

where $x = \left[x_1, \dot{x}_1, \cdots, x_1^{n_1}, x_2, \dot{x}_2, \cdots, x_2^{n_2}, \cdots \right]$.

Remark 4. The system discussed here refers to a fully-actuated system. The under- and over- actuated system control problems are solved with other techniques, which are beyond the scope of this paper and will not be discussed here.

Remark 5. For a special class of MIMO Systems of the form $x^{(n)} = f(x) + g(x)u$, where $x, f, u \in \mathbb{R}^m$, $g \in \mathbb{R}^{m \times m}$. The controller in the same form can be obtained providing that $n_1 = n_2 = \cdots = n$.

To follow the derivation for SISO systems, the MIMO system in (21) is first written into the following canonical form:

$$\begin{cases} \dot{x}_{i,1} = x_{i,2} \\ \dot{x}_{i,2} = x_{i,3} \\ \quad \vdots \\ \dot{x}_{i,n} = f_i(x) + \sum_{j=1}^{m} g_{i,j}(x)\, u_j \end{cases} , \; i = 1, 2, \cdots m \tag{22}$$

Then, following the backstepping design for the SISO system, we get:

$$\begin{cases} \dot{z}_{i,1} = -\gamma_{i,1} z_{i,1} + z_{i,2} \\ \dot{z}_{i,2} = -\gamma_{i,2} z_{i,2} + z_{i,3} \\ \quad \vdots \\ \dot{z}_{i,n_i} = \sum_{j=1}^{n_i-1} \alpha_{i,j} x_{i,j+1} + \dot{x}_{i,n} \end{cases} , \; i = 1, 2, \cdots m \tag{23}$$

Similarly, the matching conditions are given in assumption 2 as follows.

Assumption 2. The matching conditions requirement assumes that the uncertainty of f and g in Equation (21) satisfy the following:

$$\begin{cases} \left| f - \hat{f} \right| \leq F_f \\ \left| d - \hat{d} \right| \leq F_d, \; d_i = \sum_{j=1}^{n_i-1} \alpha_{i,j} x_{i,j+1} \\ \hat{g} = (I + \Delta)g, \; |\Delta| \leq G, \; 0 < \rho(G) < 1 \end{cases} \tag{24}$$

where F_f is the upper bound of uncertainty in f, F_d is the upper bound of uncertainty in d, Δ is the discrepancy between g and $\hat{g}$, G is the elementary upper bound matrix of $|\Delta|$, and $\rho(G)$ is the spectral radius of G.

Theorem 2. For a class of nonlinear MIMO systems given in form in (21), meeting the matching condition requirement in Equation (24), a nonlinear SBC is given by:

$$u = \hat{g}(x)^{-1}\left[-\hat{f}(x) - \hat{d}(x) - k \circ sgn(s)\right] \tag{25}$$

where

$$s_i = x_{i,n_i} + \sum_{j=1}^{n_i-1} \alpha_{i,j} x_{i,j} = \left(\prod_{j=1}^{n_i}\left(\frac{d}{dt} + \gamma_{i,j}\right)\right) x_i \tag{26}$$

and the adaptive gain is chosen as

$$k \geq (I - G)^{-1}\left(F_f + F_d + G\big|\hat{f}(x) + \hat{d}(x)\big| + \eta\right) \tag{27}$$

where η is an arbitrary positive vector, and $I \in \mathbb{R}^{m \times m}$ is an identity matrix, and $\circ$ denotes entrywise product. With this controller in Equation (25), the system is asymptotically stable.

Remark 6. Similar to **Remark 3** for SISO system, the control gain for MIMO system can be simply chosen as:

$$k = (I - G)^{-1}\left(F_f + F_d + G\big|\hat{f}(x) + \hat{d}(x)\big| + \eta\right) \tag{28}$$

Proof. Choose the Lyapunov function as $V = \frac{1}{2}s^T s$, where s_i is given in (26), and

$$\dot{s}_i = \dot{x}_{i,n_i} + \sum_{j=1}^{n_i-1} \alpha_{i,j}\dot{x}_{i,j} = \dot{x}_{i,n_i} + \sum_{j=1}^{n_i-1} \alpha_{i,j} x_{i,j+1} = d_i(x) + f_i(x) + \sum_{j=1}^{m} g_{i,j}(x)\, u_j \tag{29}$$

Then, the derivative of V becomes

$$\begin{aligned}
\dot{V} &= s^T \dot{s} = s^T (d(x) + f(x) + g(x)u) \\
&= s^T\left(d(x) + f(x) + g(x)\hat{g}(x)^{-1} \cdot \left[-\hat{f}(x) - \hat{d}(x) - k \circ sgn(s)\right]\right) \\
&= s^T\left(d(x) + f(x) + (I + \Delta) \cdot \left[-\hat{f}(x) - \hat{d}(x) - k \circ sgn(s)\right]\right) \\
&= s^T\left(\left[f(x) - \hat{f}(x)\right] + [d(x) - \hat{d}(x)] - \Delta\left[\hat{f}(x) + \hat{d}(x)\right] - (I + \Delta)k \circ sgn(s)\right) \\
&\leq \big|s^T\big|\left(\begin{array}{c} F_f + F_d + G\big|\hat{f}(x) + \hat{d}(x)\big| \\ -(I - G)k \end{array}\right)
\end{aligned} \tag{30}$$

If we choose $k \geq 0$ such that

$$F_f + F_d + G\big|\hat{f}(x) + \hat{d}(x)\big| - (I - G)k \leq -\eta \tag{31}$$

where $\eta \geq 0$ is an arbitrary non-negative vector. Then, $\dot{V} \leq -\eta|s|$. According to the Lyapunov stability theorem, the system is asymptotically stable. The existence of such k is guaranteed by the following Lemma. $\square$

Lemma 1. There always exists a non-negative $k \geq 0$ under the given matching condition in Equation (24), such that Equation (31) can be satisfied.

Proof. From the definition of F_f, F_d, G, and η, we know that:

$$\xi = F_f + F_d + G\big|\hat{f}(x) + \hat{d}(x)\big| + \eta > 0 \tag{32}$$

Then, the existence of k depends on whether there exists a solution to the inequality of $(I - G)k \geq \xi > 0$. Since $\rho(G) < 1$, according to Frobenius-Perron Theorem [42], there is always a non-negative solution $k \geq 0$ such that $(I - G)k \geq \xi > 0$. Hence, $k \geq 0$ always exists such that Equation (31) can be satisfied. $\square$

4. Compare SBC to SMC

4.1. Recall SMC Design

Since the comparison of these two controllers are similar for both SISO and MIMO systems, the following discussion is performed on the controller design of a SISO system. Recall SMC design [42], which started by constructing a sliding surface given in the form of:

$$s_{SMC} = \left(\frac{d}{dt} + \lambda\right)^{n-1} x = x^{(n-1)} + \sum_{j=1}^{n-1} \beta_j x^{(j-1)} \tag{33}$$

where λ is a scalar control parameter, β_j are the summation of the jth order terms in the expansion of $\left(\frac{d}{dt} + \lambda\right)^{n-1}$. The sliding mode control of the system in (1) can be given as:

$$u_{SMC} = \frac{1}{\hat{g}(x)}\left[-\hat{f}(x) - \hat{d}_{SMC}(x) - k_{SMC}sgn(s_{SMC})\right] \tag{34}$$

where

$$d_{SMC}(x) = \sum_{j=1}^{n-1} \beta_j x^{(j)} \tag{35}$$

$$k_{SMC} = \frac{1}{1-G}\left(F_f + F_{d_{SMC}} + G\left|\hat{f}(x) + \hat{d}_{SMC}(x)\right| + \eta\right) \tag{36}$$

4.2. Sliding Surface and Chatter

Comparing the sliding surface in Equation (33) to $s_{SBC} = s$ in Equation (13) of SBC design, we can see that s_{SMC} is a special case of s_{SBC} given that $\alpha_j = \beta_j$ or $\gamma_1 = \gamma_2 = \cdots \gamma_n$. From the other side, the SBC controller provides more flexibility in robust controller design by alternating the convergence speed in the successive lower orders of the systems; which can potentially improve the overall system performance. We can say SMC is a special case of SBC if we name Equations (13) and (26) sliding surfaces of the SBC.

From the design of both SBC and SMC, we know that the chatter effects exist. As it has been discussed before, chattering reduction techniques for SMC can be also used for the SBC.

4.3. Useful Nonlinear Dynamics

Besides the flexibility in designing the convergence speed, the SBC has another advantage that the useful nonlinear dynamics portion, which helps with system stability, can be leveraged to simplify the controller design. This advantage has been studied for standard BC and can similarly be used for SBC design. Note that the discussion in this section is based on a SISO system, but the results can be extended to MIMO systems.

Theorem 3. $f(x)$ is composed of useful dynamics $f_u(s)$ and the residual is $f_r(x)$ as

$$f(x) = -f_u(s) + f_r(x) \tag{37}$$

and the useful dynamic can be defined as Lipschitz condition:

$$\left|f_u(s)\right| < \lambda|s| \tag{38}$$

The controller of the revised system

$$x^{(n)} = -f_u(s) + f_r(x) + g(x)u \tag{39}$$

can be given by

$$u = \frac{1}{\hat{g}(x)}\left[-\hat{f}_r(x) - \hat{d}(x) - k\,sgn(s)\right] \tag{40}$$

then, the system is asymptotically stable.

Remark 7. Due to the change in the system structure, the first matching condition in (11) becomes $\left|f_r - \hat{f}_r\right| \le F_r$.

Proof. Choose the positive definite Lyapunov function $V = \frac{1}{2}s^2$. Its derivative becomes:

$$
\begin{aligned}
\dot{V} &= s\dot{s} = s\left(\sum_{j=1}^{n-1}\alpha_j\dot{x}_j + \dot{x}_n\right) = s\left(d(x) + \dot{x}_n\right) \\
&= s(d(x) - f_u(s) + f_r(x) + g(x)u) \\
&= -sf_u(s) + s\left(\begin{array}{c} d(x) + f_r(x) + g(x)\frac{1}{\hat{g}(x)}\cdot \\ \left[-\hat{f}(x) - \hat{d}(x) - k\,sgn(s)\right] \end{array}\right) \\
&= -sf_u(s) + s\left(\begin{array}{c} d(x) + f(x) + (1+\Delta)\cdot \\ \left[-\hat{f}(x) - \hat{d}(x) - k\,sgn(s)\right] \end{array}\right) \\
&= -sf_u(s) + s\left(\begin{array}{c} \left[f(x) - \hat{f}(x)\right] + \left[d(x) - \hat{d}(x)\right] \\ -\Delta\left[\hat{f}(x) + \hat{d}(x)\right] - (1+\Delta)k\,sgn(s) \end{array}\right) \\
&\le -\lambda s^2 + |s|\left(\begin{array}{c} F_f + F_d + G\left|\hat{f}(x) + \hat{d}(x)\right| \\ -(1-G)k \end{array}\right)
\end{aligned} \tag{41}
$$

If k is chosen according to

$$k \ge \frac{1}{1-G}\left(F_r + F_d + G\left|\hat{f}(x) + \hat{d}(x)\right| + \eta\right) \tag{42}$$

then, $\dot{V} \le -\lambda s^2 - \eta\,|s|$, which is negative definite. Hence, the system is asymptotically stable. $\quad\square$

5. Simulation Results

To validate the performance of the proposed SBC and ISBC, two cases of control of an autonomous underwater vehicle (AUV) are provided, including (1) Case I: line following cruise control (SISO system) and Case II: planar motion control (MIMO system). As shown in Figure 1, the motion of the AUV in XY plane is controlled by the thrust force F and steering torque τ. The position of the AUV is represented by the location of its center of mass (x, y), V is the speed, and φ is the heading angle.

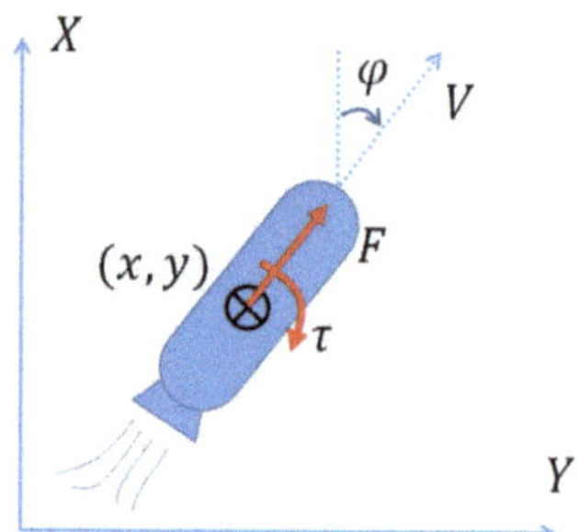

Figure 1. Sketch of an autonomous underwater vehicle.

Case I: SISO system control—line following cruise control of the AUV

This line following cruise tracking control problem simulates an application of using an AUV for sea floor mapping or searching, where the AUV scans a field with a constant speed following a zig-zag pattern. Here, we assume that speed control is handled by another controller, such that the system is an SISO system. The goal of cruise tracking control is to drive the AUV to follow a straight line at a cruise speed V_0. More specifically, if we desire that the AUV cruises along the x-axis, the goal is to drive y to zero in a finite time, while also converging to the cruise speed V_0. The associated system dynamics are given by:

$$\begin{cases} \dot{x} = V_0 cos\varphi \\ \dot{y} = V_0 sin\varphi \\ \dot{\varphi} = \omega \\ J\dot{\omega} = \tau \end{cases} \tag{43}$$

where ω is the angular rate and J is the mass moment of inertia of the AUV. Since this is a SISO system, we can follow Theorem 1 to design the controller. First, the system is converted to the canonical form by using input-output linearization:

$$\begin{cases} x_1 = y \\ x_2 = \dot{y} = V_0 sin\varphi \\ x_3 = y^{(2)} = V_0 cos\varphi\dot{\varphi} = V_0\omega cos\varphi \\ y^{(3)} = -V_0\omega^2 sin\varphi + \frac{1}{J}V_0 cos\varphi\tau \end{cases} \tag{44}$$

Then, the equivalent dynamics are given as:

$$\begin{cases} \dot{x}_1 = x_2 \\ \dot{x}_2 = x_3 \\ \dot{x}_3 = f(x) + g(x)u \end{cases} \tag{45}$$

where $f(x) = -V_0\omega sin\varphi\dot{\varphi}$, $g(x) = \frac{1}{J}V_0 cos\varphi$ and $u = \tau$.

Secondly, we define s_{SBC} and d_{SBC} as follows:

$$\begin{cases} s_{SBC} = x_3 + \alpha_2 x_2 + \alpha_1 x_1 \\ d_{SBC} = \alpha_2 x_3 + \alpha_1 x_2 \end{cases} \tag{46}$$

Correspondingly, for ISBC, we have:

$$\begin{cases} s_{ISBC} = x_3 + \alpha_2 x_2 + \alpha_1 x_1 + \alpha_0 x_0 \\ d_{ISBC} = \alpha_2 x_3 + \alpha_1 x_2 + \alpha_0 x_1 \end{cases} \tag{47}$$

and, for SMC and Integral sliding mode control (ISMC)

$$\begin{cases} s_{SMC} = x_3 + 2\beta x_2 + \beta^2 x_1 \\ d_{SMC} = 2\beta x_3 + \beta^2 x_2 \end{cases} \tag{48}$$

$$\begin{cases} s_{ISMC} = x_3 + 3\beta x_2 + 3\beta^2 x_1 + \beta^3 x_0 \\ d_{ISMC} = 3\beta x_3 + 3\beta^2 x_2 + \beta^3 x_1 \end{cases} \tag{49}$$

Then, the control laws share the same form:

$$u = \frac{1}{\hat{g}(x)}\left[-\hat{f}(x) - \hat{d}_{(*)}(x) - k\,sgn(s_{(*)})\right] \tag{50}$$

where $\hat{f}(x)$, $\hat{g}(x)$, and $\hat{d}(x)$ are nominal functions of $f(x)$, $g(x)$, and $d(x)$, respectively; and the subscript $_{(*)}$ can be one of the four choices of $\{SBC, ISBC, SMC, ISMC\}$.

The matching condition can be quantified through Monte Carlo simulation and experimental tests. The control input is given by Equation (12), where s and d are chosen according to the specific control algorithm to be used (i.e., if SBC is chosen, $s = s_{SBC}$ and $d = d_{SBC}$).

The four controllers were compared under the same testing scenario where the mass is m = 200 kg, moment of inertia is J = 450 kg·m^2, cruise speed is V_0 =4 m/s, and the vehicle is controlled to track the x-axis from the initial position of $[x_0, y_0] = [0, 5]$ m. The noise in position, angle, and angular rate are assumed to be zero-mean and Gaussian with a standard deviation of 0.1 m, 1 degree, and 0.01 degree/s, respectively. Note that all noise is assumed to be bounded within a range of $\pm 3\sigma$. A MATLAB/Simulink model is developed, and the results are shown in Figures 2 and 3.

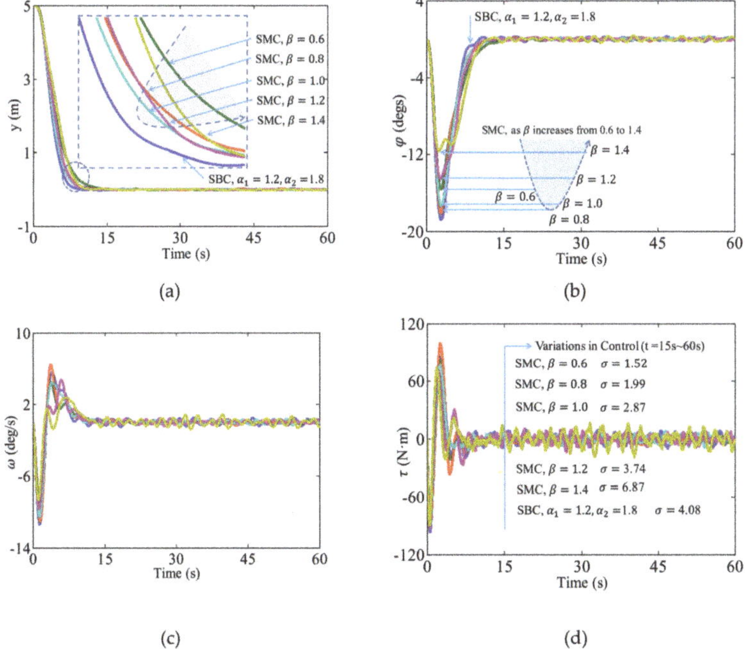

Figure 2. Performance comparison of SBC and SMC. (**a**), (**b**), (**c**) and (**d**) are the y-position, heading angle, angular rate, and control signal.

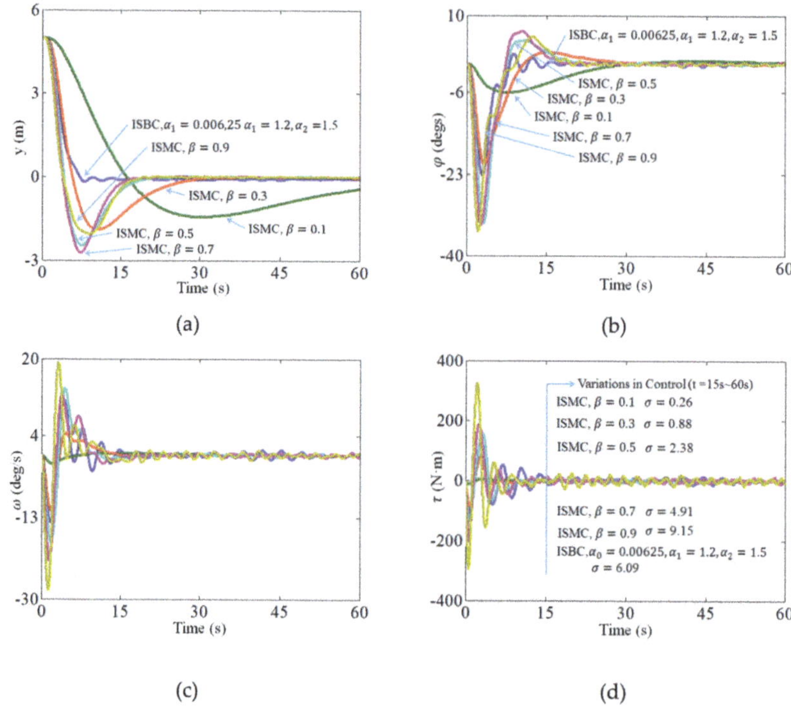

Figure 3. Performance comparison of ISBC and ISMC.

Figure 2 shows the performance comparison of the SBC and SMC controllers, where (a), (b), (c) and (d) are the y-position, heading angle, angular rate, and control signal, respectively. To provide a complete performance comparison, results for SMCs with different control parameters (i.e., $\beta = 0.6$, 0.8, 1.0, 1.2, and 1.4) are produced to compare with the proposed SBC ($\alpha_1 = 1.2$, and $\alpha_2 = 1.8$). It can be seen from Figure 2a that SBC can drive the AUV to approach zero in a shorter time compared to SMC. As the parameter β of SMC increases from 0.6 to 1.4, the performance becomes similar to the SBC before diverging from it. The rise time under these six controls are given in Table 1. It can be seen that the SBC has the shortest rise time of 6.73 s, while the quickest rise time of the five SMCs is 7.66 s.

Table 1. Rise times of SBC and SMC.

Rise Time	C0	C1	C2	C3	C4	C5
t_r	6.73 s	9.61 s	8.10 s	7.66 s	7.96 s	8.65 s

Note: C0 = SBC, C1, C2, ... , C5 = SMC with $\beta = 0.6, 0.8, ... , 1.4$.

The performance changes as β increases can be analyzed from the variations in the coefficients of s_{SMC} and d_{SMC}. Recall that $x_1 = y$ and $x_2 = \dot{y}$, where the coefficients as shown in Equation (16) correspond to the proportional and derivative gain in PID controller design. Since the two coefficients β^2 and 2β are dependent, the performance of the controller is also restricted to a certain extent due to the freedom in parameter selection being reduced from 2 to 1.

Figure 3 shows the performance comparison of the SBC and SMC controllers, where (a), (b), (c) and (d) are the y-position, heading angle, angular rate, and control signal, respectively. To provide a complete performance comparison, results of ISMCs with different control parameters (i.e., $\beta = 0.1, 0.3$, 0.5, 0.7, and 0.9) are produced to compare with the proposed SBC ($\alpha_0 = 0.0625$, $\alpha_1 = 1.2$, and $\alpha_2 = 1.5$). It can be seen from Figure 2a that ISBC can drive the AUV to reach steady state in a much shorter time of 8.9 seconds compared to ISMCs. The performance indices, including rise time t_r, settling time t_s, peak time t_p, and overshot $\sigma\%$ are summarized in Table 2. It can be seen that the ISBC has the best performance with respect to the rising time, settling time, peak time, and overshoot. Due to the dependency in the control gains, ISMCs have difficulty in achieving better performances; i.e., as β increases (from C1 to C5 in Table 2), the settling time, rising time, peak time, and overshoot first decrease then increase.

Table 2. Performance comparison between ISBC and ISMC.

System Response	C0	C1	C2	C3	C4	C5
t_r	6.28 s	15.14 s	5.32 s	3.83 s	3.52 s	3.52 s
t_s	8.90 s	> 60 s	30.53 s	21.29 s	17.01 s	18.25 s
t_s	8.01 s	30.21 s	10.64 s	7.50 s	7.15 s	8.97 s
$\sigma\%$	3.5%	28.6%	37.8%	49.5%	54.6%	40.9%

Note: C0 = ISBC, C1, C2, ... , C5 = ISMC with $\beta = 0.1, 0.3, ... , 0.9$.

It can be seen from Figure 3b,c that, during the control process, ISBC has the smallest variation in heading angle and angular rate compared to ISMCs. Figure 3d shows that all the controllers can maintain the desired states with relatively small variations in control signals with a standard deviation of less than 10 N.

From the comparison in these two experiments, we can see that the dependency in control gain selection simplifies the controller tuning, meanwhile the dependency also limits the performance of the control system.

Case II: MIMO system—planar motion control of the AUV

Planar motion control of the AUV is used to show the effectiveness of the proposed controller design for MIMO systems. The dynamics of planar motion of the AUV is given as follows:

$$\begin{cases} \dot{x} = V\cos\varphi \\ \dot{y} = V\sin\varphi \\ \dot{\varphi} = \omega \\ m\dot{V} = F - cV^2 \\ J\dot{\omega} = \tau \end{cases} \tag{51}$$

where c is the drag coefficient and the drag force is cV^2. Since this is a MIMO system, we can follow Theorem 2 to design the controller. First, the system is converted to the canonical form by using input-output linearization.

$$\text{Let } x_1 = \begin{Bmatrix} x \\ y \\ \varphi \end{Bmatrix}, \quad x_2 = \begin{Bmatrix} \dot{x} \\ \dot{y} \\ \dot{\varphi} \end{Bmatrix} = \begin{Bmatrix} V\cos\varphi \\ V\sin\varphi \\ \omega \end{Bmatrix}, \quad \text{then } \dot{x}_2 = \begin{Bmatrix} x \\ y \\ \varphi \end{Bmatrix}^{(2)} =$$

$$\begin{Bmatrix} \frac{1}{m}(F - cV^2)\cos\varphi - V\omega\sin\varphi \\ \frac{1}{m}(F - cV^2)\sin\varphi + V\omega\cos\varphi \\ \frac{1}{J}\tau \end{Bmatrix} \text{ and the equivalent dynamics is given as}$$

$$\begin{cases} \dot{x}_1 = x_2 \\ \dot{x}_2 = f(x) + g(x)u \end{cases} \tag{52}$$

where $f(x) = \left\{ \begin{array}{c} -\frac{1}{m}cV^2cos\varphi - V\omega sin\ \varphi \\ -\frac{1}{m}cV^2sin\varphi + V\omega cos\varphi \\ 0 \end{array} \right\}$, $g(x) = \left[\begin{array}{cc} \frac{1}{m}cos\varphi & 0 \\ \frac{1}{m}sin\varphi & 0 \\ 0 & \frac{1}{J} \end{array} \right]$, and $u = \left\{ \begin{array}{c} F \\ \tau \end{array} \right\}$. Secondly, we can

define s and d functions for SBC as follows

$$\begin{cases} s_{SBC} = x_2 + \alpha_1 x_1 \\ d_{SBC} = \alpha_1 x_2 \end{cases} \tag{53}$$

Correspondingly, for ISBC, we have:

$$\begin{cases} s_{ISBC} = x_2 + \alpha_1 x_1 + \alpha_0 x_0 \\ d_{ISBC} = \alpha_1 x_2 + \alpha_0 x_1 \end{cases} \tag{54}$$

and, for SMC and ISMC,

$$\begin{cases} s_{SMC} = x_3 + 2\beta x_2 + \beta^2 x_1 \\ d_{SMC} = 2\beta x_3 + \beta^2 x_2 \end{cases} \tag{55}$$

$$\begin{cases} s_{ISMC} = x_3 + 3\beta x_2 + 3\beta^2 x_1 + \beta^3 x_0 \\ d_{ISMC} = 3\beta x_3 + 3\beta^2 x_2 + \beta^3 x_1 \end{cases} \tag{56}$$

Then, the control laws share the same form

$$u = \hat{g}(x)^+ \left[-\hat{f}(x) - \hat{d}_{(*)}(x) - ksgn(s_{(*)}) \right] \tag{57}$$

where $_{(*)}$ can be one of the four choices $\{SBC,\ ISBC,\ SMC, ISMC\}$ given in Equations (53)–(56).

The matching conditions and AUV specifications are the same as those for Case I, while the speed of the AVU is treated as a state variable. The control parameters are listed in Table 3 and simulation results are shown in Figures 4 and 5. The initial conditions are given as $x(0) = -3m$, $y = -3m$, and $\varphi(0) = \frac{\pi}{2}$.

Table 3. Control parameter selection.

Control Method	Parameters
SMC	$\beta = 0.2$
SBC	$\alpha_1 = 0.2 \times [1; 0.9; 1]$
ISMC	$\beta = 0.2$
ISBC	$\alpha_1 = 0.4 \times [1.1; 1; 1]$ and $\alpha_0 = 0.04 \times [1.25; 1; 1]$

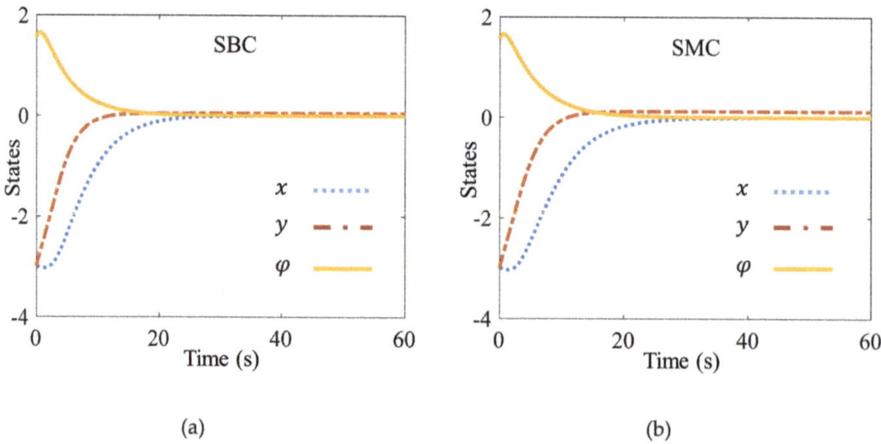

Figure 4. Performance comparison of SBC (**a**) and SMC (**b**).

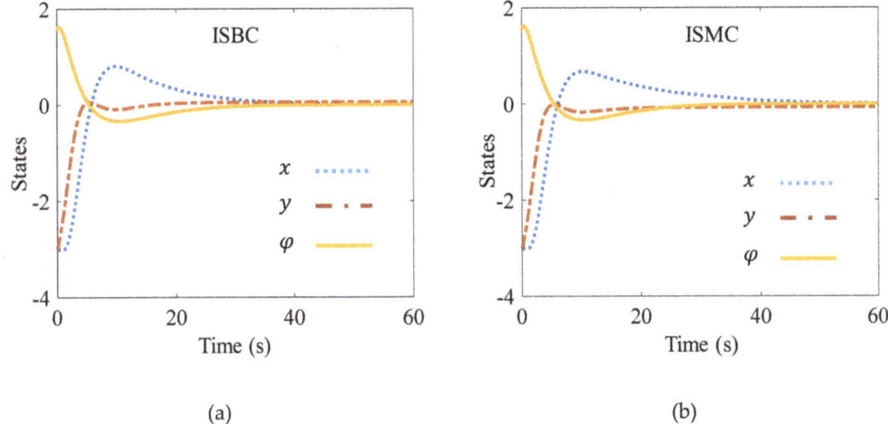

Figure 5. Performance comparison of ISBC (**a**) and ISMC (**b**).

Figure 4 shows the performance comparison of the SBC and SMC. It can be seen from that SBC can control the AUV to approach zero in a shorter time compared to SMC. The rise time is given in Table 4. Moreover, SBC allows us to adjust the performance of each state by tuning the corresponding control parameters. By comparing Figure 4a,b it can also be seen that the steady state error in y direction in SBC is much smaller than that for SMC. This is achieved by choosing a different control parameter for y direction as indicated in Table 3. Figure 5 shows the performance comparison of the ISBC and ISMC. By choosing a different control parameter for x direction in α_0, the settling time of state x is reduced from 41.22 s for ISMC to 34.72 s for ISBC.

Table 4. Rise time of SBC and SMCs.

Rise Time	SMC	SBC
$t_r\,(x, y, \varphi)$	13.01, 7.73, 11.21 (s)	11.71, 7.18, 10.25 (s)

By comparing the controller performance for the two cases, it can be seen that the proposed SBC and ISBC can achieve better performance comparing to SMC and ISMC. This improvement is achieved by flexibility in control parameter selections. Similar strategies can be used for controller parameter selection for all these four different controllers. Additionally, if $\alpha_1 = \beta\{1;1;1\}$, SBC and SMC two controllers will be equivalent if $\alpha_1 = \beta\{1;1;1\}$. Similarly, for ISMC and ISBC, if $\alpha_1 = 2\beta\{1;1;1\}$, and $\alpha_0 = \beta^2\{1;1;1\}$, these two controllers will be equivalent.

6. Conclusions

In this paper, a sliding mode control in the backstepping framework (SBC) was proposed for a class of nonlinear systems. The comparison between the SBC and sliding mode control was discussed by comparing dependency in the control gains. An example of cruise control of an unmanned underwater vehicle was given to show the effectiveness of the controller. It was found that the proposed controller is able to achieve better control performance due to its advantages in the flexibility of control gain selection and simplicity in control law formulation when considering useful nonlinearities.

J. Mar. Sci. Eng. **2019**, *7*, 452

Author Contributions: The first author H.S. contributed to conceptualization, methodology, original draft preparation, formal analysis, and editing. The second author J.I. contributed to the original draft preparation, validation, result preparation, and revision. The third author N.L. contributed to conceptualization, methodology, original draft preparation, result preparation, and final editing.

Funding: This research received no external funding.

Conflicts of Interest: The authors declare no conflict of interest.

References

1. Khalil, H. *Nonlinear Systems*, 3rd ed.; Prentice Hall: Upper Saddle River, NJ, USA, 2001.
2. Chen, M.; Ge, S.S.; Ren, B. Robust attitude control of helicopters with actuator dynamics using neural networks. *IET Control Theory Appl.* **2010**, *4*, 2837–2854. [CrossRef]
3. Ki-Seok, K.; Youdan, K. Robust backstepping control for slew maneuver using nonlinear tracking function. *IEEE Trans. Control Syst. Technol.* **2003**, *11*, 822–829. [CrossRef]
4. Mou, C.; Bin, J. Robust bounded control for uncertain flight dynamics using disturbance observer. *J. Syst. Eng. Electron.* **2014**, *25*, 640–647.
5. Azinheira, J.; Moutinho, A. Hover Control of an UAV with Backstepping Design Including Input Saturations. *IEEE Trans. Control Syst. Technol.* **2008**, *16*, 517–526. [CrossRef]
6. Chen, M.; Ge, S.; How, B.V.; Choo, Y.S. Robust adaptive position mooring control for marine vessels. *IEEE Trans. Control Syst. Technol.* **2013**, *21*, 395–409. [CrossRef]
7. Fossen, T.; Grovlen, A. Nonlinear output feedback control of dynamically positioned ships using vectorial observer backstepping. *IEEE Trans. Control Syst. Technol.* **1998**, *6*, 121–128. [CrossRef]
8. Ghommam, J.; Mnif, F.; Benali, A.; Derbel, N. Asymptotic Backstepping Stabilization of an Underactuated Surface Vessel. *IEEE Trans. Control Syst. Technol.* **2006**, *14*, 1150–1157. [CrossRef]
9. Kwan, C.; Lewis, F. Robust backstepping control of induction motors using neural networks. *IEEE Trans. Neural Netw.* **2000**, *11*, 1178–1187. [CrossRef]
10. Drid, S.; Naït-Saïd, M.-S.; Tadjine, M. Robust backstepping vector control for the doubly fed induction motor. *IET Control Theory Appl.* **2007**, *1*, 861–868. [CrossRef]
11. Lin, F.-J.; Lee, C.-C. Adaptive backstepping control for linear induction motor drive to track periodic references. *IEE Proc. Electr. Power Appl.* **2000**, *147*, 449. [CrossRef]
12. Zhao, D.; Gao, F.; Li, S.; Zhu, Q. Robust finite-time control approach for robotic manipulators. *IET Control Theory Appl.* **2010**, *4*, 1–15. [CrossRef]
13. Yoo, S.J.; Park, J.B.; Choi, Y.H. Adaptive Output Feedback Control of Flexible-Joint Robots Using Neural Networks: Dynamic Surface Design Approach. *IEEE Trans. Neural Netw.* **2008**, *19*, 1712–1726.
14. Shojaei, K.; Shahri, A. Adaptive robust time-varying control of uncertain non-holonomic robotic systems. *IET Control Theory Appl.* **2012**, *6*, 90. [CrossRef]
15. Zhou, H.; Liu, Z. Vehicle Yaw Stability-Control System Design Based on Sliding Mode and Backstepping Control Approach. *IEEE Trans. Veh. Technol.* **2010**, *59*, 3674–3678. [CrossRef]
16. Shen, P.-H.; Lin, F.-J. Intelligent backstepping sliding-mode control using RBFN for two-axis motion control system. *IEE Proc. Electr. Power Appl.* **2005**, *152*, 1321. [CrossRef]
17. Liu, Z.; Jia, X. Novel backstepping design for blended aero and reaction-jet missile autopilot. *J. Syst. Eng. Electron.* **2008**, *19*, 148–153.
18. Hua, C.; Liu, P.X.; Guan, X.; Liu, P.; Liu, P. Backstepping Control for Nonlinear Systems With Time Delays and Applications to Chemical Reactor Systems. *IEEE Trans. Ind. Electron.* **2009**, *56*, 3723–3732.
19. Mazenc, F.; Bliman, P.-A. Backstepping Design for Time-Delay Nonlinear Systems. *IEEE Trans. Autom. Control* **2006**, *51*, 149–154. [CrossRef]
20. Shiromoto, H.S.; Andrieu, V.; Prieur, C. Relaxed and Hybridized Backstepping. *IEEE Trans. Autom. Control* **2013**, *58*, 3236–3241. [CrossRef]
21. Zhang, Y.; Fidan, B.; Ioannou, P.A. Backstepping control of linear time-varying systems with known and unknown parameters. *IEEE Trans. Autom. Control* **2003**, *48*, 1908–1925. [CrossRef]
22. Cheng, C.-C.; Su, G.-L.; Chien, C.-W. Block backstepping controllers design for a class of perturbed non-linear systems with m blocks. *IET Control Theory Appl.* **2012**, *6*, 2021–2030. [CrossRef]

23. Zhou, J.; Zhang, C.; Wen, C. Robust Adaptive Output Control of Uncertain Nonlinear Plants with Unknown Backlash Nonlinearity. *IEEE Trans. Autom. Control* **2007**, *52*, 503–509. [CrossRef]
24. Wang, Z.; Ge, S.; Lee, T. Robust adaptive neural network control of uncertain nonholonomic systems with strong nonlinear drifts. *IEEE Trans. Syst. Man Cybern. Part B Cybern.* **2004**, *34*, 2048–2059. [CrossRef] [PubMed]
25. Wang, J.; Hu, J. Robust adaptive neural control for a class of uncertain non-linear time-delay systems with unknown dead-zone non-linearity. *IET Control Theory Appl.* **2011**, *5*, 1782–1795. [CrossRef]
26. Mazenc, F.; Ito, H. Lyapunov technique and backstepping for nonlinear neutral systems. *IEEE Trans. Autom. Control* **2013**, *58*, 512–517. [CrossRef]
27. Hyeongcheol, L. Robust adaptive fuzzy control by backstepping for a class of MIMO nonlinear systems. *IEEE Trans. Fuzzy Syst.* **2011**, *19*, 265–275.
28. Wang, H.; Chen, B.; Liu, X.; Liu, K.; Lin, C. Robust Adaptive Fuzzy Tracking Control for Pure-Feedback Stochastic Nonlinear Systems With Input Constraints. *IEEE Trans. Cybern.* **2013**, *43*, 2093–2104. [CrossRef]
29. Tong, S.-C.; He, X.-L.; Zhang, H.-G. A Combined Backstepping and Small-Gain Approach to Robust Adaptive Fuzzy Output Feedback Control. *IEEE Trans. Fuzzy Syst.* **2009**, *17*, 1059–1069. [CrossRef]
30. Wai, R.; Lee, J. Robust levitation control for linear maglev rail system using fuzzy neural network. *IEEE Trans. Control Syst. Technol.* **2009**, *17*, 4–14.
31. Shafiei, S.E.; Soltanpour, M.R. Robust neural network control of electrically driven robot manipulator using backstepping approach. *Int. J. Adv. Robot. Syst.* **2009**, *6*, 285–292. [CrossRef]
32. Kwan, C.; Lewis, F. Robust backstepping control of nonlinear systems using neural networks. *IEEE Trans. Syst. Man Cybern. Part A Syst. Hum.* **2000**, *30*, 753–766. [CrossRef]
33. Chen, M.; Ge, S.S.; How, B. Robust Adaptive Neural Network Control for a Class of Uncertain MIMO Nonlinear Systems with Input Nonlinearities. *IEEE Trans. Neural Netw.* **2010**, *21*, 796–812. [CrossRef] [PubMed]
34. Kuljaca, O.; Swamy, N.; Lewis, F.; Kwan, C. Design and implementation of industrial neural network controller using backstepping. *IEEE Trans. Ind. Electron.* **2003**, *50*, 193–201. [CrossRef]
35. Hsu, C.-F.; Lin, C.-M.; Lee, T.-T. Wavelet Adaptive Backstepping Control for a Class of Nonlinear Systems. *IEEE Trans. Neural Netw.* **2006**, *17*, 1175–1183.
36. Wai, R.-J.; Chang, H.-H. Backstepping Wavelet Neural Network Control for Indirect Field-Oriented Induction Motor Drive. *IEEE Trans. Neural Netw.* **2004**, *15*, 367–382. [CrossRef]
37. Hua, C.; Guan, X.; Shi, P. Robust backstepping control for a class of time delayed systems. *IEEE Trans. Autom. Control* **2005**, *50*, 894–899.
38. Ezal, K.; Pan, Z.; Kokotovic, P. Locally optimal and robust backstepping design. *IEEE Trans. Autom. Control* **2000**, *45*, 260–271. [CrossRef]
39. French, M. An analytical comparison between the nonsingular quadratic performance of robust and adaptive backstepping designs. *IEEE Trans. Autom. Control* **2002**, *47*, 670–675. [CrossRef]
40. Sabanovic, A. Variable structure systems with sliding modes in motion vontrol-a survey. *IEEE Trans. Ind. Inform.* **2011**, *7*, 212–223. [CrossRef]
41. Young, K.; Utkin, V.; Ozguner, U. A control engineer's guide to sliding mode control. *IEEE Trans. Control Syst. Technol.* **1999**, *7*, 328–342. [CrossRef]
42. Slotine, J.J.; Li, W. *Applied Nonlinear Control*; Prentice-Hall International: Upper Saddle River, NJ, USA, 1991.
43. Kachroo, P.; Tomizuka, M. Chattering reduction and error convergence in the sliding-mode control of a class of nonlinear systems. *IEEE Trans. Autom. Control* **1996**, *41*, 1063–1068. [CrossRef]
44. Fallaha, C.J.; Saad, M.; Kanaan, H.Y.; Al-Haddad, K. Sliding-mode robot control with exponential reaching law. *IEEE Trans. Ind. Electron.* **2011**, *58*, 600–610. [CrossRef]
45. Acary, V.; Brogliato, B.; Orlov, Y.V. Chattering-rree digital sliding-mode control with state observer and disturbance rejection. *IEEE Trans. Autom. Control* **2012**, *57*, 1087–1101. [CrossRef]
46. Lin, C.-K.; Liu, T.-H.; Wei, M.-Y.; Fu, L.-C.; Hsiao, C.-F. Design and implementation of a chattering-free non-linear sliding-mode controller for interior permanent magnet synchronous drive systems. *IET Electr. Power Appl.* **2012**, *6*, 332. [CrossRef]
47. Lin, F.-J.; Shen, P.-H.; Hsu, S.-P. Adaptive backstepping sliding mode control for linear induction motor drive. *IEE Proc. Electr. Power Appl.* **2002**, *149*, 184. [CrossRef]

48. Skjetne, R.; Fossen, T. On integral control in backstepping: Analysis of different techniques. In Proceedings of the 2004 American Control Conference, Boston, MA, USA, 30 June–2 July 2004; Volume 2, pp. 1899–1904.

49. Xiao, L.; Zhu, Y. Passivity-based integral sliding mode active suspension control. In Proceedings of the 19th World Congress of the International Federation of Automatic Control, Cape Town, South Africa, 24–29 August 2014.

50. Bartolini, G.; Ferrara, A.; Usai, E. Chattering avoidance by second-order sliding mode control. *IEEE Trans. Autom. Control* **1998**, *43*, 241–246. [CrossRef]

51. Kurode, S.; Spurgeon, S.K.; Bandyopadhyay, B.; Gandhi, P.S. Sliding mode control for slosh-free motion using a nonlinear sliding surface. *IEEE/ASME Trans. Mechatron.* **2013**, *18*, 714–724. [CrossRef]

52. Leung, F.; Wong, L.; Tam, P. Algorithm for eliminating chattering in sliding mode control. *Electron. Lett.* **1996**, *32*, 599. [CrossRef]

53. Kachroo, P. Existence of solutions to a class of nonlinear convergent chattering-free sliding mode control systems. *IEEE Trans. Autom. Control* **1999**, *44*, 1620–1624. [CrossRef]

Article

System Modeling and Simulation of an Unmanned Aerial Underwater Vehicle

Yuqing Chen [1], Yaowen Liu [1], Yangrui Meng [1], Shuanghe Yu [1,*] and Yan Zhuang [2]

[1]   College of Marine Electrical Engineering, Dalian Maritime University, Dalian 116026, China; chen@dlmu.edu.cn (Y.C.); lyw1120181170@sina.com (Y.L.); mi177212@163.com (Y.M.)
[2]   School of Control Science and Engineering, Dalian University of Technology, Dalian 116024, China; zhuang@dlut.edu.cn
*   Correspondence: shuanghe@dlmu.edu.cn; Tel.: +86-411-8472-6950

Received: 13 October 2019; Accepted: 25 November 2019; Published: 4 December 2019

Abstract: Unmanned Aerial Underwater Vehicles (UAUVs) with multiple propellers can operate in two distinct mediums, air and underwater, and the system modeling of the autonomous vehicles is a key issue to adapt to these different external environments. In this paper, only a single set of aerial rotors with switching propulsion abilities are designed as driving components, and then a compound multi-model method is investigated to achieve good performance of the cross-medium motion. Furthermore, some additional variables, such as water resistance, buoyancy and their corresponding moments are considered for the underwater case. In particular, a critical coefficient for air-to-water switching is presented to express these gradually changing additional variables in the cross-medium motion process. Finally, the sliding mode control method is used to reduce the altitude error and attitude error of the vehicles with external environmental disturbances. The proposed scheme is tested and the model is verified on the simulation platform.

Keywords: aerial underwater vehicle; dynamics; modeling; cross-medium

---

1. Introduction

Autonomous unmanned systems are very popular all over the world due to their broad range of applications such as surveillance, inspection, fast delivery, search and rescue, among many others [1–3]. In recent years, the rapid development of composite materials and the continuous upgrading of intelligent control technology have been of great help in the development of unmanned aerial vehicles (UAVs) and unmanned underwater vehicles (UUVs) [4,5]. More challenging, the unmanned aerial underwater vehicle (UAUV) has become a novel research field, which aims to design equipment that can fly in the air and navigate underwater. To enhance the characteristics of these unmanned systems, many scholars and enterprises have done lots of relevant work in fuselage shape, fuselage material, wing layout, mass-volume ratio, power system design, controller design and their intelligent applications and so on [6–9].

The system modeling and control of UAVs are investigated in some works. Many kinds of UAVs, such as fixed-wing, multi-rotor wing and flapping wing, have been developed and analyzed about their different structures and mathematical description method [10,11], and a lot of achievements have been made. Compared with traditional air vehicles, the control of multi-rotor vehicles are more stable and efficient. Thus, the multi-rotor vehicles have significant advantages in vertical take-off and landing (VTOL), precise hovering and aggressive maneuvers. In addition, transition flight modeling of a fixed-wing VTOL UAV is also investigated, and the model is specifically tailored for the design of a hover to forward flight and forward flight to hover transition control system [12]. Currently, there is no well accepted generic methodology that can be used for system modeling due to the fact that there are numerous challenges regarding the various configurations [13].

In addition, the system dynamics of UUVs has also been paid increasing attention by researchers. With the development of modern marine researches, many intelligent marine equipment, such as surface ships, semi-submarines, unmanned submarines and deep-sea robots, have been developed in all aspects. In particular, the underwater autonomous vehicles are facing unknown and hazardous environments, and their research and deployment have been regarded as one of significant goals and challenges by human beings. Therefore, an UUV with autonomous control should have capability to perceive its own position as well as its environment and react to unexpected or dynamic circumstances properly [14,15]. In addition, the system models of different shape configurations, such as open structure, torpedo-like and multi-thruster, are studied respectively [16,17]. Thruster fault detection and the isolation method and switching control of multi-thruster have been discussed [18,19].

UAVs and UUVs have broadened the space of operation from land to air or underwater, but both can still run only in a single medium. Unlike these unmanned systems, the UAUVs have attracted much attention and the existing literature mainly focuses on the mechanical structure design and cross-medium control [20,21]. The work in [22] focused on the modeling and trajectory tracking control of a special class of air-underwater vehicles with full torque actuation and a single thrust force directed along the vehicle's vertical axis. Furthermore, the work in [23] presented a robust switching control for stabilizing the attitude of a hybrid UAUV, and discussed the effects of buoyancy force and added inertia. A hybrid controller was designed for trajectory tracking considering the full system, including a transition strategy to assure switching between mediums. Stability analysis for the full system was provided using hybrid Lyapunov and invariance principles [24]. Based on the adaptation of typical platforms for aerial and underwater vehicles, the architecture for this kind of quadrotor-like vehicle was evaluated to allow the navigation in both environments [25].

At present, there are still many difficulties in the applications of UAUVs. Firstly, the structures of most existing vehicles are too complicated. In order to realize the movement in the air and in the water, most of them adopt variable wings or install two kinds of water-air wings directly. They are switched by mechanical equipment, and the complex mechanical mechanisms undoubtedly greatly increase the complexity of system control. Furthermore, the environment in water and in air differs greatly, and the dynamical and kinematical characteristics differ greatly, and the modeling and control methods cannot be single and are more challenging due to their cross-medium application.

The main contribution of this work is threefold:

(1) A four-rotor vehicle with adjustable arms is designed to be used in the switching operation of propulsion direction. The simple mechanism makes the vehicle more flexible and reliable. The use of only a single set of aerial rotors in both mediums greatly reduces cost, weight, and complexity compared to the other aerial underwater vehicle.

(2) Mathematical models of the UAUV are deduced, the continuous dynamics are modeled by the Newton–Euler formalism, taking into account the effects of some additional variables, such as water resistance, buoyancy and their corresponding moments, normally neglected in aerial vehicles. Furthermore, a critical coefficient for air-to-water switching is presented to model the changed mass, force and moment in the cross-medium motion process.

(3) Robust sliding mode switching controllers are designed for effectively handling the cross-medium motion and achieving precise trajectory tracking. Finally, as a proof of concept, some simulation results for the trajectory tracking control are provided for the cross-medium unmanned vehicle.

The rest of this article is organized as follows. The second section introduces the reference frames and presents the modeling preliminaries of multi-rotor vehicles. The third section investigates the dynamics of cross-medium UAUVs according to the Newton–Euler laws. Section 4 designs a sliding mode control method and analyzes the switching control problems. Section 5 simulates the unmanned system and analyzes the attitude and position regulation of the vehicle. Finally, Section 6 gives a short conclusion and future work of the UAUVs.

2. Preliminaries

In this section, the mathematical model for the UAUV is illustrated. This model is basically obtained by representing the rotor-vehicle as a non-rigid body with adjustable rotor direction evolving in six dimensional space. Our aim is to provide the hybrid dynamic models of the cross-medium vehicle working in the air, air-to-water and underwater. Thus, the system dynamics can also be decomposed into three categories, detailed in Section 3. At the same time, the Newton–Euler equations are taken into account.

2.1. Reference Frames

This section first introduces the vehicle's mechanical structure and its operation principle. Traditional hybrid vehicles have a multi-rotor structure, but the rotors are divided into two groups, which are used as power sources in the air and in the water respectively. Inspired by this model, a novel air-underwater configuration with only one set of rotors is proposed, shown in Figure 1.

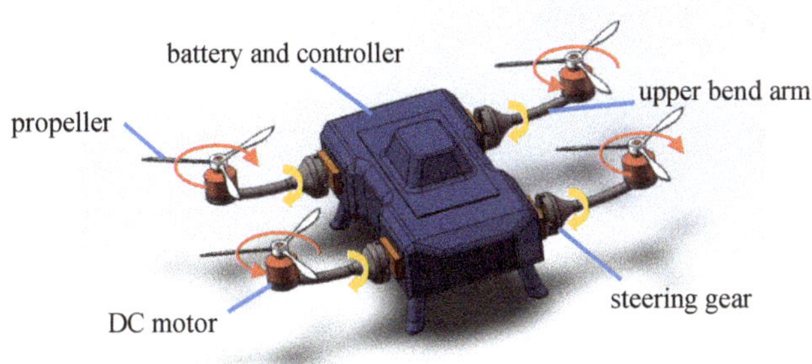

Figure 1. Configuration of the aerial underwater vehicle.

The vehicle flies in the air like a four-rotor aircraft. However, in underwater navigation, the four rotors reverse 180 degrees through the steering gear, forming upward thrust to overcome the buoyancy of the vehicle underwater, and change the motion attitude of the vehicle by controlling the thrust of the four rotors, thus changing the trajectory of the vehicle underwater. It presents simple mechanical structure and mathematical model, good maneuverability and controllability and allows hovering.

In order to accurately describe the movement of the vehicle and establish a mathematical model of the vehicle, the body frame $O_b - x_b y_b z_b$ and the inertial frame $O_e - x_e y_e z_e$ are defined as shown in Figure 2. Note that O_e is an East-North-Up (ENU) coordinate system and the axes of x_e, y_e and z_e point to the east, north and upper directions respectively. In order to express consistency for the vehicle motion, the definition of z-axis coordinates is upwards in both aerial and underwater cases in this paper. The position of the vehicle is written as $\xi = \begin{bmatrix} x & y & z \end{bmatrix}^T$ and the attitude angle of the vehicle is denoted as $\eta = \begin{bmatrix} \phi & \theta & \psi \end{bmatrix}^T$ in the inertial frame O_e. In addition, the angle's magnitude satisfies $|\eta| < \eta_L$ with the upper limit $\eta_L = \begin{bmatrix} \pi/2 & \pi/2 & 2\pi \end{bmatrix}^T$. This just means the vehicle can tilt the fuselage in motion, but the fuselage is not allowed to reverse. In the actual cases, the upper limits for the pitch and roll angles are usually less than $\pi/2$.

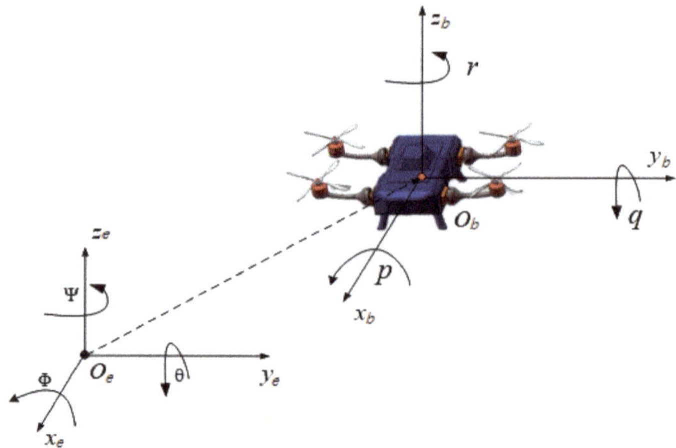

Figure 2. Frames of the aerial underwater vehicle.

This paper describes the attitude of the aerial underwater vehicle using the Euler angle, and the rotation matrices C_ϕ, C_θ and C_ψ in the angle ϕ, θ and ψ are defined as:

$$C_\phi = \begin{bmatrix} 1 & 0 & 0 \\ 0 & c_\phi & s_\phi \\ 0 & -s_\phi & c_\phi \end{bmatrix}, C_\theta = \begin{bmatrix} c_\theta & 0 & -s_\theta \\ 0 & 1 & 0 \\ s_\theta & 0 & c_\theta \end{bmatrix}, C_\psi = \begin{bmatrix} c_\psi & s_\psi & 0 \\ -s_\psi & c_\psi & 0 \\ 0 & 0 & 1 \end{bmatrix}$$

where the element $s_{\phi,\theta,\psi}$ are the sine function and $c_{\phi,\theta,\psi}$ are the cosine function of the rotation angles. Obviously, the transformation from the inertial coordinate system to the body coordinate system can be obtained as:

$$R_{eb} = C_\phi C_\theta C_\psi = \begin{bmatrix} c_\theta c_\psi & c_\theta s_\psi & -s_\theta \\ s_\theta s_\phi c_\psi - c_\phi s_\psi & s_\theta s_\phi s_\psi + c_\phi c_\psi & s_\phi c_\theta \\ s_\theta c_\phi c_\psi + s_\phi s_\psi & s_\theta c_\phi s_\psi - s_\phi c_\psi & c_\phi c_\theta \end{bmatrix} \tag{1}$$

It can be seen that R_{eb} is a direction cosine matrix expressed by the Euler angle of the unmanned vehicle.

2.2. Modeling of Multi-Rotor Vehicles

The mathematical models of four-rotor-driven vehicles are mostly described in the form of continuous differential equations. In general, it is assumed that the vehicle's center of gravity coincides with the body-fixed frame origin and each propeller rotates at the angular velocity Ω_i (where the subscript i denotes the serial number of rotor, and $i = 1, 2, 3, 4$). Consequently, a force F_i is generated along or opposite to the z-direction relative to the body frame and a reaction torque M_i is produced on the vehicle body by each rotor expressed as:

$$\begin{cases} F_i = k_T \Omega_i^2 \cdot \text{sign}(h), \\ M_i = (-1)^{i+1} k_Q \Omega_i^2. \end{cases} \quad i = 1, 2, 3, 4 \tag{2}$$

where the two positive coefficients $k_T = c_T \rho r^4 \pi$ and $k_Q = c_Q \rho r^5 \pi$ denote the aerodynamic coefficient and the drag coefficient of the rotor respectively, and their magnitudes depend on the environmental medium density ρ, the radius of the propeller r, the thrust factor c_T and torque factor c_Q of the designed rotors [26]. Furthermore, in the sign function, h denotes the vehicle altitude to the air–water interface.

The propellers represent the main source of thrust and can be designed specially in this hybrid configuration. Since the dynamic principle of the rotor structure is the same, aerial and aquatic propellers can present the same shape by optimizing the blade shape and size although the environments have different density. Therefore, the coefficients c_T, c_Q, the efficiency of engines and the energy delivered by the battery module have to be considered as the criteria to design the propeller shape (radius, width and curvature, et al.). Environmental density is directly proportional to the drag force and the added mass. Due to the large water density compared to the air density, it will generate high thrust forces with the same size and rotation speed to overcome the large movement resistance of the vehicles in the water. For convenience, the design process can be aided by the optimal propeller design software, such as OpenPro (OpenPro, Inc., Fountain valley, CA, USA), ShipPower (China state shipbuilding corporation limited, Shanghai, China).

The power system of the vehicle mainly includes rechargeable batteries, DC brushless motors and blades. Therefore, the magnitudes of the rotor's angular velocities decide the system's control inputs, which usually include the force and the torques created around a particular axis in body-fixed frame. In this section, the control signal can be defined as a vector $U = [\ U_1 \quad U_2 \quad U_3 \quad U_4\ ]^T$, and the components $U_{1\sim 4}$ denote the total thrust, the roll moment, the pitch moment and the yaw moment of the vehicle respectively. According to the dynamic principle of actuators, these control inputs can further be expressed by the following equation with the square vector of the rotor's angular velocities as:

$$U = K\left[\ \Omega_1^2 \quad \Omega_2^2 \quad \Omega_3^2 \quad \Omega_4^2\ \right]^T \tag{3}$$

where the coefficient K is a constant matrix for a symmetrical rigid multi-rotor vehicle. In addition, the arm's length is denoted as l, then the coefficient matrix K in (3) can be represented as the Kronecker product:

$$K = \begin{bmatrix} k_T \\ \sqrt{2}k_T l/2 \\ \sqrt{2}k_T l/2 \\ k_Q \end{bmatrix} \otimes \begin{bmatrix} 1 & 1 & 1 & 1 \\ 1 & -1 & -1 & 1 \\ 1 & 1 & -1 & -1 \\ 1 & -1 & 1 & -1 \end{bmatrix} \tag{4}$$

Consequently, the dynamic model of the vehicle in frame O_b can be obtained by the Newton's second law and theorem of moment of momentum as follows:

$$\begin{aligned} F_b &= m\dot{v}_b + m\omega_b \times v_b \\ M_b &= J\dot{\omega}_b + \omega_b \times J\omega_b \end{aligned} \tag{5}$$

where m denotes the mass, F_b and M_b denote the resultant force and moment acting on the multi-rotor vehicle, and $v_b = [\ u \quad v \quad w\ ]^T$ denotes the linear velocity and $\omega_b = [\ p \quad q \quad r\ ]^T$ denotes the angular velocity. It is often convenient to let the origin of the frame O_b coincide with the center of gravity, then the simplest form of the equations of motion is obtained when the body axes coincide with the principal axes of inertia. This implies that $J = diag[\ J_x \quad J_y \quad J_z\ ]^T$, and J_x, J_y, J_z are the inertia moments around x-axis, y-axis and z-axis respectively in the frame O_b. If this is not the case, the body-fixed coordinates can be rotated about their axes to obtain a diagonal inertia tensor by simply performing a principal axis transformation [27]. Then we have:

$$\begin{aligned} \dot{\xi} &= R_{eb}^T v_b \\ \omega_b &= \dot{\phi}I_1 + C_\phi\big(\dot{\theta}I_2 + C_\theta\big(\dot{\psi}I_3\big)\big) \end{aligned} \tag{6}$$

where $I_1 = \begin{bmatrix} 1 & 0 & 0 \end{bmatrix}^T$, $I_2 = \begin{bmatrix} 0 & 1 & 0 \end{bmatrix}^T$, $I_3 = \begin{bmatrix} 0 & 0 & 1 \end{bmatrix}^T$. Substituting C_ϕ, C_θ, C_ψ into the Equation (6), it also can be rewritten as:

$$\dot{\xi} = \begin{bmatrix} c_\theta c_\psi & s_\phi s_\theta c_\psi - c_\theta s_\psi & c_\phi s_\theta c_\psi + s_\phi s_\psi \\ c_\theta s_\psi & s_\phi s_\theta s_\psi + c_\phi c_\psi & c_\phi s_\theta s_\psi - s_\phi c_\psi \\ -s_\theta & s_\phi c_\theta & c_\phi c_\theta \end{bmatrix} v_b$$

$$\dot{\eta} = \begin{bmatrix} 1 & s_\phi t g_\theta & c_\phi t g_\theta \\ 0 & c_\phi & -s_\phi \\ 0 & s_\phi/c_\theta & c_\phi/c_\theta \end{bmatrix} \omega_b$$

After that, by deriving the above formula, the dynamics of the vehicle can be expressed as:

$$\ddot{\xi} = A_\xi(U_\xi + d_\xi) + B_\xi \tag{7}$$

$$\ddot{\eta} = A_\eta(U_\eta + d_\eta) \tag{8}$$

where $U_\xi = U_1$, $U_\eta = \begin{bmatrix} U_2 & U_3 & U_4 \end{bmatrix}^T$ are the control inputs, $d_\xi = \begin{bmatrix} 0 & 0 & d_z \end{bmatrix}^T$ and $d_\eta = \begin{bmatrix} d_\phi & d_\theta & d_\psi \end{bmatrix}^T$ denote the external disturbances for the force and the moment respectively, and the coefficient matrices A_ξ, B_ξ, A_η can be represented as:

$$A_\xi = \frac{1}{m} \begin{bmatrix} s_\phi s_\psi + c_\phi s_\theta c_\psi \\ -s_\phi c_\psi + c_\phi s_\theta s_\psi \\ c_\theta c_\phi \end{bmatrix}, \quad B_\xi = \begin{bmatrix} 0 & 0 & -g \end{bmatrix}^T, \quad A_\eta = \begin{bmatrix} c_\theta & c_\theta s_\phi & -c_\theta c_\phi \\ 0 & c_\theta c_\phi & -c_\theta s_\phi \\ 0 & s_\phi & c_\phi \end{bmatrix} J$$

In practical applications, the values of the position ξ and the attitude η can be measured by the sensors, such as Global Positioning Sysrtem (GPS), altimeter, gyroscope.

3. Dynamics of Cross-Medium Unmanned Aerial Underwater Vehicles (UAUVs)

By contrast with the single-medium operation environment, the cross-medium environment complicates the modeling analysis and stable continuous control of the system. The density ρ may jump aggressively from one value to another. In this case, the jump occurs when the vehicle's altitude h reaches the air-water interface, i.e., $h := 0$. In order to distinguish different external environments of the vehicle, a critical layer thickness of water surface is defined as ε, the value depends on the physical size of the vehicle and the model over-process requirements and $\varepsilon \to 0$. The mode switching for the cross-medium behavior of an aerial underwater vehicle is shown in Figure 3.

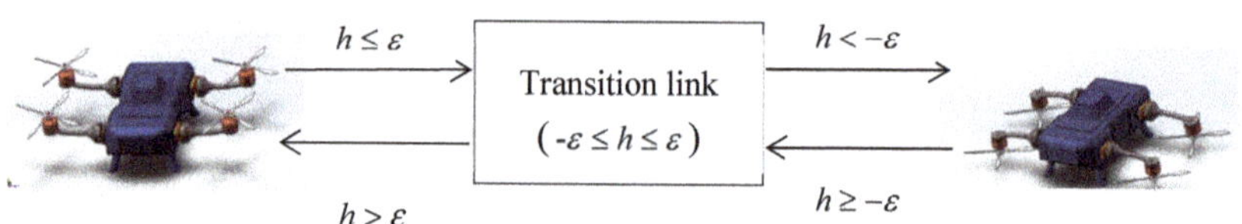

Figure 3. Mode switching for the cross-medium behavior of aerial underwater vehicle.

Therefore, operational modes of the cross-medium vehicles can be divided into three categories according to the current altitude value h, which denotes a relative altitude between the vehicle and the air–water interface.

3.1. Dynamics when $h > \varepsilon$

The aerial underwater vehicle is subjected to propeller thrust, air resistance, counter-torque generated by the propeller, and gyro moment generated by the attitude change during the movement. However, the air resistance and gyro moment can usually be ignored due to the small air density and small attitude change. Thus, only the propeller thrust, counter-torque and the gravity of the vehicle are considered during the vehicle's motion in the air.

According to the reference frame definition in Section 2, the thrust generated by the propellers is $F_m^b = \begin{bmatrix} 0 & 0 & k_T \sum_{i=1}^{4} \Omega_i^2 \end{bmatrix}$ and the direction is upward, same to the z_b axis. The motion equation of the vehicle in the inertial coordinate system can be derived as:

$$\begin{bmatrix} \ddot{x} \\ \ddot{y} \\ \ddot{z} \end{bmatrix} = \begin{bmatrix} \frac{k_T \left(c_\psi s_\theta c_\phi + s_\psi s_\phi\right)\sum_{i=1}^{4}\Omega_i^2}{m} \\ \frac{k_T \left(s_\psi s_\theta c_\phi - c_\psi s_\phi\right)\sum_{i=1}^{4}\Omega_i^2}{m} \\ \frac{k_T c_\theta c_\phi \sum_{i=1}^{4}\Omega_i^2 - mg}{m} \end{bmatrix} \tag{9}$$

The angular velocity of the vehicle in the inertial frame O_e:

$$\begin{bmatrix} \dot{\phi} \\ \dot{\theta} \\ \dot{\psi} \end{bmatrix} = \begin{bmatrix} p + p s_\phi t g_\theta + r c_\phi t g_\theta \\ q c_\phi - r s_\phi \\ q s_\phi \sec_\theta + r c_\phi \sec_\theta \end{bmatrix} \tag{10}$$

Therefore, in the actual control, the motor's speed output can be calculated. Finally, combined with the Formulas (3), (9) and (10), the motion model of the vehicle when $h > \varepsilon$ satisfies:

$$\begin{bmatrix} \ddot{x} \\ \ddot{y} \\ \ddot{z} \\ \ddot{\phi} \\ \ddot{\theta} \\ \ddot{\psi} \end{bmatrix} = \begin{bmatrix} \frac{U_1}{m}\left(c_\psi s_\theta c_\phi + s_\psi s_\phi\right) \\ \frac{U_1}{m}\left(s_\psi s_\theta c_\phi - c_\psi s_\phi\right) \\ \frac{U_1}{m}c_\theta c_\phi - g \\ \frac{\sqrt{2}l}{2}U_2 - \dot{\theta}(\omega_1 - \omega_2 + \omega_3 - \omega_4) \\ \frac{\sqrt{2}l}{2}U_3 + \dot{\phi}(\omega_1 - \omega_2 + \omega_3 - \omega_4) \\ \frac{k_Q}{k_T}U_4 \end{bmatrix} \tag{11}$$

3.2. Dynamics when $h < -\varepsilon$

The unmanned vehicle navigates in the water in a similar way to the air movement. The underwater vehicle is subjected to gravity, buoyancy, propeller thrust and resistance during movement. In addition, the moments subjected include heavy moments, floating moments, propeller thrust moments, resistance moments and gyro moments generated when the attitude changes.

Underwater vehicles are normally designed to satisfy the condition that the buoyancy force is slightly larger than the weight, which is important for safety reasons, and increasing weight or upward thrust is necessary for submersion. Therefore, in our configuration, the propeller arms of the vehicle will rotate by 180° after entering the water, so that the thrust of the propeller is reversed in an opposite direction to the z axis in the water, which is different from the convention used for some existing underwater vehicles.

In general, the motion of an underwater vehicle moving in six degrees of freedom (6DOF) at high speed will be highly nonlinear and coupled. However, in many underwater applications the vehicle will only be allowed to move at low speed, thus the Coriolis effects are smaller compared to the inertia effects of added mass, and the Coriolis effects of added mass are not considered in this paper. In addition, added mass can be seen as pressure-induced forces and moments due to a forced harmonic motion of the body which are proportional to the acceleration of the body. Since the movement of the vehicle in the water is also affected by the added mass, and the vehicle is approximately considered symmetrical, then the added mass along the three axes of the vehicle is only related to the acceleration and angular acceleration along the respective axes, and it can be expressed as:

$$m_k = m + diag\begin{bmatrix} m_{\dot{u}} & m_{\dot{v}} & m_{\dot{w}} & m_{\dot{p}} & m_{\dot{q}} & m_{\dot{r}} \end{bmatrix} \tag{12}$$

where m_k represents the mass of the object in motion, $m_{\dot{u}}, m_{\dot{v}}, m_{\dot{w}}$ represent the added mass of axial motion in water, and $m_{\dot{p}}, m_{\dot{q}}, m_{\dot{r}}$ represent the added mass of rotating motion in water.

In the inertial frame O_e, the gravity of the vehicle in the water is in the opposite direction to the z-axis, but the buoyancy direction is the same as the z-axis. Furthermore, the matrix expression of the gravity and the buoyancy are converted to the body frame O_b as:

$$\left[\begin{array}{cc} F_g^b & F_b^b \end{array}\right] = R_{eb} = \left[\begin{array}{cc} 0 & 0 \\ 0 & 0 \\ -m_k g & \rho_w g V \end{array}\right] = \left[\begin{array}{cc} m_k g s_\theta & -\rho_w g V s_\theta \\ -m_k g s_\phi c_\theta & \rho_w g V s_\phi c_\theta \\ -m_k g c_\theta c_\phi & \rho_w g V c_\theta c_\phi \end{array}\right] \tag{13}$$

where ρ_w is the density of water and V is the effective volume of the vehicle. Due to the high density of water, the resistance of the vehicle when moving in water cannot be ignored. In the frame O_b, the water flow resistance force F_{rw}^b can be expressed as:

$$F_{rw}^b = 0.5\rho_w C_{dw} S |v_b| v_b \tag{14}$$

where C_{dw} is the dimensionless drag coefficient (for most underwater robots, $C_{dw} = 0.8 \sim 1.0$), $S = diag\left[\begin{array}{ccc} S_x & S_y & S_z \end{array}\right]$ is the area in the direction of water flow velocity.

The thrust generated by the propellers is $F_{mw}^b = -F_m^b$, it's direction is contrary to the z_b axis. In summary, the external force F_w^b of the UAUV can then be represented as:

$$\begin{aligned} F_w^b &= F_g^b + F_b^b + F_{rw}^b + F_{mw}^b \\ &= \left[\begin{array}{c} (m_k g - \rho_w g V)s_\theta + 0.5\rho_w C_{dw} S_x |u| u \\ -(m_k g - \rho_w g V)s_\phi c_\theta + 0.5\rho_w C_{dw} S_x |v| v \\ -(m_k g - \rho_w g V)c_\theta c_\phi + 0.5\rho_w C_{dw} S_x |w| w - k_T \sum_{i=1}^4 \Omega_i^2 \end{array}\right] \end{aligned} \tag{15}$$

The aerial underwater vehicle changes its attitude in the water by its torque. Since the origin of the body coordinate system is located at the center of gravity of the vehicle, the gravity does not generate the moment. The buoyancy center coordinates are assumed to be located on the z_b axis in the frame O_b, and the buoyancy center coordinates are $\left[x_f, y_f, z_f\right]^T$ in the frame O_e. Then the buoyancy moment satisfies:

$$M_b^b = -F_b^b \left[\begin{array}{c} z_f s_\phi c_\theta - y_f c_\phi c_\theta \\ x_f c_\phi c_\theta + z_f s_\theta \\ -y_f s_\theta - x_f s_\phi c_\theta \end{array}\right] \tag{16}$$

where F_b^b is the buoyancy of the vehicle underwater in the body coordinate system.

The thrust torque generated by the propeller in the water is similar to that in the air, but the thrust torque generated by the propeller in the water is opposite to that in the air due to the 180° rotation of the propeller, i.e., $M_{mw}^b = -M_m^b$ show in the 2~4th rows of the Formula (4). In addition, the change of the moving attitude in the underwater vehicle will produce the gyro moment as:

$$M_{gyrow}^b = -\left[\begin{array}{c} -J_{rw} q(\Omega_1 - \Omega_2 + \Omega_3 - \Omega_4) \\ J_{rw} p(\Omega_1 - \Omega_2 + \Omega_3 - \Omega_4) \\ 0 \end{array}\right] \tag{17}$$

where J_{rw} is the moment of inertia of the underwater motor and propeller.

If the aerial underwater vehicle changes its attitude in the water, it will also be affected by the resistance moment, which is proportional to the square of the angular velocity of the aerial underwater vehicle as $M_{rw}^b = \left[\begin{array}{ccc} k_x |p| p & k_y |q| q & k_z |r| r \end{array}\right]^T$, where k_x, k_y and k_z are the resistance moment coefficients around the x_b, y_b and z_b axes in the body coordinate system. In summary, the external moment of the vehicle in the water can then be represented as:

$$M_w^b = M_{mw}^b + M_b^b + M_{rw}^b + M_{gyrow}^b$$
$$= \begin{bmatrix} -\frac{\sqrt{2}}{2}k_TL(\Omega_1^2 - \Omega_2^2 - \Omega_3^2 + \Omega_4^2) - k_x|p|p - F_b^b(z_1c_\theta s_\phi - y_1c_\theta c_\phi) + J_{rw}q(\Omega_1 - \Omega_2 + \Omega_3 - \Omega_4) \\ -\frac{\sqrt{2}}{2}k_TL(\Omega_1^2 + \Omega_2^2 - \Omega_3^2 - \Omega_4^2) - k_y|q|q - F_b^b(x_1c_\theta c_\phi - z_1s_\theta) + J_{rw}p(\Omega_1 - \Omega_2 + \Omega_3 - \Omega_4) \\ k_Q(\Omega_1^2 - \Omega_2^2 + \Omega_3^2 - \Omega_4^2) - k_z|r|r - F_b^b(-y_1s_\theta - x_1c_\theta s_\phi) \end{bmatrix} \tag{18}$$

If the aerial underwater vehicle fuselage is symmetrical about each coordinate system, the inertia matrix of its movement in the water can be set to $J_w = diag\begin{bmatrix} J_{xxw} & J_{yyw} & J_{zzw} \end{bmatrix}$. Similarly, the translational and rotational dynamics of the UAUV in water can be obtained by the Formulas (5), (15) and (18) as:

$$\begin{bmatrix} \dot{u} \\ \dot{v} \\ \dot{w} \\ \dot{p} \\ \dot{q} \\ \dot{r} \end{bmatrix} = \begin{bmatrix} \frac{1}{m_k}[(m_kg - \rho_w gV)s_\theta + 0.5\rho_w C_{dw}S_x|u|u] - wq + vr \\ \frac{1}{m_k}[-(m_kg - \rho_w gV)s_\phi + 0.5\rho_w C_{dw}S_y|v|v] - ur + wp \\ \frac{1}{m_k}[-(m_kg - \rho_w gV)c_\theta s_\phi + 0.5\rho_w C_{dw}S_z|w|w - k_T\sum_{i=1}^{4}\Omega_i^2] \\ \frac{1}{J_{xxw}}\sum M_{xw}^b + \frac{J_{yyw}-J_{zzw}}{J_{xxw}}qr \\ \frac{1}{J_{yyw}}\sum M_{yw}^b + \frac{J_{zzw}-J_{xxw}}{J_{xxw}}rp \\ \frac{1}{J_{zzw}}\sum M_{zw}^b + \frac{J_{xxw}-J_{yyw}}{J_{zzw}}pq \end{bmatrix} \tag{19}$$

3.3. Dynamics when $-\varepsilon \leq h \leq \varepsilon$

For water and air cross-medium unmanned vehicles, the key processes include air-water transition (surface landing) and water-air transition (surface takeoff). How to achieve a stable and reliable conversion between the water and the air is significant. The unmanned vehicle can splash into water, this method has certain advantages in both the water entering time and the falling distance. It is a suitable way for the air–water transition of the amphibious cross-medium vehicle.

The unmanned vehicle is in the semi-submersible dynamic process when $-\varepsilon \leq h \leq \varepsilon$. This kind of vehicle resolves the problem of sailing across the different mediums. The additional variables, such as water resistance, buoyancy, resistance moment, buoyancy moment and gyro moment, cannot be neglected. They are modeled for the completely underwater vehicle in Section 3.2. However, the values of these variables are related to the vehicle altitude, which are not same to the underwater case.

In order to model the dynamics of the UAUV in the environmental transition stage, a critical coefficient of the additional variables is defined as:

$$k_s = \frac{1}{2}(1 - h/\varepsilon), \quad -\varepsilon \leq h \leq \varepsilon \tag{20}$$

Thus, the critical coefficient k is inverse proportional to the altitude h, and obviously its magnitude satisfies $k_s \in [0, 1]$. If the vehicle decreases to below the air-water interface, then k_s increases and the additional variables' function will augment gradually to the underwater case, and vice versa. Similarly, the dynamics can be deduced from the Formula (19) as:

$$\begin{bmatrix} \dot{u} \\ \dot{v} \\ \dot{w} \\ \dot{p} \\ \dot{q} \\ \dot{r} \end{bmatrix} = \begin{bmatrix} \frac{1}{m_k}[(m_kg - \rho_w gV)s_\theta + 0.5k_s\rho_w C_{dw}S_x|u|u] - wq + vr \\ \frac{1}{m_k}[-(m_kg - \rho_w gV)s_\phi + 0.5k_s\rho_w C_{dw}S_y|v|v] - ur + wp \\ \frac{1}{m_k}[-(m_kg - \rho_w gV)c_\theta s_\phi + 0.5k_s\rho_w C_{dw}S_z|w|w - k_T\sum_{i=1}^{4}\Omega_i^2] \\ \frac{1}{J_{xxw}}\sum k_sM_{xw}^b + \frac{J_{yyw}-J_{zzw}}{J_{xxw}}qr \\ \frac{1}{J_{yyw}}\sum k_sM_{yw}^b + \frac{J_{zzw}-J_{xxw}}{J_{xxw}}rp \\ \frac{1}{J_{zzw}}\sum k_sM_{zw}^b + \frac{J_{xxw}-J_{yyw}}{J_{zzw}}pq \end{bmatrix} \tag{21}$$

4. Controller Design

In this section, a non-linear control scheme is proposed, which provides the integrated control method of the air and underwater dynamics of the UAUV. The aerial underwater vehicle is a typical under-actuated system, and it faces a nonlinear and strong coupling environment. The advantages of sliding mode control are obvious, which is used to design the altitude controller and attitude controller of this unmanned system. Figure 3 shows the integrated control block diagram for the UAUV.

The deviations of altitude and attitude are the inputs of the SMC controller, and the controller outputs are the system inputs $U_{1\sim4}$. According to altitude h of the current vehicle distance from air-water interface, the dynamics of the multi-model system discussed in Section 3 is determined, and the angular velocities of the vehicle motor are calculated. Thus, a closed-loop nonlinear multi-model feedback control scheme is formed. The system outputs of the designed scheme mainly include the position and the attitude angles which can be measured directly.

In the above scheme depicted in Figure 4, the position loop is designed to attenuate the tracking position error which is expressed as $\xi_e = \begin{bmatrix} x_e & y_e & z_e \end{bmatrix}^T$. In addition, we substitute the position error ξ_e into the Equation (7) and neglect the external disturbance temporarily, the following resultant error equation is deduced as:

$$\ddot{\xi}_e = A_\xi U_1 + B_\xi \tag{22}$$

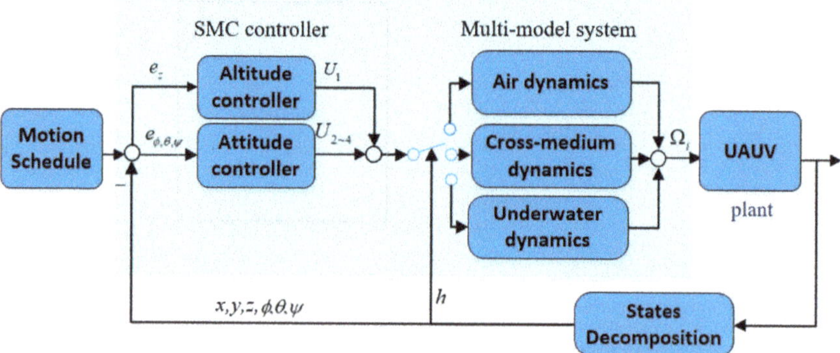

Figure 4. Aerial underwater vehicle control block diagram.

Thus, the altitude sliding manifold of the vehicle can be defined as:

$$\sigma_z = k_1^z z_e + k_2^z \dot{z}_e \tag{23}$$

where two coefficients k_1^z and k_2^z are constants. According to the SMC method, the system states are desired to remain on the manifold defined as $\sigma_z = 0$, and the error dynamics can be written as $k_1^z \dot{\xi}_e + k_2^z \ddot{\xi}_e = 0$. After that, we substitute the Equation (22) into this altitude second-order equation and get the equivalent control formula as:

$$U_1^{eq} = -\left(1/k_2^z\right) A_\xi^{-1} \left(k_1^z \dot{z}_e + k_2^z B_\xi\right) \tag{24}$$

So long as the system dynamics are accurately modeled, U_1^{eq} can maintain the altitude in the absence of external disturbances.

However, some unknown disturbances have to be considered in the real system, and the dynamics (22) can be represented as:

$$\ddot{\xi}_e = A_\xi(U_1 + d_\xi) + B_\xi \tag{25}$$

where d_ξ denotes the external disturbance mentioned before. To keep the system states still on the defined sliding mode manifold, a feedback control law of the system is redesigned as:

$$U_1 = -\left(1/k_2^z\right)A_\xi^{-1}\left(k_1^z \dot{z}_e + k_2^z B_\xi\right) - \lambda_z \cdot sat(\sigma_z) \tag{26}$$

where λ_z is a positive constant and $sat(\cdot)$ is a continuous saturation function defined as:

$$sat(x) = \begin{cases} -1 & x \leq -\delta \\ x/\delta & |x| < \delta \\ 1 & x \geq \delta \end{cases}, \quad \delta > 0. \tag{27}$$

where δ is the thickness of the boundary layer. For our sliding mode method, when the control error enters the given boundary layer, it will slip into the vicinity of zero in the boundary layer. Thus the saturation function is used to smooth out the control discontinuity around zero to reduce undesired chattering caused by imperfection switching of the discontinuous term. The performance of robustness is dependent on the thickness of the boundary layer, that means a thicker boundary layer may not be helpful to the robustness of the system.

In addition, the desired attitude can be calculated from the position errors, and the attitude error is obtained. After that, we consider the attitude loop used for attenuating the tracking attitude error $\eta_e = \begin{bmatrix} \phi_e & \theta_e & \psi_e \end{bmatrix}^T$, Substituting the error η_e into (8) and neglecting the environmental disturbance d_η, we get $\ddot{\eta}_e = A_\eta U_\eta$. Assume λ_η is a positive constant, the corresponding control law obtained can be given by:

$$U_\eta = -\left(k_1^\eta/k_2^\eta\right)A_\eta^{-1}\dot{\eta}_e - \lambda_\eta \cdot sat(\sigma_\eta) \tag{28}$$

where k_1^η, k_2^η are constant control parameters.

5. Simulations

In this section, the dynamics and the proposed control method are verified on a simulation platform of an autonomous UAUV with 4 rotors. The numerical parameters used in the simulation are a mixture of previous, similar projects and measurements of the real platform. The mass of the simulation vehicle is 1.2 kg, and the length of its arm is 0.19 m, with a small body-volume of 10^{-3} m^3, the maximum power of the motor is 202.8 W and the maximum rotational speed of the motor is 8500 rpm respectively. The detailed parameters used in the simulator are shown in Appendix A. In order to express clearly the work flow of the system, a flow diagram of the motion control process of the vehicle is presented in Figure 5.

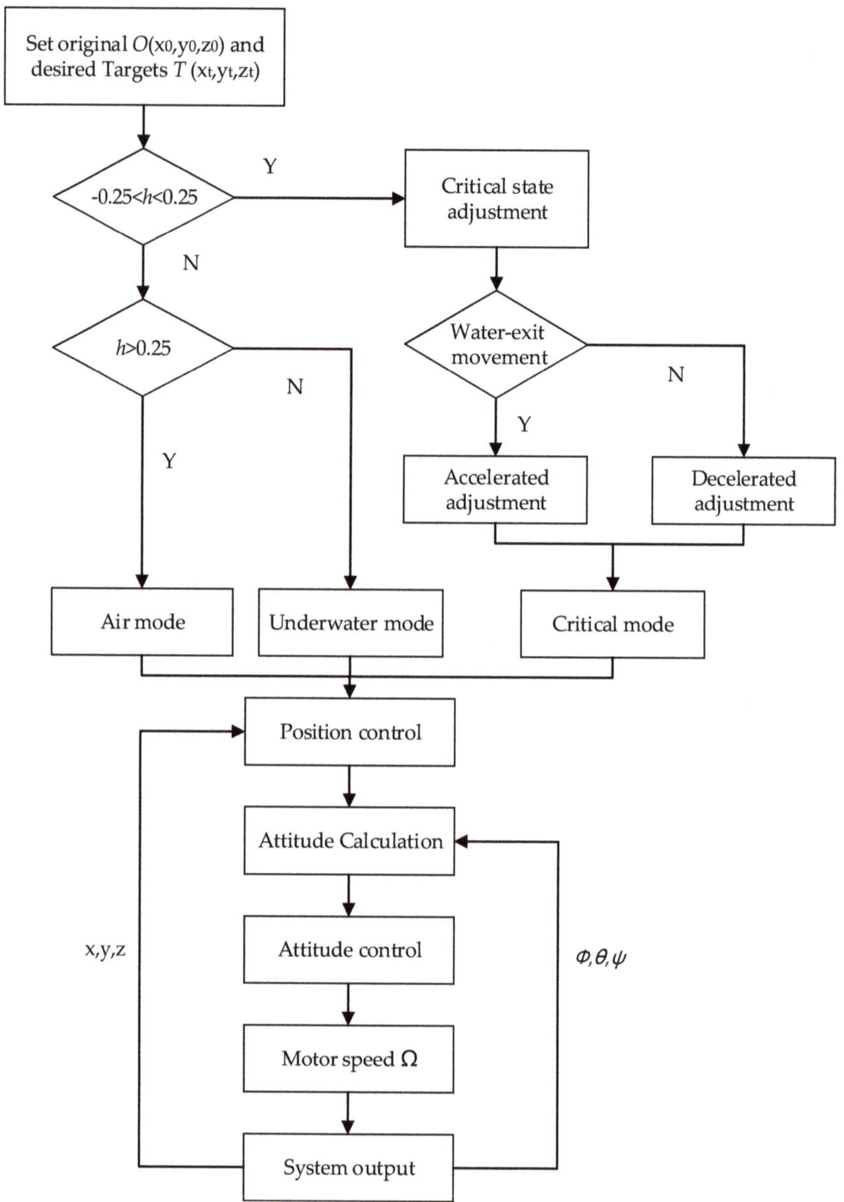

Figure 5. Diagram of the motion control process of the unmanned aerial underwater vehicle (UAUV).

At first we simulated the air position response and the underwater position response of the vehicle in the air and in the water single-medium environments in Sections 5.1 and 5.2, respectively. Then, to verify the cross-medium characteristics of the unmanned vehicle, a cross-media control motion was performed in Section 5.3. Consequently, the vehicle navigation behavior is mainly divided into three cases: the air case, the underwater case and the transition case. In our platform, a critical layer thickness ε of the air–water interface is set to 0.25 m. Then for the case $-0.25 < h < 0.25$, the vehicle is in the transition mode. The air–water motion triggers the deceleration adjustment of the four rotors, and the rotors will reverse 180 degrees by the steering gears to change the thrust directions. Otherwise, the water–air motion triggers the acceleration adjustment of the four rotors and reverses the rotor directions similarly. In addition, for the case $h > 0.25$ and $h < -0.25$, the vehicle operates in the air mode and in the underwater mode respectively. Furthermore, the sliding mode controllers are used so as to control the vehicle's position and attitude to infinitely close to the desired targets, and the desired attitude is calculated from the position errors of the vehicle, then the motor speeds are derived to drive the vehicle.

5.1. Air Position Response

In this case, we tested a straight-line flight control of the vehicle in the air. Set the original position $O(0, 0, 0)$ of the vehicle on the water surface and the target position at the inertia coordinates $T(5, 5, 5)$ in the air, and the route is a straight way from O to T. Consequently, the position response curve and attitude response curves can be obtained, shown in Figures 6 and 7, respectively.

It can be seen that the aerial underwater vehicle can reach the specified position in a short time. Furthermore, this illustrates that although there is an overshoot of about 10% during its movement, the specified position can be reached within 5 s, and the actual path followed by the vehicle is very close to the desired straight-line path. In addition, the peak values of roll angle and pitch angle is 12.6° and 11.5° respectively, while the adjustment of yaw angle is relatively small.

The control inputs U_i ($i = 1,2,3,4$) of the vehicle are shown in the following Figure 8. In the initial acceleration stage, the upward thrust force U_1 increases rapidly to about 40 N, then falls back gradually and stabilized on 12 N which exactly counteracts its own gravity and maintained hovering state after vehicle reaches the target point. Furthermore, the moments U_{2-4} of the three coordinate axes are also effective within the initial 5 s, which drive the vehicle to move to the target point by attitude adjustment during its ascending process. The outputs of the actuators for the system, i.e., the rotational speeds of the vehicle motors, are also shown in Figure 9.

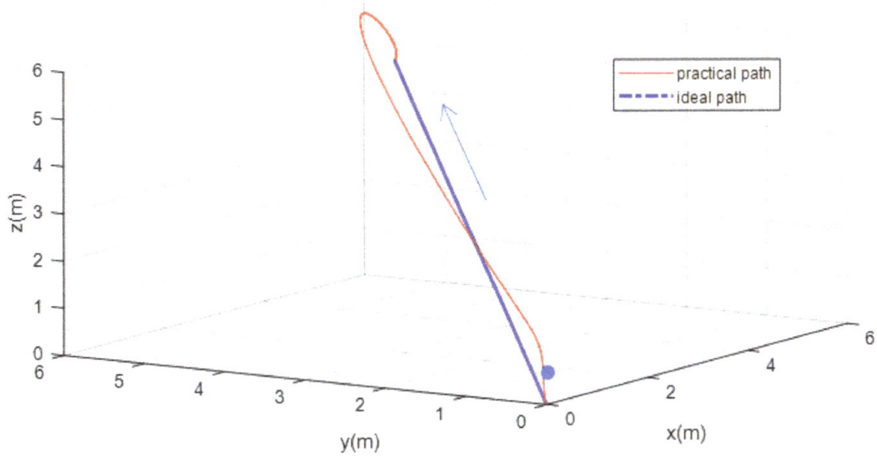

Figure 6. Three-dimensional position graphic of the aerial underwater vehicle in the air.

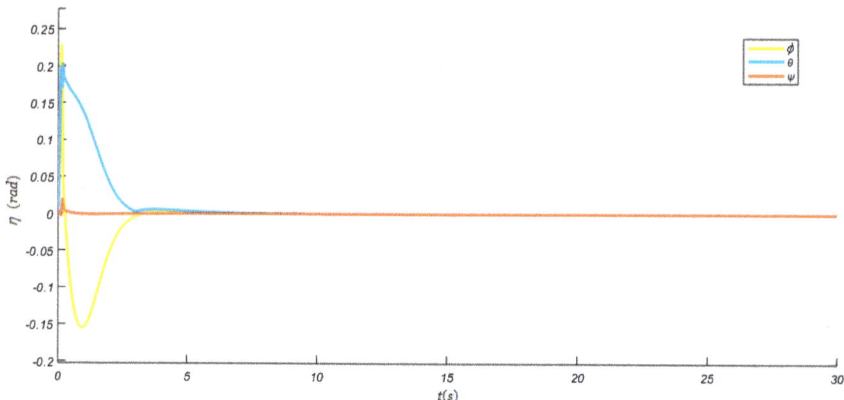

Figure 7. Attitude response curves of the aerial underwater vehicle in the air.

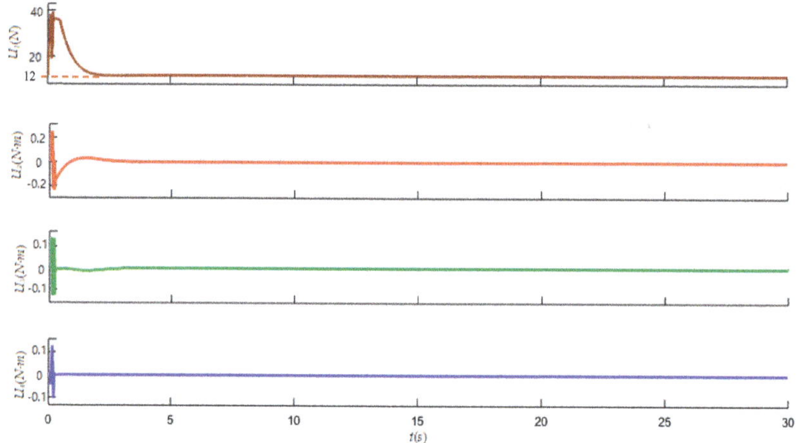

Figure 8. Control inputs of the vehicle (Thrust force U_1 and moments U_{2-4}).

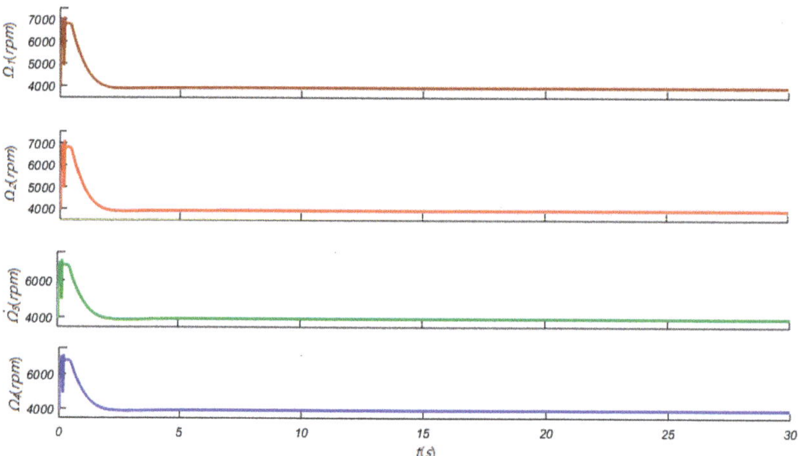

Figure 9. Rotational speed outputs of the vehicle motors.

It can be seen that the speeds of the motors increase rapidly to about 6800–7000 rpm, then decrease gradually to less than 4000 rpm and remain unchanged, which corresponds to the hovering state of the UAV. The speed differences of the four motors are not obvious because the upward thrust of the system is relatively large and the rotation moments are relatively small. When the UAV is in the vertical take-off and landing motion and hovering state, the four motors have the same speed.

5.2. Underwater Position Response

In this case, we tested a flight motion of the vehicle in the air. Set the original position $O(0, 0, 0)$ of the vehicle on the water surface and the underwater target position at the inertia coordinates $T(-1, -1, -1)$, and the route is a straight way from O to T. Consequently, the position response curve and attitude response curves can be obtained, as shown in Figures 10 and 11, respectively. It can be seen that the aerial underwater vehicle can reach the set point steadily in the water. Because the water density is 1000 kg/m^3, which is about 1000 times larger than the air density, it can produce high thrust with reduced motor speed. Furthermore, the peak values of roll angle and pitch angle are 16° and 24.2° respectively, while the adjustment of yaw angle is relatively small during the vehicle's movement in water as shown in Figure 11.

The control inputs U_i (i = 1,2,3,4) of the vehicle are shown in the following Figure 12. In the initial acceleration stage, the upward thrust force U_1 increases rapidly to about 30 N, then falls back gradually and stabilizes on 3.5 N which is used to counteract the effects of its own gravity, buoyancy and water resistance. Furthermore, the moments U_{2-4} of the three coordinate axes are also effective within the initial 5 s, which drive the vehicle to move to the target point by attitude adjustment during its descending process. The outputs of the actuators for the system, i.e., the rotational speeds of the vehicle motors are also shown in the Figure 13. It can be seen that the speeds of the motors increase rapidly to about 1410 rpm, then decrease gradually to less than 500 rpm and remained unchanged. However, the motor speeds are slower (one eighth) than those in the air.

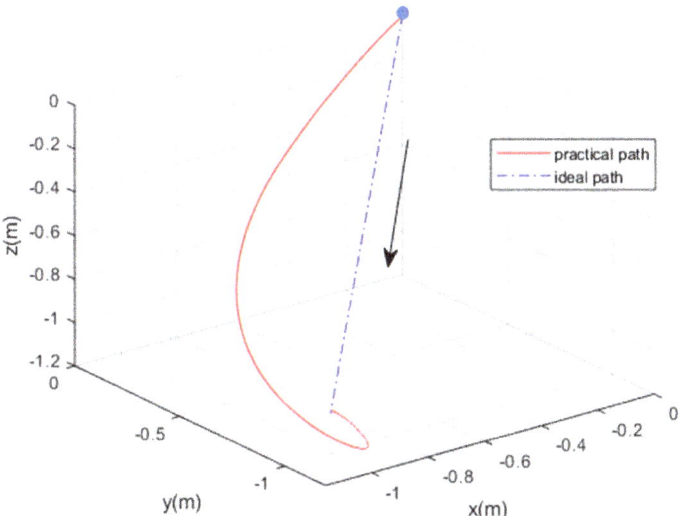

Figure 10. Three-dimensional position graphic of the aerial underwater vehicle underwater.

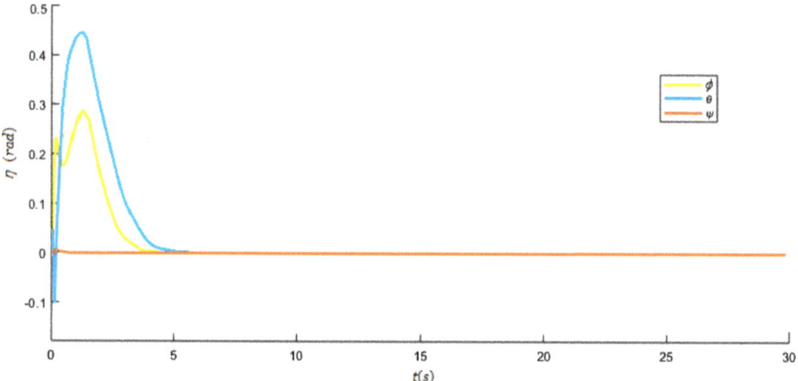

Figure 11. Attitude response curves of the vehicle underwater.

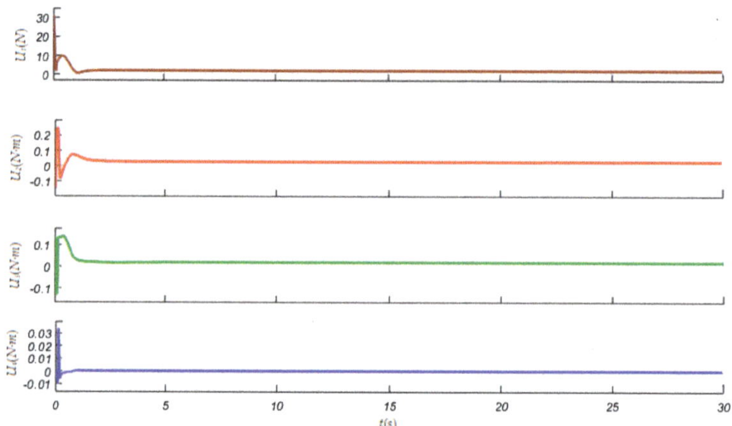

Figure 12. Control inputs of the vehicle (Thrust force U_1 and moments U_{2-4}).

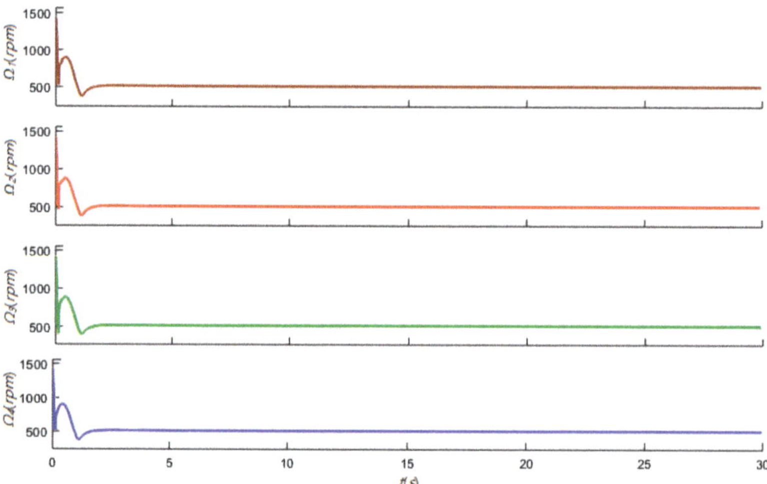

Figure 13. Rotational speed outputs of the vehicle motors.

5.3. Aerial Underwater Vehicle Cross-Media Response

In this case, we test a cross-medium motion near the air-water interface, and we set the desired motion trajectory: $x_d = \sin(0.5t)$, $y_d = \sin(0.25t)$, $z_d = \sin(0.25t) + 0.25 \cdot sign(z)$ when the altitude of the vehicle satisfies $|z| \geq 0.25$. Otherwise, the aerial underwater vehicle is in vertical lift transition. As can be seen from the simulation result of the three-dimensional cross-medium trajectory shown in Figure 14, the vehicle first flies an arc in the air, then lands and enters the water from the point O, and dives through an arc trajectory, returning to the vicinity of the starting point. Moreover, point O is the incidence point on the water surface, and point P_a and point P_b are the vehicle state adjustment points respectively, and the vehicle moves vertically when its altitude is within the range $[-0.25, 0.25]$. Therefore, the vehicle can travel across the medium and track the predetermined trajectory along the arrow directions.

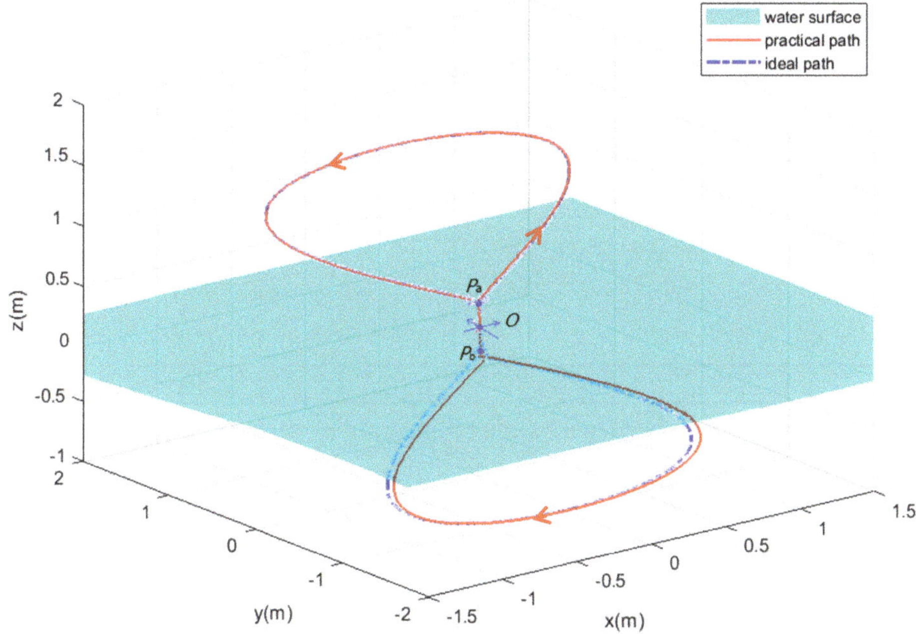

Figure 14. Three-dimensional cross-medium trajectory of the UAUV.

In addition, Figure 15 shows that the attitude angles are adjusted to be small when $t \leq 13$ s at initial stage. However, the roll angle and pitch angle fluctuate slightly at $t = 13$ s and $t = 35$ s, which corresponded to the vehicle's critical state adjustments of water-entering and water-exiting respectively. However, at $t = 22.5$ s, these two attitude angles suddenly increase to 0.4–0.5 rad and last for about 12.5 s, which corresponds to the vehicle's motion in water.

The control inputs U_i ($i = 1,2,3,4$) of the vehicle are shown in the following Figure 16. In the initial stage, the upward thrust force U_1 is adjusted and stabilized on 12 N until $t = 13$ s. Then, the thrust force decreases to near zero and the vehicle's arms are in the state of steering adjustment, and then the force gradually increases to about 16N at $t = 20$ s and lasts for a period of time. Thereafter, the reverse thrust force adjustment is made at $t = 35$ s, and the final effluent operation is completed. Furthermore, the moments U_{2-4} of the three coordinate axes increase at about $t = 20$ s and the maximum is not more than 1 Nm, which drives the vehicle to adjust its attitude during the underwater motion. The rotational speeds of the vehicle motors are also shown in Figure 17. The motor speeds are approximately 4020 rpm and 560 rpm for the air condition and underwater condition respectively. In addition, all the rotational speeds of the motor have to undergo a process of first deceleration and then acceleration in the process of critical state adjustment.

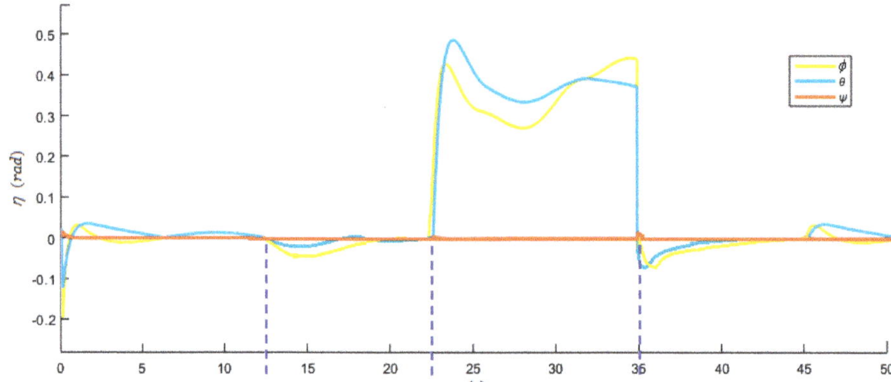

Figure 15. Cross-medium attitude response curves of the aerial underwater vehicle.

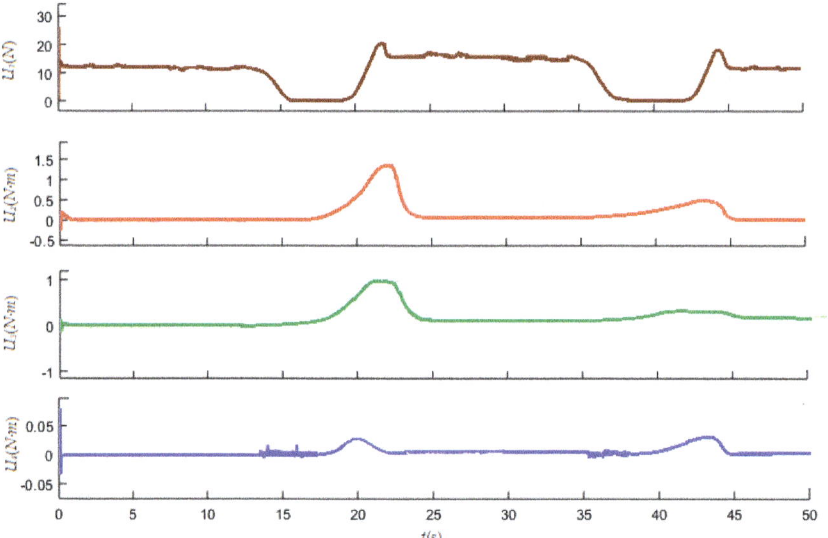

Figure 16. Cross-medium control inputs of the vehicle (thrust force U_1 and moments U_{2-4}).

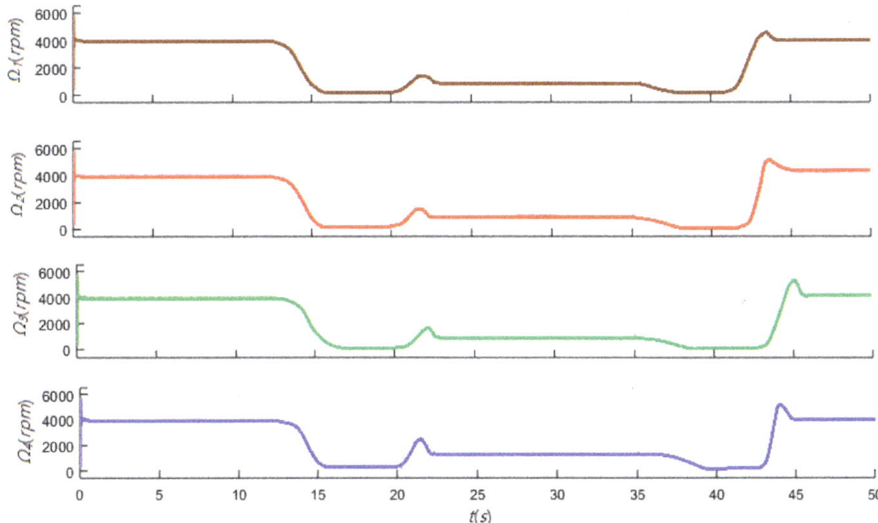

Figure 17. Rotational speed outputs of the vehicle motors.

6. Conclusions

This study evaluates the cross-medium motion of a quadrotor-like UAUV. The goal is to model the vehicle and simulate the autonomous control process both in the air and in the underwater environments. A hybrid vehicle configuration with only 4 rotors is proposed, and the thrust direction changes by the steering gears installed at the arm junctions. The simple mechanism makes the vehicle more flexible and reliable. The use of a single set of rotors in both mediums greatly reduces cost, weight, and complexity compared to other aerial underwater vehicle concepts. Without losing generality, the body frame and the inertial frame are defined to describe the position and attitude of the vehicle, and the transformation between these two frames is expressed as a direction cosine matrix of the vehicle's Euler angles.

The thrust force and the reaction torque generated by each rotor are proportional to the square of the propeller's angular velocities. Furthermore, the aerodynamic coefficient and the drag coefficient depend on the medium density, the radius of the propeller, the thrust and the torque coefficient. In our configuration, the aerial and aquatic propellers present the same shape, and the large water density generates high thrust forces with the same size and rotation speed to overcome the large movement resistance of the vehicles in the water. After that, the Quaternary control force and torque of the vehicle are defined. The system dynamics of the vehicle in the body frame are summarized and reconstructed as a second-order system based on Newton's second law and the theorem of the moment of momentum.

For the cross-medium motion the of aerial underwater vehicle, a critical layer thickness of the water surface is first defined and the dynamics model switches according to the vehicle's altitude h to the air–water interface. In our configuration, the thrust of the propeller is downward in the air to counteract the forces of gravity and acceleration, and is upward to submerge underwater. For the air environments, the air resistance and gyro moment are usually ignored to simplify the vehicle's dynamics expression due to the small air density and small attitude change. However, the buoyancy, the resistance and the corresponding resistance moments and gyro moments generated when the attitude changes have to be considered in the dynamics due to the large water density. Besides, the added mass is considered and its magnitude is related to the acceleration and angular acceleration along the respective axes. Then the external force and moment are derived and represented. In particular, for the air–water and water–air transitions, a critical coefficient is presented to model the changed mass, force and moment in the cross-medium motion process. In addition, the sliding mode control method is used to design the altitude controller and attitude controller of this unmanned system, and a continuous saturation function is performed to reduce undesired chattering caused by imperfection switching of the method. Finally, the deduced dynamics and proposed control scheme are tested on the simulation platform of an autonomous UAUV in different scenarios.

In future work, it is important to apply our modeling and control method to the real cross-medium experimental platform, and study how to improve the vehicle's maneuverability and stability, as well as investigating the influence of wind, wave, current and other disturbances on the system.

Author Contributions: Individual contributions include, conceptualization, Y.C.; methodology, Y.C. and Y.L.; software, Y.M.; validation, Y.L. and Y.M.; formal analysis, Y.C.; investigation, Y.C.; resources, Y.L.; writing—original draft preparation, Y.L.; writing—review and editing, S.Y. and Y.Z.; visualization, Y.M.; supervision, Y.Z.; project administration, Y.C.; funding acquisition, S.Y.

Funding: This work was supported in part by the Fundamental Research Funds for the Central Universities of China under Grand 3,132,019,318 and in part by the Natural Science Foundation of Liaoning Province under Grand 20,180,520,036.

Acknowledgments: The author would like to thank Haomiao Yu, Hui Li, Zhipeng Shen and Chen Guo from Dalian maritime University for editorial review. Two anonymous reviewers are gratefully acknowledged for comments which improved the manuscript.

Conflicts of Interest: The authors declare no conflict of interest.

Appendix A

This appendix presents the UAUV parameters of the simulation as following in Table A1, which includes the symbol of the quantities used in this work and their given numerical values.

Table A1. UAUV parameters used in simulation.

Symbol	Quantity	Numerical Value
m	Mass of platform	1.2 kg
r	Radius of the propeller	0.15 m
l	Length of arm	0.19 m
K_Q	Propeller torque coefficient in air	1.126×10^{-4}
c_T	Propeller lift coefficient in air	3.5×10^{-2}
c_{Tw}	Propeller lift coefficient in water	4.97×10^{-6}
K_{Qw}	Propeller torque coefficient in water	2.012×10^{-5}
g	acceleration of gravity	9.81 m/s^2
J_r	Motor and propeller moment of inertia	8.61×10^{-4} kg/m^2
ρ	Air density	1.29 kg/m^3
S_x	X direction area	1.05×10^{-2} m^2
S_y	Y direction area	1.96×10^{-2} m^2
S_z	Z direction area	4.2×10^{-2} m^2
J_x	Rotational inertia around the x-axis	2.365×10^{-2} kg/m^2
J_y	Rotational inertia around the y-axis	1.318×10^{-2} kg/m^2
J_z	Rotational inertia around the z-axis	1.318×10^{-2} kg/m^2
ρ_w	Water density	1000 kg/m^3
(x_f, y_f, z_f)	Buoyancy center coordinates	$(0, 0, -0.02)$
C_{dw}	No dimension resistance coefficient in water	0.9
k_x	Resistance coefficient of rotation around the x-axis in water	0.8
k_y	Resistance coefficient of rotation around the y-axis in water	1
k_z	Resistance coefficient of rotation around the z-axis in water	0.8
V	Volume	2×10^{-3} m^3

References

1. Hildreth, E. Capturing Images of a Game by An Unmanned Autonomous Vehicle. U.S. Patent 15/397,286, 4 April 2019.

2. Moud, H.I.; Shojae, A.; Flood, I. Current and Future Applications of Unmanned Surface, Underwater and Ground Vehicles in Construction. In Proceedings of the Construction Research Congress, New Orleans, LA, USA, 2–4 April 2018; pp. 106–115.

3. Kaub, L.; Seruge, C.; Chopra, S.D.; Glen, J.M.G.; Teodorescu, M. Developing an autonomous unmanned aerial system to estimate field terrain corrections for gravity measurements. *Lead. Edge* **2018**, *37*, 584–591. [CrossRef]

4. Zhong, Y.; Wang, X.; Xu, Y.; Wang, S.; Jia, T.; Hu, X.; Zhao, J.; Wei, L.; Zhang, L. Mini-UAV-Borne Hyperspectral Remote Sensing: From Observation and Processing to Applications. *IEEE Geosci. Remote Sens. Mag.* **2018**, *6*, 46–62. [CrossRef]

5. Soriano, T.; Pham, H.A.; Ngo, V.H. Analysis of coordination modes for multi-UUV based on Model Driven Architecture. In Proceedings of the 2018 12th France-Japan and 10th Europe-Asia Congress on Mechatronics, Tsu, Japan, 10–12 September 2018; IEEE: Piscataway, NJ, USA, 2018; pp. 189–194.

6. Morishima, S.; Hatano, M.; Sugeta, K.; Kozasa, T.; Momose, T.C.; Tani, T.; Ojika, H.; Matsuhashi, M.; Azuma, T. Design Method, Design Device, and Design Program for Cross-Sectional Shape of Fuselage. U.S. Patent 16/099,468, 21 March 2019.

7. Zhu, Z.; Guo, H.; Ma, J. Aerodynamic layout optimization design of a barrel-launched UAV wing considering control capability of multiple control surfaces. *Aerosp. Sci. Technol.* **2019**, *93*. [CrossRef]

8. Chen, Y.; Zhang, G.; Zhuang, Y.; Hu, H. Autonomous Flight Control for Multi-Rotor UAVs Flying at Low Altitude. *IEEE Access* **2019**, *7*, 42614–42625. [CrossRef]

9. Basri, M.A.M. Robust backstepping controller design with a fuzzy compensator for autonomous hovering quadrotor UAV. *Iran. J. Sci. Technol. Trans. Electr. Eng.* **2018**, *42*, 379–391. [CrossRef]

10. Owen, M.; Beard, R.W.; McLain, T.W. Implementing dubins airplane paths on fixed-wing uavs. In *Handbook of Unmanned Aerial Vehicles*; Springer: Dordrecht, The Netherlands, 2014; pp. 1677–1701. [CrossRef]

11. Nguyen, H.V.; Chesser, M.; Koh, L.P.; Rezatofighi, S.H.; Ranasinghe, D.C. TrackerBots: Autonomous unmanned aerial vehicle for real-time localization and tracking of multiple radio-tagged animals. *J. Field Robot.* **2019**, *36*, 617–635. [CrossRef]

12. Yuksek, B.; Vuruskan, A.; Ozdemir, U.; Yukselen, M.A.; Inlhan, G. Transition flight modeling of a fixed-wing VTOL UAV. *J. Intell. Robot. Syst.* **2016**, *84*, 83–105. [CrossRef]

13. Kamal, A.M.; Serrano, A.R. Design methodology for hybrid (VTOL+ Fixed Wing) unmanned aerial vehicles. *Aeronaut. Aerosp. Open Access J.* **2018**, *2*, 165–176. [CrossRef]

14. MahmoudZadeh, S.; Powers, D.M.W.; Zadeh, R.B. Introduction to autonomy and applications. In *Autonomy and Unmanned Vehicles*; Springer: Singapore, 2019; pp. 1–15. ISBN 978-981-13-2244-0.

15. He, Y.; Zhu, L.; Sun, G.; Dong, M. Study on formation control system for underwater spherical multi-robot. *Microsyst. Technol.* **2019**, *25*, 1455–1466. [CrossRef]

16. Bian, J.; Xiang, J. QUUV: A quadrotor-like unmanned underwater vehicle with thrusts configured as X shape. *Appl. Ocean Res.* **2018**, *78*, 201–211. [CrossRef]

17. Aminur, R.B.A.M.; Hemakumar, B.; Prasad, M.P.R. Robotic Fish Locomotion & Propulsion in Marine Environment: A Survey. In Proceedings of the 2018 2nd International Conference on Power, Energy and Environment: Towards Smart Technology (ICEPE), Haryana, India, 2 July 2018; IEEE: Piscataway, NJ, USA, 2018; pp. 1–6.

18. Kemp, M. Underwater Thruster Fault Detection and Isolation. In Proceedings of the AIAA Scitech 2019 Forum, San Diego, CA, USA, 7–11 January 2019; p. 1959. [CrossRef]

19. Jin, S.; Kim, J.; Kim, J.; Seo, T. Six-degree-of-freedom hovering control of an underwater robotic platform with four tilting thrusters via selective switching control. *IEEE/ASME Trans. Mechatron.* **2015**, *20*, 2370–2378. [CrossRef]

20. Alzu'bi, H.; Akinsanya, O.; Kaja, N.; Mansour, L.; Rawashdeh, O. Evaluation of an aerial quadcopter power-plant for underwater operation. In Proceedings of the 2015 10th International Symposium on Mechatronics and its Applications (ISMA), Sharjah, UAE, 8–10 December 2015; IEEE: Piscataway, NJ, USA, 2015; pp. 1–4. [CrossRef]

21. Maia, M.M.; Mercado, D.A.; Diez, F.J. Design and implementation of multirotor aerial-underwater vehicles with experimental results. In Proceedings of the 2017 IEEE/RSJ International Conference on Intelligent Robots and Systems (IROS), Vancouver, BC, Canada, 24–28 September 2017; IEEE: Piscataway, NJ, USA, 2017; pp. 961–966.

22. Mercado, D.; Maia, M.; Diez, F.J. Aerial-underwater systems, a new paradigm in unmanned vehicles. *J. Intell. Robot. Syst.* **2019**, *95*, 229–238. [CrossRef]
23. Neto, A.A.; Mozelli, L.A.; Drews, P.L.J.; Campos, M.F.M. Attitude control for an hybrid unmanned aerial underwater vehicle: A robust switched strategy with global stability. In Proceedings of the 2015 IEEE International Conference on Robotics and Automation (ICRA), Seattle, WA, USA, 26–30 May 2015; IEEE: Piscataway, NJ, USA, 2015; pp. 395–400.
24. Ravell, D.A.M.; Maia, M.M.; Diez, F.J. Modeling and control of unmanned aerial/underwater vehicles using hybrid control. *Control Eng. Pract.* **2018**, *76*, 112–122. [CrossRef]
25. Drews, P.L.J.; Neto, A.A.; Campos, M.F.M. Hybrid unmanned aerial underwater vehicle: Modeling and simulation. In Proceedings of the 2014 IEEE/RSJ International Conference on Intelligent Robots and Systems, Chicago, IL, USA, 14–18 September 2014; IEEE: Piscataway, NJ, USA, 2014; pp. 4637–4642.
26. Shi, D.; Dai, X.; Zhang, X.; Quan, Q. A practical performance evaluation method for electric multicopters. *IEEE/ASME Trans. Mechatron.* **2017**, *3*, 1337–1348. [CrossRef]
27. Fossen, T.I. *Guidance and Control of Ocean Vehicles*; Wiley: New York, NY, USA, 1996; pp. 28–29.

Journal of
Marine Science and Engineering

Article

A Novel Obstacle Localization Method for an Underwater Robot Based on the Flow Field

Xinghua Lin [1], Jianguo Wu [1,*] and Qing Qin [2]

[1] School of Mechanical Engineering, Hebei University of Technology, Tianjin 300401, China; sslxh2009@126.com

[2] China Automotive Technology and Research Center Co., Ltd, Tianjin 300300, China; sxyqqq@sina.com

* Correspondence: wujianguo@hebut.edu.cn

Received: 2 October 2019; Accepted: 21 November 2019; Published: 29 November 2019

Abstract: Because the underwater environment is complex, autonomous underwater vehicles (AUVs) have difficulty locating their surroundings autonomously. In order to improve the adaptive ability of AUVs, this paper presents a novel obstacle localization strategy based on the flow features. Like fish, the strategy uses the flow field information directly to locate the object obstacles. Two different localization methods are provided and compared. The first method, which is named the Method of Spatial Distribution (MSD), is based on the spatial distribution of the flow field. The second method, which is named the Method of Amplitude Variation (MAV), is provided by the amplitude variation of the flow field. The flow field around spherical targets is obtained by a numerical method, and both methods use the parallel velocity component on the virtual lateral line. During the study, different target numbers, detective ratios, spacing ratios, and flow velocities are taken into account. It is demonstrated that both methods are able to locate object obstacles. However, the prediction accuracy of MAV is higher than that of MSD. That implies that MAV is more robust than MSD. These new findings indicate that the object obstacles can be directly located based on the flow field information and robust flow sensing is perhaps not based on the spatial distribution of the flow field but rather, on its fluctuation range.

Keywords: artificial lateral system; flow sensing; object obstacle avoidance; underwater robot

---

1. Introduction

The autonomous underwater vehicle (AUV) is an important piece of equipment that is used to explore the ocean. However, the adaptive ability of the AUV is poor because the underwater environment is complex. In order to improve its environmental adaptability, studies have mainly focused on its control strategy [1–3]. However, the AUV is still not as adaptable as fish. One main reason for this is that the AUV has difficulty sensing its surroundings autonomously. Conventionally, electromagnetic sensors, sonars, and vision sensors are applied for perception. They can offer numerous advantages to help AUV localization, but their usage is also limited for a variety of reasons. For example, the sonars locate object obstacles based on the emission of sound waves, a process which has high energy consumption. Moreover, their medium to low resolution and high cost is the main limitation of their usage in underwater localization. In addition, the electromagnetic waves attenuate faster in seawater [4]. The vision sensors are overly dependent on light conditions, resulting in a greater power requirement and a limited operating scope [5–7]. Despite the exploitation of electromagnetic sensors, sonars, and vision sensors for successful object obstacle avoidance by underwater robots, the new method has obvious room for development because of the limitations.

Recently, some studies have proven that flow sensing can provide useful information about an AUV's surroundings [8,9]. Flow sensing has no relationship with illumination and noise conditions, and it can provide exteroceptive information in extreme environments [8,10,11]. The research on flow sensing based on lateral line mechanism can be divided into two parts. One is the study of ichthyology, and the other one is hydrodynamics.

Ichthyology mainly focuses on the structure of lateral organs and the behavior of fish after being stimulated. Based on the biologists' studies of the lateral line system [12–14], some researchers have designed many artificial lateral line sensors. Such designs employ a variety of sensing principles, such as piezoresistive [15], capacitive [16], thermal [17], magnetic [18], piezoelectric [19], and optical techniques [20]. Hydrodynamics mainly focuses on the characteristics of the flow field around the target [21–24]. Research methods mainly include theoretical analysis and numerical simulations. Among the theoretical analysis methods, the potential flow theory has been used widely [25–27]. However, the potential flow theory is based on the assumption of idealized fluid, and the details of the flow field cannot be described accurately. With the development of computer technology, computational fluid dynamics (CFD) has become the main hydrodynamics method [28–33]. Based on these studies, we determined the structures of neuromasts and the flow fields around different targets. Because there is a wide academic gap between the ichthyology and hydrodynamics, we cannot determine the mechanism of the lateral line system used by the fish. For example, the specific stimuli information obtained by fish and the sensing processes based on this information are unknown. If we cannot build a bridge between ichthyology and hydrodynamics, the AUV will not be able to use lateral line sensors (LLS) to sense the environment [34].

In recent years, flow sensing has been introduced into the detection of objects. However, the most common method is the detection and localization of the dipole source [35–37]. One main method is the detection of Karman vortex streets (KVS) [38,39]. Another technique is the study of multisignal fusion and detection algorithms, which are based on the neural network scheme and other algorithms [40,41]. However, the excessive computing and storage resources and the difficulty of using these approaches for system level implementation mean that they cannot be used in engineering.

In this paper, we present a new flow sensing method for target localization. The relationship between the flow features and object location is investigated. The complete localization technique only uses flow velocity data as input parameters, and this technique has never been reported. This novel obstacle localization strategy uses the spatial distribution and amplitude variation of the flow field as flow features. An extensive comparative analysis between the two methods is carried out. The evaluation aims to study the differences between the methods—the Method of Spatial Distribution (MSD) and the Method of Amplitude Variation (MAV). The ultimate goal is to find a simple and effective method to locate the object obstacles based on the flow features.

The paper is organized as follows: Section 2 describes the background of the system model and the numerical method used to obtain the flow field, and the validation of numerical method is also described in this section. The novel object localization strategy is described in Section 3, including descriptions of MSD and MAV. Additionally, when the detective ratio changes, a comparison between two methods is provided. Section 4 presents the case of two tandem targets and the evaluation results of the two methods when the spacing ratio and flow velocity change. Finally, concluding remarks and directions for future work are provided in Section 5.

2. Implementation of the Test Problem

2.1. System Model

The lateral line and the targets are located in a Cartesian coordinate. As depicted in Figure 1, the diameter of the target is H. In order to avoid the influence of the computational domain on the flow field around the targets, the size of the computation field is 10 times larger than the target. Because the two targets are arranged in tandem and in order to ensure the full development of the wake field, the size of computation field along the flow direction is 30 times larger than the target. Thus, the computational domain Ω is $30H \times 10H \times 10H$. In the case of two targets, L is the distance between them. The lateral line is under the center of the targets, and the interval between two LLS is $0.01H$. The influence of the LLS is negligible. The sensors are located at (x_i, y_i, z_i), and their range is shown in Equation (1):

$$\left\{ \begin{array}{l} -12H \leq x_i \leq 12H; \\ y_i = -D; \\ z_i = 0. \end{array} \right. \qquad , \qquad (1)$$

where H is the diameter of the target and D is the detective distance.

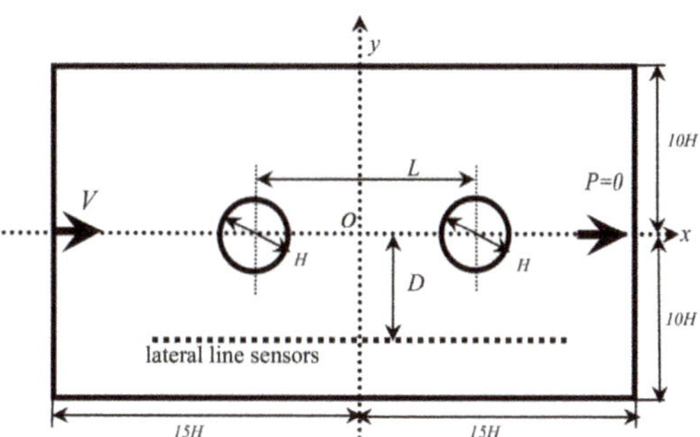

Figure 1. Schematic diagram of the computational domain in the XOY plane.

At the inlet boundary, a uniform flow is prescribed to be V. The outlet condition is a pressure outlet, and the initial pressure is zero. A no-slip boundary is set as the surface of the targets, and a symmetry condition is used on the lateral boundaries. In order to ensure that the dynamic characteristics between the fluid and targets can be embodied accurately, the computational domain is defined by $\Omega = \Omega_i + \Omega_e$. The domain Ω_i is the vicinity of targets, whose size is $24H \times 8H \times 8H$. The lateral lines used in this paper are located in this domain. Its grids are presented by employing an unstructured, triangular, and refined grid, whose size is $\Delta\chi$. Restricted by the computer resources, a grid increasing function is used to mesh the domain Ω_e by a hexahedral grid. The surfaces of domain Ω_i are used as the source, with an increasing ratio of 1.2 and a maximum size of 0.05H.

To ensure the solution's convergence, grid independence is studied. Three cases of the mesh dependency test are provided. When the residual is smaller than 1×10^{-6}, the solution can be seen as convergent. The minimum sizes $\Delta\chi_{min}$ of the three cases are 0.001H, 0.005H, and 0.01H, respectively. The drag coefficient C_d and its percentage changes are shown in Table 1. It is observed that C_d decreases while $\Delta\chi$ becomes smaller. However, there is no significant change between Grid-1 and Grid-2. Therefore, the mesh characteristics of Grid-2 were used in the investigations presented.

Table 1. Analysis of grid independence.

Case	$\Delta\chi_{min}$	Number of Elements	C_d	Percentage Changes/%
Grid-1	0.001H	7.94×10^6	0.1418	0.77
Grid-2	0.005H	1.33×10^6	0.1429	3.99
Grid-3	0.01H	0.47×10^6	0.1486	\

2.2. Numerical Method

The two-step Taylor-characteristic-based Galerkin method (TCBG) is used to solve the Navier–Stokes equation and the continuity equation, written in the Eulerian form as Equation (2) and Equation (3) [42]. The momentum equation must be split in the classical method, but that is not required in the current method. The momentum–pressure Poisson equation method is used to segregate the pressure from the calculation of velocity. The preferable accuracy of this method with less numerical dissipation was proven by Bao et al. [42].

$$\left(\frac{\partial u_i}{\partial t} + u_j \frac{\partial u_i}{\partial x_i}\right) = -\frac{\partial p}{\partial x_i} + \frac{1}{Re}\frac{\partial \tau_{ij}}{\partial x_j},$$

(2)

$$\frac{\partial u_i}{\partial x_i} = 0$$

(3)

where u_i is the i-component velocity, u_j is the convective velocity, t_i is the time, ρ is the water density, p is the pressure, and τ_{ij} is the deviatoric stress that is given by Equation (4):

$$\tau_{ij} = \mu\left(\frac{\partial u_i}{\partial x_i} + \frac{\partial u_j}{\partial x_j}\right),$$

(4)

where μ is the viscosity constant.

Based on the TCBG method, the discretizing process has three steps. Firstly, we can get $u_i^{n+1/2}$ from Equation (5), and then the relationship of p^{n+1} and u_i^{n+1} can be obtained from Equation (7). Finally, Equation (6) can be solved to get u_i^{n+1}:

$$u_i^{n+1/2} = u_i^n - \frac{\Delta t}{2}(u_j^n\frac{\partial u_i^n}{\partial x_j} + \frac{\partial p^n}{\partial x_i} - \frac{1}{Re}\frac{\partial \tau_{ij}^n}{\partial x_j}) + \frac{\Delta t^2}{8}u_k^n\frac{\partial}{\partial x_k}(u_j^n\frac{\partial u_i^n}{\partial x_i} + 2(\frac{\partial p^n}{\partial x_j} - \frac{1}{Re}\frac{\partial \tau_{ij}^n}{\partial x_j})),$$

(5)

$$u_i^{n+1} = u_i^n - \Delta t(u_j^{n+1/2}\frac{\partial u_i^n}{\partial x_j} + \frac{\partial p^{n+1}}{\partial x_i} - \frac{1}{Re}\frac{\partial \tau_{ij}^{n+1/2}}{\partial x_j}) + \frac{\Delta t^2}{2}u_k^{n+1/2}\frac{\partial}{\partial x_k}(u_j^{n+1/2}\frac{\partial u_i^n}{\partial x_i} + \frac{\partial p^{n+1}}{\partial x_i} - \frac{1}{Re}\frac{\partial \tau_{ij}^{n+1/2}}{\partial x_j}))$$

(6)

$$\frac{\partial^2 p^{n+1}}{\partial x_i \partial x_i} = \frac{1}{\Delta t}\frac{\partial}{\partial x_i}(u_i^n - u_i^{n+1}) - \frac{\partial}{\partial x_i}(u_j^{n+1/2}\frac{\partial u_i^{n+1/2}}{\partial x_j} - \frac{1}{Re}\frac{\partial \tau_{ij}^{n+1/2}}{\partial x_j})$$

(7)

where n, $n+1/2$, and $n+1$ denote the time points of t_n, $t_{n+1/2}$, and t_{n+1}, respectively.

2.3. Numerical Validation

In this section, the accuracy of the numerical algorithm is demonstrated and validated by theoretical and previous results.

In the analysis of theoretical validation, the velocity potential ϕ meets the Laplace equation, as shown in Equation (8). Because of its properties, the complex flow can be decomposed into several simple flows:

$$\frac{\partial^2 \phi}{\partial x_i^2} + \frac{\partial^2 \phi}{\partial y_i^2} + \frac{\partial^2 \phi}{\partial z_i^2} = 0,$$

(8)

In uniform flow, the flow field around a sphere can be considered to be a mixed flow of uniform flow and dipole flow. The interest velocity potential ϕ can be expressed as Equation (9), the flow velocity near the target can be shown as Equation (10), and the parallel velocity component $v(x)_{x,//}$ can be expressed as Equation (11) [43]. If we do not consider the influence of gravity and the component along the y-direction, the parallel velocity component $v(x)_{x,//}$ meets the requirements of the potential flow theory. For validation of the proposed method, flow past a single sphere is computed. When the detective distance is equal to $2H$, the parallel velocity on the lateral line is as shown in Figure 2. As seen in Figure 2, it is found that the numerical result shows satisfactory agreement with the theoretical result.

$$\phi = \phi_1 + \phi_2 = Vx_i + \frac{H^3(V \cdot r)}{2\|r\|^3} \cdot \omega$$

(9)

$$v(r) = V - \frac{H^3\big(3(V \cdot r) \cdot r - \|r\|^2 V\big)}{2\|r\|^5} \cdot \omega \tag{10}$$

$$v(x)_{x,//} = V - \frac{H^3 V \big(2(x_i - x_o)^2 - D^2\big)}{2\big\|(x_i - x_o)^2 + D^2\big\|^{5/2}} \cdot \omega \tag{11}$$

where V is the flow velocity, H is the spherical diameter, r is the Euclidean distance between the sensor and the target, $\|\cdot\|$ represents the vector's Euclidean norm, and ω is a dimensionless coefficient, which was 1.143 in this paper.

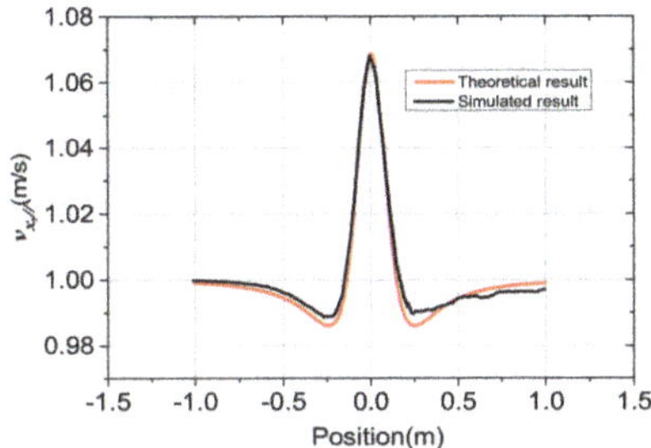

Figure 2. Comparison of the simulated parallel velocity and the theoretical result.

Moreover, Zhao et al. [44] studied the flow past a single square cylinder. The same case was studied by the above computational algorithm in this paper. The Strouhal number of square targets with different postures θ was computed. The comparison results are shown in Figure 3, and the numerical results are in accordance with previous reports. The computational algorithm is adequate to solve the flow field around the underwater target, which is confirmed by reasonable agreement between our numerical results and the previous data.

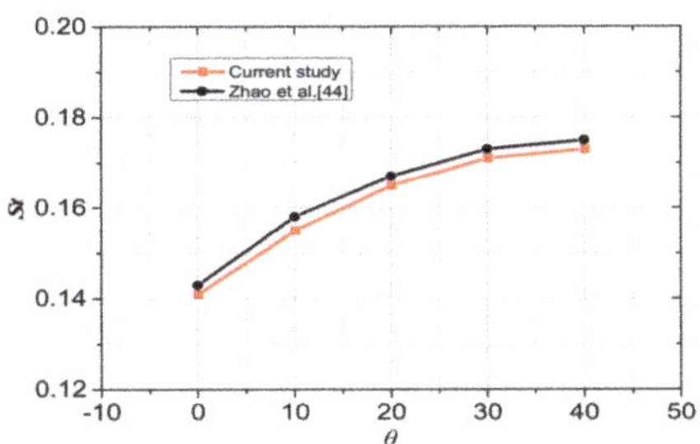

Figure 3. Comparison of simulated results and previous results.

3. Location Strategies

3.1. Method of Spatial Distribution (MSD)

As shown in Figure 2, we determined the parallel velocity distribution on the lateral line. It is clear that there are three extreme points on the curve. Using the operation of partial derivatives on the parallel component, the results are shown in Equation (12), and the abscissa of three extreme points are shown in Equation (13).

$$v(x)'_{x,//} = \frac{3H^3 V (x_i - x_o)\big(3D^2 - 2(x_i - x_o)^2\big)}{2\big\|(x_i - x_o)^2 + D^2\big\|^{7/2}} \cdot \omega, \tag{12}$$

$$\begin{cases} x_1 = x_0 - \frac{\sqrt{6}}{2}D; \\ x_2 = x_0; \\ x_3 = x_0 + \frac{\sqrt{6}}{2}D \end{cases} \tag{13}$$

As shown in Figure 2, the abscissa of the spherical center and the maximum point are equal, and the spatial variation of two minimum points is $\sqrt{6}D$, where D is equal to $\left\|y_i - y_0\right\|$. The results show that the position of a target can be estimated based on the spatial variation of the extremum points. The method is similar to the model presented by Dagamseh et al. [35], which was proven in the study of dipole source localization.

Because the spherical target is symmetrical, in order to discuss it conveniently, the velocity distribution hereinafter is in the XOY plane. As seen in Figure 4, the velocity in the area away from the target is similar to the inlet velocity, resulting in the positions of minimum points being non-obvious. From Figure 5, it is clear that the estimated error ε_1, which is shown in Equation (14), increases when the detective ratio increases. In other words, as the detective ratio increases, we cannot locate the target based on the MSD accurately. Therefore, the other more robust method is needed.

$$\varepsilon_i = \frac{|D - D_{est}|}{D} \times 100\%, \tag{14}$$

where D is the actual detective distance and D_{est} is the estimated detective distance. The i values are 1 and 2, and they represent the estimated errors of the previous method and new method, respectively.

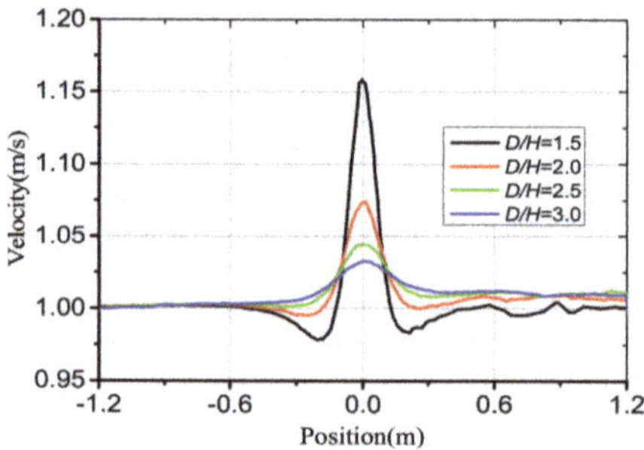

Figure 4. Parallel velocity on the lateral line at different detective ratios.

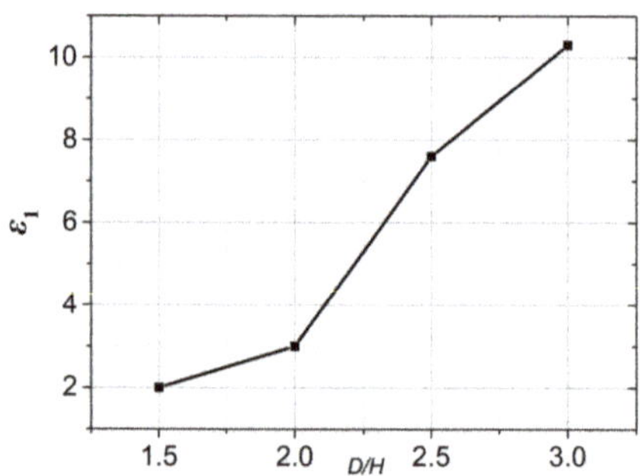

Figure 5. Estimated error at different detective ratios.

3.2. Method of Amplitude Variation (MAV)

Another interesting observation from Figure 4 is that the amplitude variation of velocity $\Delta_{max}v$ decreases when the detective ratio D/H increases. The amplitude variation $\Delta_{max}v$ is plotted in Figure 6 as a function of the detective ratio. As seen in Figure 6, the connection can be approximated by an exponent regression. The regression equation is shown in Equation (15) and the equation analog effect is good. This implies that the localization of targets can be approximated by the amplitude variation of extreme points in the velocity distribution.

$$\begin{cases} |\Delta_{\max}v| = m/(D/H)^n \\ |\Delta_{\max}v| = |v_{\max} - v_{\min}| \end{cases},\tag{15}$$

where m and n are the regression coefficients. $v_{\max}$ and $v_{\min}$ are the maximum velocity and the minimum velocity on the lateral line.

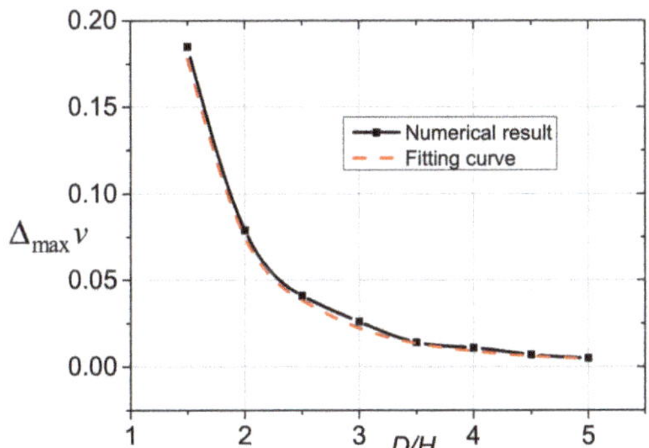

Figure 6. Velocity drop amplitude at different detective ratios and the fitting cure.

According to Equation (11), we can get $v_{\max}$ and $v_{\min}$, which are shown in Equation (16). The amplitude $\Delta_{\max}v$ of the velocity drop based on the potential flow theory is shown in Equation (17), whose form is similar to that of Equation (15). Compared with Equation (15), the coefficients m and n are $0.6V$ and 3.0, respectively, in this paper. As seen in Equation (17), the regression coefficients depend on the flow velocity.

$$\begin{cases} v_{\max} = V + \frac{V}{2}(D/H)^{-3} \cdot \omega \\ v_{\min} = V - \frac{\sqrt{10}V}{125}(D/H)^{-3} \cdot \omega \end{cases},\tag{16}$$

$$\Delta_{\max}v = \frac{\frac{(125 + 2\sqrt{10}) \cdot \omega \cdot V}{250}}{(D/H)^3}\tag{17}$$

In order to prove that the method is effective at different detective ratios, the velocity curves on more lateral lines were computed, and the detective ratio is estimated based on the MAV, as shown in Table 2. We can see that the estimated error of the MAV is less than 1%, and it is smaller than that of MSD obviously. This implies MAV is more robust than MSD.

Table 2. Estimated detective ratios compared with the actual values.

D/H	1.25	1.75	2.25	2.75	3.25
D_{est}/H	1.26	1.74	2.23	2.73	3.28
ε_2	0.80	0.63	0.89	0.73	0.92

4. Discussion of Tandem Targets

4.1. Influence of the Spacing Ratio

In this section, the case of double targets arrayed in tandem is presented. The influence of the spacing ratio L/H is taken into consideration. It should be pointed out that the amplitude of the velocity curve decreases when the spacing ratio is decreased, as shown in Figure 7. When the spacing ratio becomes bigger, the flow field is more stable. As seen in Figure 7, when the value of L/H is 4 and 6, the positions of two minimum points are not obvious. This implies that the estimated error will become bigger based on the MSD, as shown in Figure 8.

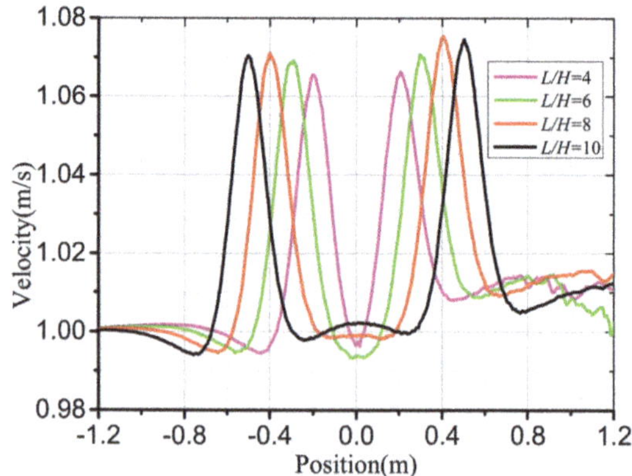

Figure 7. Parallel velocity on the lateral line at different spacing ratios.

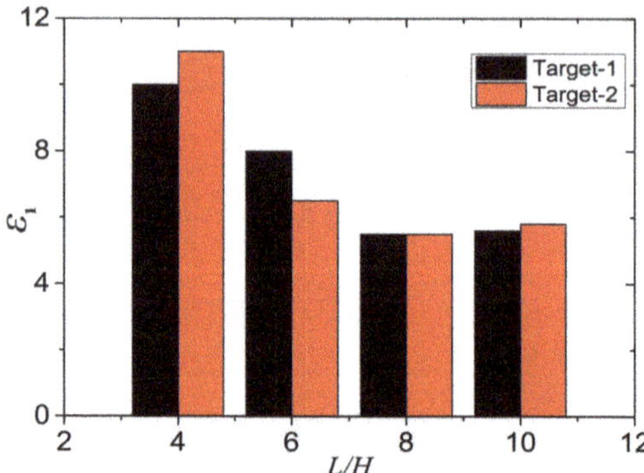

Figure 8. Parallel velocity on the lateral line at different spacing ratios.

We chose the minimum velocity in the upstream region of the target to calculate the velocity drop amplitude in MAV. Additionally, it is interesting that both the estimated errors are less than 1%. This means that MAV is effective for the case of double targets, and the spacing ratio has no influence on its prediction accuracy.

4.2. Influence of the Flow Velocity

In order to determine the influence of the flow velocity on the estimated results, the cases of different flow velocity are computed. Because the velocity range of most AUV is from 0.5 to 2.0 m/s, this range is used for the research considered. For all the cases in this section, the detective ratio and the spacing ratio are set as 2.0 and 10.0, respectively. The parallel velocity curve is treated with the dimensionless method, as shown in Equation (18):

$$V_{norm} = \frac{v}{V},$$

(18)

where V_{norm} is the dimensionless velocity, v is the velocity on the lateral line, and V is the inlet velocity.

As seen in Figure 9, the dimensionless velocity curves are plotted for various flow velocities. In this range, the change of velocity is not sufficient to change the velocity distribution on the lateral line. When the spacing ratio is enlarged to 10.0, the shear layers formulated from the upstream target are stable in the gap between the targets. Therefore, the flow patterns are similar to each other.

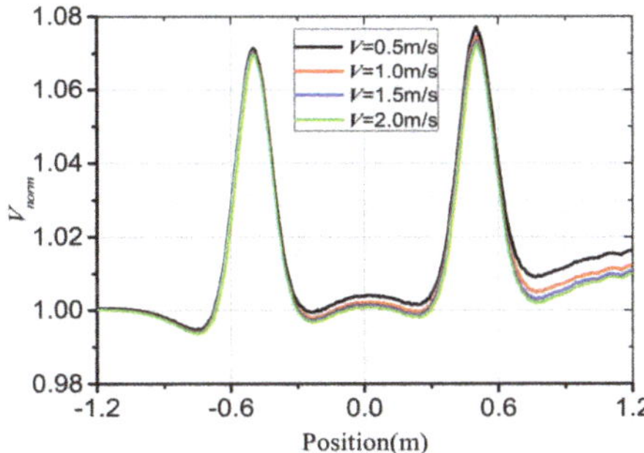

Figure 9. Dimensionless velocity curves on the lateral lines with different inlet velocities.

It can be seen from Figure 9 that the dimensionless velocity curves are similar to each other in the range of flow velocities. This implies that the flow velocity has little influence on the estimated errors of both methods. The estimated errors of both methods for target-2 are shown in Table 3. As seen in Table 3, the estimated errors of MAV with different velocities are smaller.

Table 3. Estimated errors of both methods for target-2.

V (m/s)	0.5	1.0	1.5	2.0
ε_1	5.846	6.495	6.279	6.306
ε_2	0.573	0.746	0.685	0.652

5. Conclusions

We presented a new strategy for underwater target localization using the flow features including MSD and MAV. The two methods were compared in different cases. The following conclusions were obtained from the results.

1. The amplitude variation of extreme points on the parallel velocity curve can be used to estimate the locations of targets. For the cases of flow past one and two targets in a uniform flow field, the estimated error of MAV was less than 1%, which was evidently smaller than that of MSD.
2. Changes in the detective ratio and spacing ratio had obvious influences on the estimated error of MSD, but they had little influence on the results of MAV.
3. Robust flow sensing is not based on the spatial distribution of the flow field but rather, on the fluctuation range.

Future work will focus on other target shapes and flow patterns, such as irregular targets in turbulent oscillatory and pulsing flows. We hope that this research will investigate the connections among various target shapes and flow patterns and that the full range of practical applications of this new object location strategy will be established.

J. Mar. Sci. Eng. **2019**, *7*, 437

Author Contributions: Conceptualization, X.L. and J.W.; methodology, X.L.; software, X.L. and Q.Q.; validation, J.W., X.L. and Q.Q.; formal analysis, X.L. and Q.Q.; investigation, X.L. and J.W.; resources, J.W.; data curation, X.L.; writing—original draft preparation, X.X.; writing—review and editing, J.W. and Q.Q.; visualization, X.L. and Q.Q.; supervision, J.W.; project administration, X.L.; funding acquisition, J.W.

Funding: This research was funded by NATIONAL NATURAL AND SCIENCE FOUNDATION OF HEBEI, grant number E2018202259.

Acknowledgments: The authors would like to thank Xiaoming Wang and Haitao Liu, who provided insight and expertise that greatly assisted the writing of this manuscript.

Conflicts of Interest: The authors declare no conflict of interest. The funders had no role in the design of the study; in the collection, analyses, or interpretation of data; in the writing of the manuscript, or in the decision to publish the results.

References

1. Li, J.H.; Kim, M.G.; Kang, H.; Lee, M.J.; Cho, G.R. UUV Simulation Modeling and its Control Method: Simulation and Experimental Studies. *J. Mar. Sci. Eng.* **2019**, *7*, 89. [CrossRef]

2. Davide, C.; Marco, B.; Gabriele, B.; Massimo, C.; Andrea, R.; Enrica, Z.; Lucia, M.; Paola, C. A Novel Gesture-Based Language for Underwater Human-Robot Interaction. *J. Mar. Sci. Eng.* **2018**, *6*, 91.

3. Corina, B.; Matthew, W.; Dunnigan, M.; Petillot, P. Coupled and Decoupled Force/Motion Controllers for an Underwater Vehicle-Manipulator System. *J. Mar. Sci. Eng.* **2018**, *6*, 96.

4. Szyrowski, T.; Khan, A.; Pemberton, R.; Sharma, S.K.; Singh, Y.; Polvara, R. Range Extension for Electromagnetic Detection of Subsea Power and Telecommunication Cables. *J. Mar. Sci. Eng. Technol.* **2019**, 1–8. [CrossRef]

5. Emberton, S.; Chittka, L.; Cavallaro, A. Underwater Image and Video Dehazing with Pure Haze Region Segmentation. *Comput. Vis. Image Underst.* **2017**, *168*, 145–156. [CrossRef]

6. Li, C.; Guo, J. Underwater Image Enhancement by Dehazing and Color Correction. *J. Electron. Imaging* **2015**, *24*, 033023. [CrossRef]

7. Zhang, J.X. Principle of Biomimetic Detection Based on Flow Field Information of Underwater Moving Object. Master's Thesis, Beijing Institute of Technology, Beijing, China, 2015.

8. Muhammad, N.; Toming, G.; Tuhtan, J.A.; Musall, M.; Kruusmaa, M. Underwater Map-based Localization Using Flow Features. *Auton. Robot.* **2017**, *41*, 417–436. [CrossRef]

9. Zheng, R.; Kamran, M. A Model of the Lateral Line of Fish for Vortex Sensing. *Bioinspir. Biomim.* **2012**, *7*, 036016.

10. Coombs, S.; Patton, P. Lateral Line Stimulation Patterns and Prey Orienting Behavior in the Lake Michigan Mottled Sculpin (*Cottus bairdi*). *J. Comput. Physiol. A* **2009**, *195*, 279–297. [CrossRef]

11. Mchenry, M.J.; Feitl, K.E.; Strother, J.A.; Trump, W.J.V. Larval Zebrafish Rapidly Sense the Water Flow of a Predator's Strike. *Biol. Lett.* **2009**, *5*, 477–479. [CrossRef]

12. Quinzio, S.; Fabrezi, M. The Lateral Line System in Anuran Tadpoles: Neuromast Morphology, Arrangement, and Innervation. *Anat. Rec.* **2014**, *297*, 1508–1522. [CrossRef] [PubMed]

13. Schaumburg, F.; Kler, P.A.; Berli, C.L.A. Numerical Prototyping of Lateral Flow Biosensors. *Sens. Actuators B Chem.* **2018**, *259*, 1099–1107. [CrossRef]

14. Tuhtan, J.A.; Fuentes-Pérez, J.F.; Strokina, N.; Toming, G.; Musall, M.; Noack, M.; Kruusmaa, M. Design and Application of a Fish-shaped Lateral Line Probe for Flow Measurement. *Rev. Sci. Instrum.* **2016**, *87*, 045110. [CrossRef] [PubMed]

15. Dusek, J.E.; Triantafyllou, M.S.; Lang, J.H. Piezoresistive Foam Sensor Arrays for Marine Applications. *Sens. Actuators A Phys.* **2016**, *248*, 173–183. [CrossRef]

16. Delamare, J.; Sanders, R.G.P.; Krijnen, G.J.M. 3D Printed Biomimetic Whisker-based Sensor with Co-planar Capacitive Sensing. In Proceedings of the 2016 IEEE International Conference on Sensors, Orlando, FL, USA, 30 October–3 November 2016.

17. Zhu, P.; Ma, B.; Jiang, C.; Deng, J.; Wang, Y. Improved Sensitivity of Micro Thermal Sensor for Underwater Wall Shear Stress Measurement. *Microsyst. Technol.* **2015**, *21*, 785–789. [CrossRef]

18. Djapic, V.; Dong, W.J.; Bulsara, A.; Anderson, G. Challenges in Underwater Navigation: Exploring Magnetic Sensors Anomaly Sensing and Navigation. In Proceedings of the 2016 IEEE International Conference on Sensors Applications Symposium, Zadar, Croatia, 13–15 April 2015.

19. Salmanpour, M.S.; Sharif, K.Z.; Aliabadi, M. Impact Damage Localisation with Piezoelectric Sensors under Operational and Environmental Conditions. *Sensors* **2017**, *17*, E1178. [CrossRef]

20. Campos, R.; Gracias, N.; Ridao, P. Underwater Multi-vehicle Trajectory Alignment and Mapping Using Acoustic and Optical Constraints. *Sensors* **2016**, *16*, 387. [CrossRef]

21. Mou, B.; He, B.J.; Zhao, D.X.; Chau, K.W. Numerical Simulation of the Effects of Building Dimensional Variation on Wind Pressure Distribution. *Eng. Appl. Comput. Fluid* **2019**, *11*, 293–309. [CrossRef]

22. Ghalandari, M.; Koohshahi, E.M.; Mohamadian, F.; Shamshirband, S.; Chau, K.W. Numerical Simulation of Nanofluid Flow inside a Root Canal. *Eng. Appl. Comput. Fluid* **2019**, *13*, 254–264. [CrossRef]

23. Jorge, P.; Beatriz, D.P.; Jesús, F.O.; Eduardo, B.M. A CFD Study on the Fluctuating Flow Field across a Parallel Triangular Array with One Tube Oscillating Transversely. *J. Fluid Struct.* **2018**, *76*, 411–430.

24. Manoochehr, F.M.; Mohammad, T.S.; Mostafa, R. Numerical Simulation of the Hydraulic Performance of Triangular and Trapezoidal Gabion Weirs in Free Flow Condition. *Flow Meas. Instrum.* **2018**, *62*, 93–104.

25. Hassan, E.S. Mathematical Description of the Stimuli to the Lateral Line System of Fish Derived from a Three-dimensional Flow Field Analysis. *Biol. Cybern.* **1992**, *6*, 443–452. [CrossRef]

26. Humphrey, J.A. Drag Force Acting on a Neuromast in the Fish Lateral Line Trunk Canal II: Analytical Modelling of Parameter Dependencies. *J. R. Soc. Interface* **2009**, *6*, 641–653. [CrossRef] [PubMed]

27. Kumar, P.S.; Vendhan, C.P.; Krishnankutty, P. Study of Water Wave Diffraction around Cylinders Using a Finite-element Model of Fully Nonlinear Potential Flow Theory. *Ships Offsh. Struct.* **2017**, *12*, 276–289. [CrossRef]

28. Zhang, J.X.; Fan, X.; Liang, D.F.; Liu, H. Numerical Investigation of Nonlinear Wave Passing through Finite Circular Array of Slender Cylinders. *Eng. Appl. Comput. Fluid* **2019**, *13*, 102–116. [CrossRef]

29. Song, X.W.; Qi, Y.C.; Zhang, M.X.; Zhang, G.G.; Zhan, W.C. Application and Optimization of Drag Reduction Characteristics on the Flow around a Partial Grooved Cylinder by Using the Response Surface Method. *Eng. Appl. Comput. Fluid* **2019**, *13*, 158–176. [CrossRef]

30. Quezada, M.; Tamburrino, A.; Nino, Y. Numerical Simulation of Scour around Circular Piles Due to Unsteady Currents and Oscillatory Flows. *Eng. Appl. Comput. Fluid* **2018**, *12*, 354–374. [CrossRef]

31. Barbier, C.; Humphrey, J.A.C. Drag Force Acting on a Neuromast in the Fish Lateral Line Trunk Canal. I. Numerical Modelling of External–Internal Flow Coupling. *J. R. Soc. Interface* **2009**, *6*, 627–640. [CrossRef]

32. Zhou, H.; Hu, T.J.; Low, K.H.; Shen, L.C.; Ma, Z.W.; Wang, G.M.; Xu, H.J. Bio-inspired Flow Sensing and Prediction for Fish-like Undulating Locomotion: A CFD-aided Approach. *J. Bionic Eng.* **2015**, *12*, 406–417. [CrossRef]

33. Singh, Y.; Bhattacharyya, S.K.; Idichandy, V.G. CFD Approach to Modelling, Hydrodynamic Analysis and Motion Characteristics of a Laboratory Underwater Glider with Experimental Results. *J. Ocean Eng. Sci.* **2017**, *2*, 90–119. [CrossRef]

34. Tao, J.; Yu, X. Hair Flow Sensors: From Bio-inspiration to Bio-mimicking a Review. *Smart Mater. Struct.* **2012**, *21*, 113001. [CrossRef]

35. Abdulsadda, A.T.; Tan, X. Nonlinear Estimation-based Dipole Source Localization for Artificial Lateral Line Systems. *Bioinspir. Biomim.* **2013**, *8*, 026005. [CrossRef] [PubMed]

36. Abdulsadda, A.T.; Tan, X. An Artificial Lateral Line System Using IPMC Sensor Arrays. *Int. J. Smart Nano Mater.* **2012**, *3*, 226–242. [CrossRef]

37. Dagamseh, A.M.K.; Lammerink, T.S.J.; Kolster, M.L.; Bruinink, C.M.; Wiegerink, R.J.; Krijnen, G.J.M. Dipole-source Localization Using Biomimetic Flow-sensor Arrays Positioned as Lateral-line System. *Sens. Actuators A Phys.* **2010**, *162*, 355–360. [CrossRef]

38. Salumae, T.; Kruusmaa, M. Flow-relative Control of an Underwater Robot. *Proc. R. Soc. A Math Phys.* **2013**, *469*, 1–19. [CrossRef]

39. Venturelli, R.; Akanyeti, O.; Visentin, F.; Jezov, J.; Chambers, L.; Toming, G.; Brown, J.; Kruusmaa, M.; Megill, W.M.; Fiorini, P. Hydrodynamic Pressure Sensing with an Artificial Lateral Line in Steady and Unsteady Flows. *Bioinsper. Biomim.* **2012**, *7*, 036004. [CrossRef]

40. Fuentes-Perez, J.F.; Uhtan, J.A.; Baeza, R.C.; Musall, M.; Toming, G.; Muhammad, N. Current Velocity Estimation Using a Lateral Line Probe. *Ecol. Eng.* **2015**, *85*, 296–300. [CrossRef]

41. Yang, Y.; Nguyen, N.; Chen, N.; Lockwood, M.; Tucker, C.; Hu, H.; Jones, D. Artificial Lateral Line with Biomimetic Neuromasts to Emulate Fish Sensing. *Bioinspir. Biomim.* **2010**, *5*, 16001. [CrossRef]

42. Bao, Y.; Zhou, D.; Zhao, Y.J. A Two Step Taylor-characteristic-based Galerkin Method for Incompressible Flows and Its Application to Flow over Triangular Cylinder with Different Incidence Angles. *Int. J. Numer. Methods Fluids* **2010**, *62*, 1181–1208. [CrossRef]
43. Lamb, H. *Hydrodynamics*; Cambridge University Press: Cambridge, UK, 1932.
44. Zhao, X.Z.; Cheng, D.; Zhang, D.K.; Hu, Z.J. Numerical Study of Low-Reynolds-number Flow Past Two Tandem Square Cylinders with Varying Incident Angles of the Downstream One Using a CIP-based Model. *Ocean Eng.* **2016**, *121*, 414–421. [CrossRef]

Article

Extracting Typhoon Disaster Information from VGI Based on Machine Learning

Jiang Yu [1], Qiansheng Zhao [1,*] and Cheng Siong Chin [2]

[1]   School of Geodesy and Geomatics, Wuhan University, Wuhan 430000, China; yujiang_sgg@whu.edu.cn
[2]   Faculty of Science, Agriculture, and Engineering, Newcastle University in Singapore,
     Singapore 567739, Singapore; cheng.chin@ncl.ac.uk
*   Correspondence: qshzhao@whu.edu.cn

Received: 20 August 2019; Accepted: 9 September 2019; Published: 12 September 2019

Abstract: The southeastern coast of China suffers many typhoon disasters every year, causing huge casualties and economic losses. In addition, collecting statistics on typhoon disaster situations is hard work for the government. At the same time, near-real-time disaster-related information can be obtained on developed social media platforms like Twitter and Weibo. Many cases have proved that citizens are able to organize themselves promptly on the spot, and begin to share disaster information when a disaster strikes, producing massive VGI (volunteered geographic information) about the disaster situation, which could be valuable for disaster response if this VGI could be exploited efficiently and properly. However, this social media information has features such as large quantity, high noise, and unofficial modes of expression that make it difficult to obtain useful information. In order to solve this problem, we first designed a new classification system based on the characteristics of social medial data like Sina Weibo data, and made a microblogging dataset of typhoon damage with according category labels. Secondly, we used this social medial dataset to train the deep learning model, and constructed a typhoon disaster mining model based on a deep learning network, which could automatically extract information about the disaster situation. The model is different from the general classification system in that it automatically selected microblogs related to disasters from a large number of microblog data, and further subdivided them into different types of disasters to facilitate subsequent emergency response and loss estimation. The advantages of the model included a wide application range, high reliability, strong pertinence and fast speed. The research results of this thesis provide a new approach to typhoon disaster assessment in the southeastern coastal areas of China, and provide the necessary information for the authoritative information acquisition channel.

Keywords: typhoon disaster; deep learning; VGI; text classification

---

1. Introduction

1.1. Background

Volunteer geographic information [1], which is also interpreted by domestic scholars as "spontaneous geographic information" [2], is similar to neogeography [3], or crowd sourcing geographic data [4,5], which refers to the phenomenon of public participation in contributing geographic information data [6] and is an important feature of "new geography" [3]. There are many VGI data sources, including both structured spatial data platforms (such as Open Street Map, or OSM for short) and unstructured social network data platforms with explicit or implicit location information (such as Twitter, Facebook, Sina Weibo, Tencent Weibo, etc.). Due to the characteristics of instantaneity and interaction of VGI, VGI played an important role in the "4.20" Lushan strong earthquake, the Haiti earthquake, and the Philippines typhoon Haiyan in 2012. The VGI played an important role that could not be achieved by traditional methods from disaster awareness, information identification, and information classification to disaster determination [7–9].

The southeastern coast of China is struck by typhoons every year. Secondary disasters caused by typhoons, such as heavy rainfall, storms, floods, debris flows, and landslides have a great impact on the eastern part of China [10]. During and after typhoon disasters, the relevant media track and report, and the affected people also interact with the typhoon-related information through various social networks (such as Weibo). According to the data, since 2012, during the period of each typhoon with a serious impact, the netizens in the affected areas published more than 100,000 microblog messages, including locations, winds, rainfalls, secondary disasters, rescue, and other related disaster information, with strong instantaneity and interaction. Compared to the disadvantages of using traditional means to obtain updates on the disaster situation during typhoon disasters, it is of great significance to use VGI to assist typhoon disaster situation assessment. In view of this, this paper established a classification system that meets the needs of typhoon emergency response based on the characteristics of microblog data, and constructed a generic typhoon disaster information automatic acquisition model using a neural network method. The model can automatically select microblogs related to disasters from a large number of microblog messages, and further subdivide them into different types of disasters, which is different from the general classification system, so as to facilitate subsequent emergency response and loss estimation. The advantages of the model included a wide application range, high reliability, strong pertinence, and fast speed.

1.2. Analysis of Existing Studies

Generally speaking, there are three ways to use social media [11]. One is to regard social media as a huge sensor through which we can collect a lot of information [12] about typhoon disasters. The difficulty lies in that the sensor is too sensitive. It not only collects information about the typhoon disaster, but also contains a lot of irrelevant information. And author et al. [13,14] mainly discussed how to extract effective information. The second is to make full use of the social media's communication function. Emergency agencies use social media to publish corresponding emergency messages so that they can be received by the people who really need them, as it may be difficult to receive such information in time through other channels [15,16]. Thirdly, the analysis tools of social media, such as hot spot analysis, can be used to analyze the characteristics of the disaster [17].

From the perspective of research methods, many studies have focused on classifying and visualizing social media data in the order of pre-disaster, disaster, and post-disaster to verify the relationship between the number of tweets and the disaster process [11,18]. We implemented an algorithm based on K nearest neighbor (KNN) for extracting information from VGI which resulted in about 70% of microblogs classified correctly [19], which was not enough. A further approach used is to verify the coincidence of the typhoon trajectory and the number of tweets in conjunction with their time and location [13,20–22]. To date, not much research has been done to further analyze each twitter's contents, and most studies have stayed at statistical analysis of the amount of data from a specific time or place, as described above. Clearly, the extraction of social media data is still a difficult point for researchers. Nevertheless, some papers have analyzed social media content, although the analysis has mainly stayed at the level of sentiment analysis, using SVM [20]. Although this is also an exploration, objectively speaking, staying at the level of sentiment analysis has had little effect on the emergency response. On the contrary, some studies have used relatively unique methods to carry out research which has been of great significance for reference, such as forecasting the areas of power outages caused by typhoon impact [23], and the idea of weighting different emergency strategies with social media information to ensure multi-sectoral collaboration [15]. Author et al. [24] focused on identifying informative tweets posted during disasters, naive Bayesian classifier and the classification model based on neural network were designed respectively, and their classification effects were compared. The results showed that the deep neural network, especially Convolutional Neural Network (CNN), were more effective in identifying informative tweets.

To sum up, there is great potential in the study of the application of social media data in typhoon emergency response. Nevertheless, its research is still in the exploratory stage, and there is no optimal method that can be consistently applied. To date, most research is still at a relatively superficial stage, and few papers have focused on deeper mining, and this is the main work of our study. In addition, the lack of a unified dataset and classification method is also a major problem hindering this research. Some researchers have attempted this work abroad [11,25,26], but it has not yet been seen in China. Therefore, the classification strategy used with the microblog data and the convolutional neural network used to classify the microblog data one by one are both somewhat original, and the classification effect was able to reach a good level. The specific classification results may have a great effect on subsequent emergency work.

2. Methodology

This paper first studied the characteristics of Weibo data related to typhoon disasters, and then established a classification system of typhoon disaster information which met the characteristics of Weibo data. On this basis, we established the corresponding dataset. Following the general process of text classification using a neural network [27], the model's construction and the training process were then completed. Finally, the model was verified by the verification set to verify the effect of the model. The final result was an extraction model able to gather typhoon disaster information from Weibo data, and the model also fit other typhoons well.

2.1. Design of Classification

Sina Weibo is an important source of typhoon disaster information, but as a public social platform, it contains a lot of useless information; even the useful information is generally lacking in professionalism and pertinence. Therefore, it was necessary to design a suitable classification method, which would not only separate the useful information from the useless information, but more importantly, link the colloquial expression about the disaster with the disaster situation. If the word description of the classification system was too professional, it might have collected little or no disaster information. For example, when marine fisheries are damaged, there are specific categories like aquaculture affected area, aquaculture disaster area, aquaculture ruin area, etc. It is conceivable that the possibility of such professional terminology appearing in Sina Weibo is extremely low. However, if the classification design is too casual, then the value of the collected data will be debatable. After reading and analyzing a certain amount of Weibo information, and considering it comprehensively, this paper divided the Weibo disaster information into the following categories:

1. Building: Weibo content mainly describing damage to buildings, such as water flooding into the house, buildings destroyed, billboards blown off, etc;
2. Green plants: Weibo content mainly describing damage to trees, green belts, etc. in the city;
3. Transportation: Weibo content mainly describing road flooding caused by the typhoon, poor traffic, etc.;
4. Water and electricity: Weibo content mainly describing water being cut off and power cuts caused by typhoons;
5. Other: Data related to typhoon disaster information, but not explicitly related to the above categories;
6. Useless: Data that were not related to the above categories.

2.2. Text Representation

The study used the method of word embedding [28] for text representation. In this paper, we used the embedding layer of Tensorflow to complete word embedding. The embedding layer, as a part of the deep learning model, trains with the model and updates the word embedding vector. After word embedding, the vector corresponding to each word and punctuation was of equal length. The text was converted into the form shown in Figure 1, and each microblog text was converted into a two-dimensional matrix in which each line was a vector form of a word or punctuation symbol constituting the text. The number of rows in the two-dimensional matrix was the number of words and punctuation the text contained. For the convenience of subsequent processing, each microblog text was artificially changed into equal length. Thus, the final result was a batch of two-dimensional matrices of the same size as shown in Figure 2. The actual processing was stored in a three-dimensional array when the data were finally entered into the model.

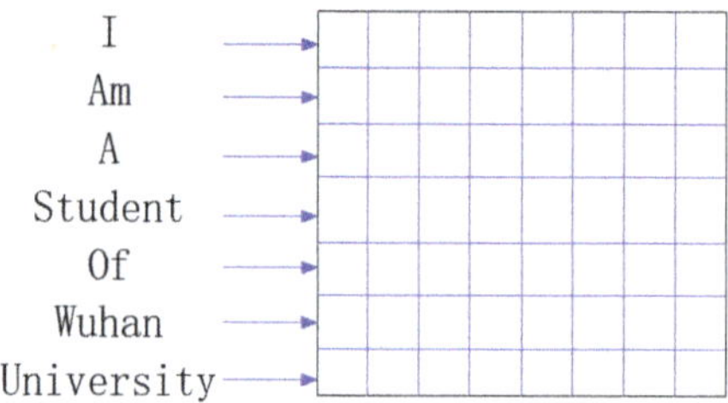

Figure 1. Word embedding.

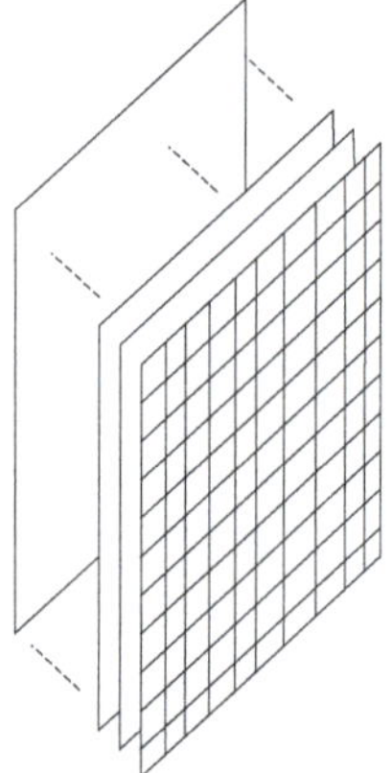

Figure 2. Two-dimensional matrices of the same size.

2.3. Model Construction

2.3.1. Structure of Model

We use the basic model structure from the research of Reference [29,30]; the structure of the model is shown in Figure 3 below. The left side the figure shows the input layer, with each input entering one piece of texts to build a two-dimensional matrix. Feature extraction was performed on the convolutional layer using 128 different filters with a height of 5, and 128 feature vectors were obtained. The kernel of max pooling was used at the pooling layer to get the strongest part of each feature and combine them. These features were then integrated through the fully connected layer, and finally input into the Softmax classifier to obtain classification results.

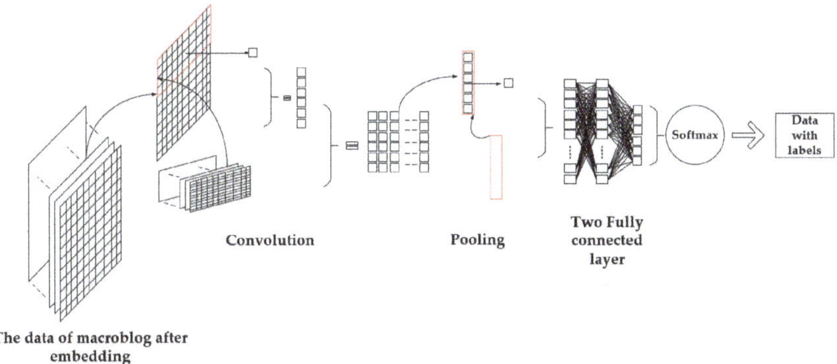

Figure 3. Neural network structure.

2.3.2. The Loss Function

The above describes the specific structure of the convolutional neural network used in this study, from the input layer to the output layer. The mainframe of the entire network has been established. However, a very important aspect that has been missed so far is the evaluation of the classification results. There is no doubt that the quality of the classification results is the most concerning issue. In addition, the training of neural networks is aimed at continuously improving the accuracy of classification. If there is no good method to assess the results of the classification, the training of the neural network cannot be carried out, let alone the model be finally used to help us classify. The concept of the loss function was proposed to solve this problem. The loss function reflects the quality of the classification by measuring the closeness between the classified result and the expected output. There are many forms of loss function, and in our study we use cross entropy as a loss function. There are three main features of the cross entropy: (1) the cross entropy of two identical functions is zero; (2) The smaller the cross entropy of the two functions, the more similar the two functions are, and the larger, the opposite; (3) Cross entropy can measure the difference between two random distributions with values greater than zero [22].

3. Case Study

In order to verify the feasibility and effect of the proposed method, this paper selected the relevant microblog texts of Typhoon Anemone for a case study. Anemone landed in Zhejiang on August 8, 2012, affecting five provinces (cities), namely Zhejiang, Shanghai, Jiangsu, Anhui, and Jiangxi. Anemone was strong and had a long stay over the mainland. It had caused tremendous damage to the affected area. The specific statistics are shown in the Table 1. Anemone was a typical typhoon landing on the coast of China, and it had a large social impact, resulting in lots of microblog texts, giving it a certain representativeness as a research object. The Typhoon Anemone database contained a total of 22,317 microblog texts. Some texts only had description about the typhoon disaster, and some data included both content and position information of the people who posted on Weibo.

Table 1. Statistics of losses caused by typhoon anemone.

Provinces (Cities)	People Affected	Transferred	Died	Houses Collapsed	Damaged
Zhejiang	7,010,000	1,546,000	0	5100	15,000
Shanghai	361,000	311,000	2	50	700
Jiangsu	662,000	126,000	1	600	2400
Anhui	1,576,000	163,000	0	1500	13,000

3.1. Data Preprocessing

For the smooth progress of the experiment, data preprocessing was required. It was mainly divided into two parts: dataset production and text representation. The text representation has been described in detail in the previous section, and will not be repeated here. This section mainly introduces the process of making the dataset.

Dataset production mainly consists of two parts: one is classification and screening, and the other is unified formatting. The data format specified in this article was: label + tab + content, with each line separated by a new line. These messages translated into English looked like this:

(1)   /Green plants/tab/After the typhoon, the trees on the side of the road fell down and the traffic lights were broken./

(2)   /Water and electricity/tab/#typhoon# Go ahead, it's all the sound of the wind and the wind..., And it's still out of power./

The difficulty in dataset production is screening and classification (data labeling), which was the most time-consuming part of the whole experiment. After trying manual screening and search keyword screening, we combined our previous experience to write a program to achieve the coarse classification of Weibo data. The algorithm flow used was as follows in Figure 4:

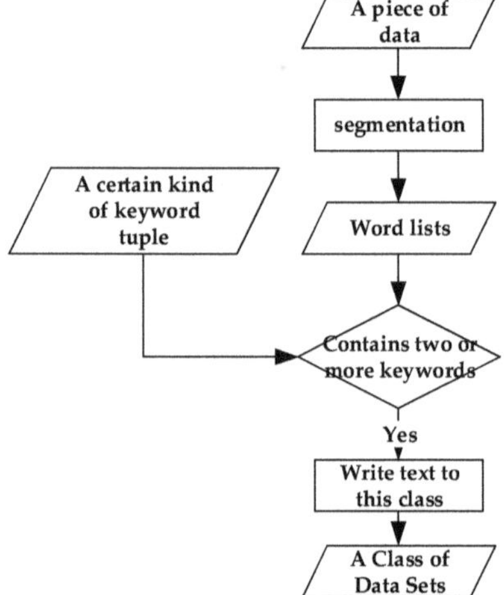

Figure 4. Pre-classification process.

Through the above procedures, most of the classification work was completed, only needing manual refinement of the already classified data and marking of other disasters at the same time.

The datasets obtained in this paper were as follows: 335 disasters in construction; 378 disasters in traffic; 525 disasters in green plants; 399 disasters in hydropower; 435 in other disasters; and 1419 in useless categories, with 3491 data in total.

3.2. Training and Verification

This section mainly introduces the implementation ideas and processes of the program, and does not discuss the specific implementation of the function. The program can be roughly divided into three parts: one is the generation of the dictionary and one-hot vector, the second is the configuration of the CNN model, and the last is the training and verification of the model in combination with the first two steps.

3.2.1. Generation of the Dictionary and the One-Hot Vector

The generation of the dictionary and the one-hot vector was the preparation work required before word embedding. The generation process was as follows in Figure 5:

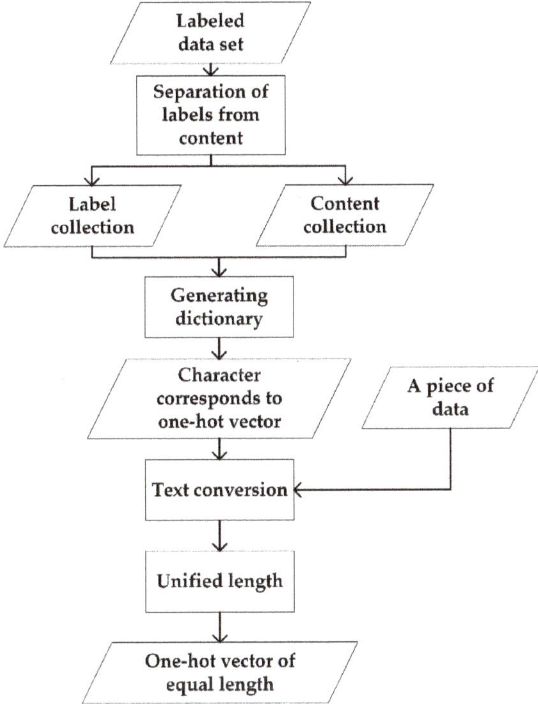

Figure 5. Generation of the one-hot vector.

The created dataset file was read, and the label and content saved separately in two lists. The two lists of tags and content were then processed separately to generate their respective dictionaries. According to the generated dictionary, the index form of the one-hot vector of each character in the dictionary was obtained. After the correspondence between the label, the character and its one-hot vector was established, each piece of data could be processed according to the corresponding relationship, and converted from the text format to the corresponding one-hot vector form. All the one-hot vectors of all sentences were then unified to the same length by adding 0.

3.2.2. Construction of the CNN Model

The construction of the CNN model included setting parameters, implementation of each layer structure, and connection.

After setting the parameters, the appropriate Tensorflow function was selected to implement the function of the corresponding layer, and the output of the previous function was then used as the input of the next function to achieve connection between the layers. Finally, the loss function and optimizer were added.

3.2.3. Training and Testing

The training process of the model was as follows in Figure 6:

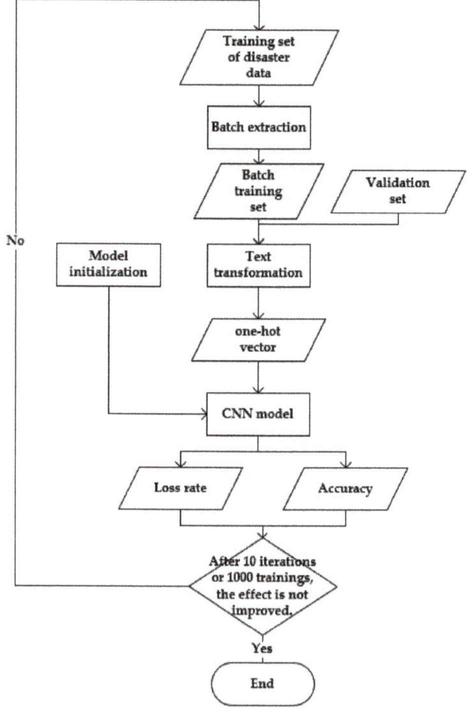

Figure 6. Training process.

Figure 6 shows the training process. First, the training set and the verification set were prepared, then the data were extracted from the training set according to the set batch size. Next, these data and the verification set were embedded, then imported into the initialized CNN model. The loss rate and accuracy of the model were then output, and then the parameters were adjusted to reduce losses. This cycle was repeated until the entire training set had been used to complete 10 rounds trainings, or there was no improvement in long-term. The process of testing and verification was similar, except that the process of initializing the model needed to be changed into importing the existing model, and then the accuracy and the loss rate were directly output.

3.3. Discussion

According to the experimental procedure described above, six experiments were carried out. First, 50% of the dataset was randomly extracted for training and verification of the model, and then 60%, 70%, and 100% of the dataset. Each time, data were randomly extracted from the dataset as a training set, a test set, and a verification set at a ratio of 7:2:1. The rest of this section is mainly divided into three subsections to explain the experimental results. First, the results are analyzed in detail with 70% of the data training and test results. The experimental results of the different-sized datasets are then compared and analyzed based on the above analysis.

3.3.1. Description of Training Results

The columns in Table 2 in represent the iteration round (Epoch), the training batch (Iter), the loss rate(Train Loss) and accuracy (Train Acc) on the training set, and the loss rate (Val Loss) and accuracy (Val Acc) on the verification set. For the convenience of analysis, these data have been visualized in Figures 7 and 8.

Table 2. Training result.

Epoch	Iter	Train Loss	Train Acc	Val Loss	Val Acc
	0	1.8	12.50%	1.8	15.16%
1	50	1.5	50.00%	1.6	41.39%
	100	1.6	37.50%	1.5	41.39%
2	150	1	56.25%	1.1	61.48%
	200	0.55	87.50%	0.84	74.59%
3	250	0.97	62.50%	0.78	73.36%
	300	0.7	75.00%	0.7	76.64%
4	350	0.78	68.75%	0.65	78.69%
	400	0.29	87.50%	0.65	78.28%
5	450	0.33	93.75%	0.64	78.28%
	500	0.61	68.75%	0.63	78.28%
6	550	0.12	100.00%	0.65	79.51%
	600	0.14	100.00%	0.63	78.69%
7	650	0.097	93.75%	0.61	79.92%
	700	0.37	87.50%	0.63	79.92%
	750	0.086	100.00%	0.68	79.92%
8	800	0.051	100.00%	0.66	80.33%
	850	0.064	100.00%	0.65	80.74%
9	900	0.28	93.75%	0.73	78.69%
	950	0.088	100.00%	0.67	79.51%
10	1000	0.014	100.00%	0.75	79.10%
	1050	0.0161	100.00%	0.76	80.74%

After analysis, the reason for the abnormal oscillation of the loss rate and accuracy curve of the training set was that the data batch was too small, but this situation did not affect the training effect of the model, so no further analysis was needed. Ignoring the abnormal oscillation of the curve, we concluded that the loss of the model dropped rapidly and then stabilized, and the accuracy first rose rapidly and then stabilized. The above conclusions were true for both the training set and the verification set. Specifically, the loss rate of the training set finally stabilized at around 0, and the accuracy finally stabilized at 100%. As far as verification is concerned, the loss rate finally stabilized at around 0.6 and the accuracy was maintained at 80%.

Figure 7. Loss rate and number of trainings.

Figure 8. Accuracy and number of trainings.

3.3.2. Description of Test Results

As shown in Table 3, the test results consisted mainly of two parts. Table 3 gives the various indicator values of the classification effect of each category, as well as the overall classification effect, which is convenient for the overall analysis model. Table 4 shows the classification confusion matrix. Each type can be analyzed in more depth through the confusion matrix.

Table 3. Test result.

Test Loss	0.62		Test Acc	80.29%
Class Name	**Precision**	**Recall**	**F1-Score**	**Number of Entries in the Category**
Building	0.93	0.71	0.80	55
Green plants	0.80	0.86	0.83	78
Transportation	0.87	0.70	0.78	57
Water and electricity	0.75	0.94	0.84	54
Other	0.77	0.44	0.56	54
Useless	0.79	0.90	0.84	189
Mean/sum	0.81	0.80	0.80	487

Table 4. Confusion matrix.

Class Name	Building	Green Plants	Transportation	Water and Electricity	Other	Useless
Building	39	1	1	3	0	11
Green plants	1	67	0	2	2	6
Transportation	0	5	40	0	1	11
Water and electricity	0	0	0	51	0	3
Other	0	4	3	8	24	15
Useless	2	7	2	4	4	170

As shown in Table 3, there were two indicators—that is, the loss rate and the accuracy rate—which were 0.62 and 80.29% respectively. As analyzed in the previous section, there was no substantial difference between the verification process and the training process. Therefore, the loss rate and the accuracy rate on the test set were consistent with the analysis of the verification set; the loss rate was finally maintained at 0.6 and the accuracy was maintained at 80%.

There were three new indicators added, namely "precision", "recall", and "F1-score". Precision indicates the accuracy of the prediction, which can be defined by the following formula: $p_i = a_i/n_i$, where n_i represents the number of data in the i-th class after the model classification is completed, and a_i represents the number of data pieces that belong to the class in the n_i data. For example, after the classification was completed, there were a total of 100 pieces of data ($n_i = 100$) classified into the construction disaster category. Compared with the original data label, it was found that there were 90 pieces of data that belonged to the construction disaster category ($a_i = 90$); thus, for the construction class, the accuracy of the classification was $p_i = a_i/n_i = 90\%$. Recall represents the recall rate, which represents the probability that the tagged data have been correctly classified. In the same way, for example, in the construction class, assuming that there were 120 construction-labeled items in the training set data, if there were 90 construction-labeled items still in the construction category after the classification, then the recall rate would be 75%. F1-score is the harmonic mean of the accuracy and recall rate, which can comprehensively reflect the effect of classification. Next, we analyzed the prediction effect by category:

Building category: accuracy rate of 93%, recall rate of 71%, and F1-score of 0.8. The classification accuracy of the building category was very high, reaching 93%. However, the recall rate was relatively low, only 71%, which meant that nearly 30% of the data belonging to the building category were misclassified by the model to other types. Combining the accuracy and the recall rate, we made the inference that the model gave too much weight to certain special features of the building class. These features generally belonged to data of building category, but not all building data had these characteristics or the features were not obvious enough, so those data may not have been classified into the building class correctly. This is why the building category had a high accuracy rate and a low recall rate. As for why some features were given too high a weight, two conjectures can be made. One is that the sample types used for training were not rich enough, so that some feature models that also belonged to the building category could not be learned. Second, the features of model learning were not abstract enough to distinguish the data belonging to the building category.

Green plants: accuracy rate of 80%, recall rate of 86%, F1-score of 0.83. Compared with the construction category, the accuracy of green plants dropped a lot to only 80%, but the recall rate increased a lot, reaching 86%. The final F1-score was also higher than the building category. According to the classification results of green plants, we speculated that there were many data for model training, and the model learned a lot about the features of green plants, so the recall rate of the model was relatively high. However, due to the limitation of the performance of the model, the characteristics learned were not deep enough, so that the distinguishing degree from other categories was not high enough, and thus there were also many misclassifications and the accuracy was reduced.

Transportation: accuracy rate of 87%, the recall rate of 70%, F1-score of 0.78. The classification situation of traffic was very similar to the building category; the supposed reasons are the same, and will not be repeated.

Hydropower: accuracy rate of 75%, the recall rate of 94%, the F1-score of 0.84. The accuracy of hydropower was ordinary, but the recall rate was high. The reason is likely similar to the green plants class.

Other category: 77% accuracy, 44% recall, F1-score of 0.56. The other class had average accuracy, but the recall rate was very low, not even reaching 50%. That is to say, more than half of the data marked as other classes were classified into the remaining categories. Two possibilities were considered. One follows the above ideas: that is, that the sample types in this category were not rich enough. The second was misclassification when making labels. Because the definition of the other class was not clear, and those data related to typhoon disasters but not obviously belonging to the above four categories will be classified into other categories, this category is very subjective. It may be that some of the data in author's opinion should not belong to the above four categories, but actually they do; the model correctly marked the data of some classification errors by comparing the characteristics, so that the recall rate of the other category was relatively low. The second possibility seems more likely in this case.

Useless class: accuracy rate of 79%, recall rate of 90%, F1-score of 0.84. Compared with the previous categories, it was seen that the final classification results of the useless class were superior not only in accuracy and recall rate, but its F1-score was also the highest. This was actually quite unexpected, because unlike the previous categories, any content data will be classified as useless as long as it is not related to the disaster. The authors originally thought that because of this arbitrariness, it would be difficult to classify useless classes, because arbitrariness means that the data characteristics of useless classes may also be diverse. If analyzed carefully, the model should not be able to achieve such a good effect by completely extracting the characteristics of useless classes, perhaps by means of reverse thinking. That is, if the last extracted feature of the sentence is not similar to the first five categories, then it is classified into the useless class, so the classification effect of the useless class actually depends on the classification effect of the first five categories.

Through the analysis, the classification effect of each type of data could be more clearly seen, as shown in Table 3. The rows and columns of the confusion matrix are arranged in the order of construction, green plants, transportation, water and electricity, others, and uselessness. Each row shows the distribution of the data with the corresponding label for that row, and each column shows the distribution of the labels of the data classified as that class. For example, the first line shows the distribution of the data with the building label after the classification was completed. Of these, 39 were classified as buildings, 1 was classified as green plants, 1 was classified as traffic, 3 were classified as hydropower, 0 were classified into other categories, and 11 were classified as useless. A total of 55 data: 39/55 = 0.71 is the recall rate, which can be found from the distribution of data with building tags. The building features learned by the model were generally consistent with the test set, but there were also some building data that were classified into useless classes. These should be the data of the building category with less obvious features, and therefore features not too similar to the building classes in the model, but since the data were labeled as a building, its features were not similar to those of other categories. Thus, they were classified as useless, that is, in the category where features were not identified as being disaster-related at all. Some readers may question why these data would be classified into the useless class rather than the other class, since the other class is at least related to disasters. The suggested reasons are as follows. From the above data, the prediction accuracy of the other class reached 77%, which was similar to the accuracy of the remaining categories, indicating that the features of the other category had been learned in the model. Therefore, it was found that the features of these data were not the same as those of the other class, and so they were not classified into the other class. Note the data in the first column, which represents the actual labels of the data that were finally classified into the building category. In other words, 39 of the data points classified as buildings were indeed in the building class, one was actually in green plants, and the other two were useless. A total of 42 data, 39/45 = 0.94 is the accuracy rate. It was seen that the differences between the characteristics of buildings and the other types of features learned by the model were still relatively large, so misclassification was rare.

The specific meanings of the rows and columns of the confusion matrix have been carefully described above. In fact, most of the information in the confusion matrix has been reflected in the analysis of the accuracy and recall rate of the previous part. Only by analyzing the confusion matrix, we can find out which classes of features in the model are not clearly separated, so it is easy to misclassify each other. For example, data of other class can be easily classified into useless class, which is hard to see by accuracy and recall.

This study has carried out a detailed analysis of the results obtained by using the 70% dataset, and obtained the relationship between the loss rate and the correct rate of the model and the number of training in experiment. And through the experiment in test set, it is proved that the prediction accuracy of the model can indeed reach 80%, and the effect and causes of the classification are analyzed.

In the following, the experimental results of different size datasets are compared horizontally to find the relationship between the dataset size and the classification effect.

3.3.3. Comparison of Results of Datasets with Different Sizes

The results of one experiment have been thoroughly analyzed in the previous section. The final results of the other datasets were similar and will not be repeated. In this section, we mainly compare the effect of different size datasets on the accuracy of the model. Based on the results of the test set for each experiment, the statistical table is as follows:

The above Table 5 shows the classification effect corresponding to the datasets of different sizes. It can be seen that the accuracy rate increased with the increase of the dataset at the beginning, but after increasing to about 80%, there was a tendency to stabilize. As the dataset continued to increase, the accuracy of the model did not seem to increase accordingly. The number 80% has appeared many times in this paper; whether using a single training set or a different training set, the accuracy of the model could not easily break the 80% limit. Thus, 80% was the precision limit of the model in this paper.

Table 5. The relationship between the size of the dataset and the accuracy.

Size of the Dataset	0.5	0.6	0.7	0.8	0.9	1.0
Accuracy	70.61%	74.70%	80.29%	78.64%	77.39%	79.66%

3.3.4. Actual Forecasting Effect

In order to further test the generalization ability of the model, a verification experiment was carried out. We selected some Weibo data from another typhoon to see if the model can correctly classify it. The experimental process was as follows in Figure 9:

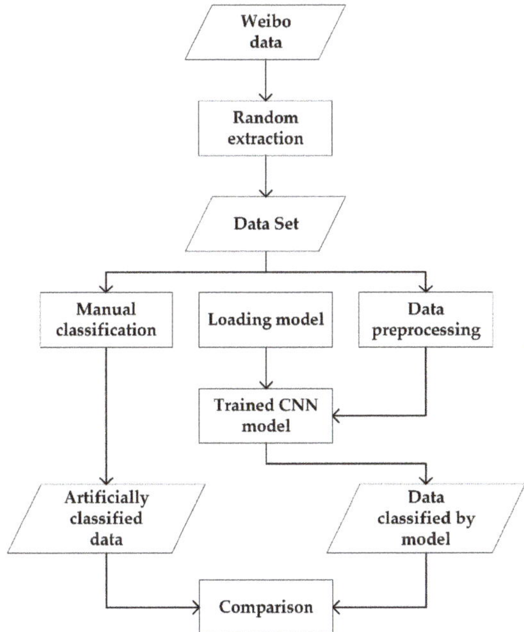

Figure 9. Verification experiment.

This study randomly selected 105 items of typhoon-related data. It was first manually classified, then the model is used for prediction, and finally, the two classification results are compared. After comparison, it was found that among the 105 data, there were 21 inconsistencies between the model prediction and the manual classification and the accuracy rate was 80%, which was consistent with the level on the test set. It is also worth mentioning that when there was a contradiction between manual classification and model prediction in the comparison process, most of the time it was indeed a prediction error of the model, but in some cases, the authors believe that the prediction of the model was more in line with the meaning of the data. With this in mind, the accuracy of the model could be slightly higher than 80%.

4. Conclusions

Starting from VGI, this paper used deep learning tools to process and classify microblog data collected after a typhoon occurred, and to extract information related to typhoon disasters to help prevent and mitigate disasters. The paper introduced the design of a classification system and the structure and function of a convolutional neural network model. The model was then programmed and implemented, and it was trained and verified by the dataset we designed. A typhoon disaster mining model with universality was obtained, which achieved 80% classification accuracy and has a certain practical value.

The biggest advantage of this model is that it is convenient and fast, our group finished the classification work in a week, and the same work could be done in two seconds using the model, which greatly reduces the labor and shortens the time required for classification. It has a high practical value in disaster situations where time and manpower are scarce. The disadvantage is that the accuracy of the model is not high enough, and the highest precision is currently around 80%. However, the accuracy is actually relative; the Sina Weibo data itself had some ambiguity. Unlike some texts classified in the past, such as sentiment analysis (positive and negative) or news classification, there were almost no difficulties and misjudgments for manual marking, this was not the case with Weibo data. In the authors' experience of manually marking more than 3000 pieces of data, many data expressions were ambiguous and the mark-up took a lot of effort. Despite this, there were still some pieces of data that were considered inappropriately when looking back at the previous mark. Therefore, the extraction of Weibo data is different from general text classification. In other words, whether the accuracy of the model in this paper should be measured by general accuracy is still open to question.

There are basically three ways to improve: First, continue to increase the dataset; because the current number of data was in the thousands, which is a relatively small dataset, 80% may have been just because the dataset number was not enough. The second is to adjust the design of the classification system; because the current categories were just my own ideas, the boundaries between classes may not have been clear enough, especially in the other category. More scientific design would not only help to reduce the mistakes of manual classification and label making, but would also help to use the data. The third is to optimize the structure of the model to give it stronger learning ability.

Author Contributions: Y.J. and Q.Z. contributed to the work equally and should be regarded as co-first authors. Contributions from each author in this work can be described by: Literature search, Q.Z., Y.J.; Methodology, Y.J., Q.Z.; Software, Y.J.; Validation, Y.J., Q.Z. and C.S.C.; Data collection, Y.J., Data analysis, Y.J., Q.Z.; Data interpretation, Y.J., Q.Z.; Writing–Original Draft Preparation, Y.J.; Writing–Review & Editing, Q.Z. and C.S.C.; Supervision, Q.Z.; Project Administration, Q.Z.; funding acquisition, Q.Z.

Funding: This work was supported by the National Key Research and Development Program of China (Grant No. 2017YFC1405300).

Acknowledgments: This work was supported by the National Key Research and Development Program of China (Grant No. 2017YFC1405300). The authors would like to express their sincere thanks to the editors and anonymous reviewers for their valuable comments and insightful feedback.

Conflicts of Interest: The authors declare no conflict of interest. The funders had no role in the design of the study; in the collection, analyses, or interpretation of data; in the writing of the manuscript, or in the decision to publish the results.

References

1. Goodchild, M.F. Citizens as sensors: The world of volunteered geography. *GeoJournal* **2007**, *69*, 211–221. [CrossRef]
2. Li, D.R.; Qian, X.L. A Brief Introduction of Data Management for Volunteered Geographic Information. *Geom. Inf. Sci. Wuhan Univ.* **2010**, *35*, 379–383.
3. Turner, A. The role of angularity in route choice: an analysis of motorcycle courier GPS traces. In Proceedings of the Spatial Information Theory, Aber Wrac'h, France, 21–25 September 2009; pp. 489–504.
4. Heipke, C. Crowd Sourcing Geospatial Data. *ISPRS J. Photogramm. Remote Sens.* **2010**, *65*, 550–557. [CrossRef]
5. Starbird, K. Digital Volunteerism During Disaster: Crowdsourcing Information Processing. In Proceedings of the Human Factors in Computing Systems, Vancouver, BC, Canada, 7–12 May 2011; pp. 7–12.
6. Shan, J.; Qin, K.; Huang, C.; Hu, X.; Yu, Y.; Hu, Q.; Lin, Z.; Chen, J.P.; Jia, T. Methods of Crowd Sourcing Geographic Data Processing and Analysis. *Geom. Inf. Sci. Wuhan Univ.* **2014**, *39*, 390–396.
7. Yates, D.; Paquette, S. Emergency knowledge management and social media technologies: A case study of the 2010 Haitian earthquake. *Int. J. Inf. Manag.* **2011**, *31*, 6–13. [CrossRef]
8. Camponovo, M.E.; Freundschuh, S.M. Assessing uncertainty in VGI for emergency response. *Cartogr. Geogr. Inf. Sci.* **2014**, *41*, 440–455. [CrossRef]
9. Liu, S.B. Crisis Crowdsourcing Framework: Designing Strategic Configurations of Crowdsourcing for the Emergency Management Domain. *Comput. Supported Cooper. Work* **2014**, *23*, 389–443. [CrossRef]
10. Niu, H.Y.; Liu, M.; Lu, M.; Quan, R.S.; Zhang, L.J.; Wang, J.J. Risk Assessment of Typhoon Disasters in China Coastal Area during Last 20 Years. *Sci. Geogr. Sin.* **2011**, *31*, 764–768.
11. Wang, L.H.; Hovy, E.; Dredze, M. The Hurricane Sandy Twitter Corpus. In Proceedings of the AAAI Workshop on the World Wide Web and Public Health Intelligence, Quebec, QC, Canada, 27 July 2014; pp. 20–24.
12. Qu, Y.; Huang, C.; Zhang, P. Microblogging after a Major Disaster in China: A Case Study of the 2010 Yushu Earthquake. In Proceedings of the 2011 ACM Conference on Computer Supported Cooperative Work, CSCW 2011, Hangzhou, China, 19–23 March 2011.
13. Yury, K.; Haohui, C.; Esteban, M. Performance of Social Network Sensors during Hurricane Sandy. *PLoS ONE* **2015**, *10*, e0117288.
14. Wang, Y.D.; Wang, T.; Ye, X.Y. Using Social Media for Emergency Response and Urban Sustainability: A Case Study of the 2012 Beijing Rainstorm. *Sustainability* **2015**, *8*, 142–143. [CrossRef]
15. Wang, D.; Qi, C.; Wang, H. Improving emergency response collaboration and resource allocation by task network mapping and analysis. *Saf. Sci.* **2014**, *70*, 9–18. [CrossRef]
16. Lerman, K.; Ghosh, R. Information Contagion: An Empirical Study of the Spread of News on Digg and Twitter Social Networks. *Comput. Sci.* **2010**, *52*, 166–176.
17. Lazo, J.K.; Bostrom, A.; Morss, R.E. Factors Affecting Hurricane Evacuation Intentions. *Risk Anal.* **2015**, *35*, 1837. [CrossRef] [PubMed]
18. Dittus, M.; Quattrone, G.; Capra, L. Mass Participation during Emergency Response: Event-centric Crowdsourcing in Humanitarian Mapping. In Proceedings of the Acm Conference on Computer Supported Cooperative Work & Social Computing, Portland, OR, USA, 25 February–1 March 2017.
19. Zhao, Q.S.; Chen, Z.; Liu, C.; Luo, N.X. Extracting and classifying typhoon disaster information based on volunteered geographic information from Chinese Sina microblog. *Concurr. Comput. Pract. Exp.* **2019**, *31*, e4910. [CrossRef]
20. Neppalli, V.K.; Caragea, C.; Squicciarini, A.; Stehle, S. Sentiment analysis during Hurricane Sandy in emergency response. *Int. J. Disaster Risk Reduct.* **2017**, *21*, 213–222. [CrossRef]
21. Neppalli, V.K.; Caragea, C.; Caragea, D.; Medeiros, M.C.; Tapia, A.H.; Halse, S.E. Predicting tweet retweetability during hurricane disasters. *Int. J. Inf. Syst. Crisis Response Manage.* **2016**, *8*, 32–50. [CrossRef]
22. Kogan, M.; Palen, L.; Anderson, K.M. Think Local, Retweet Global: Retweeting by the Geographically-Vulnerable during Hurricane Sandy. In Proceedings of the 18th ACM Conference on Computer Supported Cooperative Work & Social Computing, Vancouver, BC, Canada, 14–18 March 2015.
23. Guikema, S.D.; Nateghi, R.; Quiring, S.M. Predicting Hurricane Power Outages to Support Storm Response Planning. *IEEE Access* **2017**, *2*, 1364–1373. [CrossRef]

24. Neppalli, V.K.; Caragea, C.; Caragea, D. Deep Neural Networks versus Naive Bayes Classifiers for Identifying Informative Tweets during Disasters. In Proceedings of the Information Systems for Crisis Response and Management Asia Pacific Conference, Rochester, NY, USA, 20–23 May 2018.
25. Chew, C.; Eysenbach, G. Pandemics in the Age of Twitter: Content Analysis of Tweets during the 2009 H1N1 Outbreak. *PLoS ONE* **2010**, *5*, e14118. [CrossRef] [PubMed]
26. Imran, M.; Diaz, F.; Elbassuoni, S. Practical Extraction of Disaster-Relevant Information from Social Media. In Proceedings of the 22nd International Conference on World Wide Web, Rio de Janeiro, Brazil, 13–17 May 2013.
27. Michael, A. Nielsen, Neural Networks and Deep Learning, Determination Press 2015. Available online: http://neuralnetworksanddeeplearning.com (accessed on 22 July 2019).
28. Mikolov, T.; Sutskever, I.; Chen, K.; Corrado, G.; Dean, J. Distributed representations of words and phrases and their compositionality. In Proceedings of the International Conference on Neural Information Processing Systems, Lake Tahoe, Nevada, 8–13 December 2013; pp. 3111–3119.
29. Kim, Y. Convolutional Neural Networks for Sentence Classification. *arXiv* **2014**, arXiv:1408.5882.
30. Zhang, X.; Zhao, J.; Lecun, Y. Character-level Convolutional Networks for Text Classification. In Proceedings of the 28th International Conference on Neural Information Processing Systems, Montreal, QC, Canada, 7–12 December 2015; pp. 649–657.

Journal of
Marine Science and Engineering

Article

Received Signal Strength Indication (RSSI) of 2.4 GHz and 5 GHz Wireless Local Area Network Systems Projected over Land and Sea for Near-Shore Maritime Robot Operations

Brennan Yamamoto *, Allison Wong, Peter Joseph Agcanas, Kai Jones, Dominic Gaspar, Raymond Andrade and A Zachary Trimble

Renewable Energy, Industrial Automation, Precision Engineering Laboratory, University of Hawai'i at Mānoa, 2540 Dole Street, Honolulu, HI 96822, USA
* Correspondence: brennane@hawaii.edu

Received: 27 June 2019; Accepted: 24 August 2019; Published: 27 August 2019

Abstract: The effect of the maritime environment on radio frequency (RF) propagation is not well understood. In this work, we study the propagation of ad hoc 2.4 GHz and 5 GHz wireless local area network systems typically used for near-shore operation of unmanned surface vehicles. In previous work, maritime RF propagation performance is evaluated by collecting RSSI data over water and comparing it against existing propagation models. However, the multivariate effect of the maritime environment on RF propagation means that these single-domain studies cannot distinguish between factors unique to the maritime environment and factors that exist in typical terrestrial RF systems. In this work, we isolate the effect of the maritime environment by collecting RSSI data over land and over seawater at two different frequencies and two different ground station antenna heights with the same physical system in essentially the same location. Results show that our 2.4 GHz, 2 m antenna height system received a 2 to 3 dBm path loss when transitioning from over-land to over-seawater (equivalent to a 25 to 40% reduction in range); but increasing the frequency and antenna height to 5 GHz, 5 m respectively resulted in no meaningful path loss under the same conditions; this reduction in path loss by varying frequency and antenna height has not been demonstrated in previous work. In addition, we studied the change in ground reflectivity coefficient, R, when transitioning from over-land to over-seawater. Results show that R remained relatively constant, $-0.49 \leq R \leq -0.45$, for all of the over-land experiments; however, R demonstrated a frequency dependence during the over-seawater experiments, ranging from $-0.39 \leq R \leq -0.33$ at 2.4 GHz, and $-0.51 \leq R \leq -0.50$ at 5 GHz.

Keywords: RSSI; WLAN; airmax; path loss; free space; two-ray

1. Introduction

RF-based wireless communication is a vital tool for many industries. However, despite its prolific use for terrestrial applications, the effect of the maritime environment on RF propagation is still not well-characterized. An excellent listing of various factors that can affect RF propagation in the maritime environment is compiled in [1]. These factors include weather, sea conditions and sea state, atmospheric effects, transmitter mobility, and the specifications of the wireless system employed. Clearly, RF propagation in a maritime environment is a highly multivariate problem.

In this paper, we study the real-world propagation of an RF system appropriate for near-shore unmanned surface vehicle operations. Example applications for near-shore surface vehicles include maritime search-and-rescue, disaster response, coral reef surveying, etc. These RF systems are typically deployed "ad hoc", require high bandwidths (>20 Mb/s), low antenna heights (<5 m), and a strong receive signal strength (>−80 dBm) to ensure a reliable connection and to overpower ambient RF interference. For this reason, we chose the Ubiquiti Airmax wireless standard, which is designed to meet the needs of outdoor wireless local area networks (WLAN). We evaluate two systems, one system based on the 2.4 GHz frequency band, and one based on the 5 GHz frequency band; both of which are nearly ubiquitously legal to operate unlicensed worldwide. Previous research in maritime RF propagation compares measured received signal strength (RSSI) data to various loss models to evaluate the efficacy of an RF system in the maritime environment; however, these studies are typically only limited to RSSI data collection over water with a specific RF system configuration that cannot be directly compared to similar experiments over land by other authors. Since the effect of the maritime environment on RF propagation is so multivariate, these single-domain studies make it impossible to understand how RF propagation is impacted due to factors unique to the maritime environment versus factors related to a typical deployment of the RF system over land. To address this degeneracy, we isolate the effects of transmission over seawater on RF propagation in a maritime environment by collecting RSSI data from an ad hoc WLAN over land. Then, using the same system in essentially the same location, collect RSSI data over seawater. The differences in the RSSI data between the over-land and over-seawater deployments in the same maritime environment, allows us to isolate the factors related to differences in the RF system, and the factors related to a typical RF deployment over land. A short summary of the previous research most relevant to this work is listed below.

In Ref. [1], a survey on previous references is maritime RF propagation was compiled; this information informed the previously mentioned list of factors affecting RF propagation in maritime environments. Most references in this work compared empirically gathered RF propagation data over water to various existing RF propagation models. These include the free space path loss (FSPL) model, irregular terrain model (ITM), international telecommunication union (ITU) family of models, log-normal or log-distance model, two ray ground reflection (two-ray) mode, and three-ray ground reflection (three-ray) models. Some references proposed modified versions of these existing models based on their analyses.

In Ref. [2], RF propagation data was collected in the 5 GHz frequency band, with antenna heights of $h_t = 1.7$ m and $h_r = 9.8$ m over an antenna separation distance of 10 km. This study also investigated line-of-sight and non-line-of-sight measurements intended for operations near urban environments. This data was compared to the FSPL and two-ray propagation models. Results show that the two-ray model fits the data well for large scale path loss in line-of-sight condition. As distance increases, the data demonstrated a higher rate of signal attenuation.

In Ref. [3], RF propagation data was collected in the 2.4 GHz frequency band, with antenna heights of $h_t = 3$ m, $h_r = 4.5$ m, over an antenna separation distance of 2 km. This data was compared to the two-ray and ITU-R P.1546 propagation models. Results show that the two-ray model fit the data well, but the ITU-R P.1546 model did not. Weather conditions and wave fluctuations were also considered to analyze the received power difference in detail.

In Ref. [4], RF propagation data was collected in the 5 GHz frequency band, with antenna heights of $h_t = 5.45$ m, $h_r = \{1.9, 2.7, 9.8\}$ m, over an antenna separation distance of 3 km. This study also investigated line-of-sight and non-line-of-sight measurements intended for operations near urban environments. This data was compared to the log-normal propagation models. Results show that the log-normal propagation model fit the data well, with a path-loss exponent between 2.7 and 5.6. In addition, it was shown that small-scale measurements could be approximated by the Type I extreme value distribution function with location parameter and scale parameter values of −15.7 dB and 3.3 dB, respectively.

In Ref. [5], RF propagation data was collected in the 5 GHz frequency band, with antenna heights of $h_t = 3.5$ m, $h_r = \{7.6, 10, 20\}$ m, over an antenna separation distance of 10 km. This data was compared to the FSPL, two-ray, and three-ray propagation models. Results showed that the two-ray propagation model fits the data when inside the crossover distance of the two-ray model. Outside of the crossover distance, the three-ray path loss model, which considers a refracted wave the evaporation duct and the reflective wave considered in the two-ray model, is a better model.

In Ref. [6], RF propagation data was collected in the 1.95 GHz frequency band, with antenna heights of $h_t = 22$ m, $h_r = 2.5$ m, over an antenna separation distance of 10 km. This data was compared to the FSPL, two-ray, and three-ray propagation models. This data was compared to the FSPL model. Results demonstrated a path loss of 40 dB/decade, which does not match the prediction made by the FSPL model. This was not noted in Ref. [6], but this result supports the general consensus in the literature that the FSPL model (which drops off at 20 dB/decade) is only valid for antenna separation distances less than the cutoff distance. The 40 dB/decade path loss matches the prediction made by the two-ray model for the ranges studied. In addition, [6] also noted that the mean standard deviation of the maritime data was 10.3 dB, which is notably lager than the 8 dB mean standard deviation typical for systems over land.

In Ref. [7] RF propagation data was collected in the 968 MHz, 3.5 GHz, and 5.8 GHz frequency bands, with antenna heights of $h_t = \{12, 14\}$ m, $h_r = \{2.5, 5, 50\}$ m, over an antenna separation distance of 5 km. This data was compared to the free-space and two-ray propagation models. Results showed that the two-ray propagation model best fit the data, except when in near proximity of the antennas, where the model was understandably broken due to the antenna gain aperture not being modeled in the near field, as well as additional reflections due to stiffs and high cliffs in the proximity of the shore.

In Ref. [8] RF propagation data was collected in the 5 GHz frequency band, with antenna heights of $h_t = \{4, 76, 185\}$ m, $h_r = 8$ m, over antenna separation distances of $d = \{7, 12, 20\}$ km. The unusually high antenna heights considered were the focus of this work. Results showed that placing the transmitter higher fits the free-space model more as there is more line of sight and diminished effect of the multipath fading. By contrast, the two-ray model fit the lower antenna heights better, even in no line of sight scenarios.

In Ref. [9] RF propagation data was collected in the 412 MHz frequency band, with antenna heights of $h_t = 15$ m, $h_r = 15$ m, over antenna separation distances of $d = \{33, 48\}$ km. This study is the most similar reference to the work in our paper, as it recorded data over land and seawater. The measured data over the sea demonstrated a mean path loss delta of 10 dB when transitioning from land to sea. Other interesting phenomena observed were: a 1.1 dB path loss for every meter of tide increase, which is consistent with the ITU-R P.526-7 path loss model, and a 1 to 2 dB of path loss during calm sea states, but 3 to 6 dB path loss in sea state 4 (approximately).

Based on these references, the most common propagation models for the study of maritime RF propagation are the FSPL model, and two-ray model. In reality, the FSPL is contained within the approximate two-ray model, as will be discussed in this paper. For the near-shore, low-antenna height, operational environments we studied, these models are the most relevant and are utilized for the analysis in this paper.

2. Materials and Methods

2.1. Path Loss Models

In this work, we aim to characterize the relationship between receive signal strength indication (RSSI) and distance between the transmitting and receiving antenna. The transmitting power of an access point is typically described in units of dBm, which is the ratio of the power in the signal referenced to 1 mW in a log10 scale. Conversion equations between the mW and dBm scales used in this work are given in Equation (1).

$$\begin{cases} P\,[\text{dBm}] = 10 \cdot \log_{10}\left(\frac{P\,[\text{mW}]}{1\,[\text{mW}]}\right) \\ P\,[\text{mW}] = 10^{\frac{(P\,[\text{dBm}])}{10}} \end{cases} \tag{1}$$

One convenient method of describing the radiated power of a transmitting access point is the equivalent isotropically radiated power (EIRP), which represents the power an isotropic antenna shall emit to produce the power observed in the direction of maximum antenna gain. The logarithmic formulation for EIRP is given in Equation (2).

$$EIRP = P_t + G_t - L_t \tag{2}$$

where P_t is the access point transmitting power, G_t is the transmitting isotropic antenna gain, and L_t is the cable loss between the access point and antenna. In this work, we investigate the RSSI vs. range of a 2.412 GHz and 5.240 GHz WLAN system. The EIRP of these two systems were matched to ensure that the two systems would demonstrate reasonably similar RSSI vs. range results. Note that physically matching the EIRP of the two systems is not necessarily sufficient to normalize the range of the two different frequencies; however, it will generally ensure that the two systems produce RSSI within the same order of magnitude.

In this work, we investigate two different loss models; the free-space path loss (FSPL), and the two-ray ground reflection (two-ray) models. The FSPL model is based on the Friss transmission formula [10]. This is given in Equation (3) [11].

$$\frac{P_r}{P_t} = \frac{G_t G_r \lambda^2}{P_l (4\pi d)^2} \tag{3}$$

where P_r is the receiving power in mW, P_t is the transmitting power in mW, G_t is the transmitting antenna gain in mW, G_r is the receiving antenna gain in mW, λ is the wavelength in m, P_l are generalized path losses in mW, and d is the distance between the two antennas in m. The path loss of any transmission model is defined as the ratio of the transmitting power, P_t, over the receiving power, P_r, i.e., the reciprocal of Equation (3) in the case of the FSPL model. Because the loss ratio is typically very large, it is conventionally described logarithmically in dBm. Solving Equation (3) for P_r, and describing the result logarithmically in dBm results in the FSPL model shown in Equation (4).

$$P_r \, [\text{dBm}] = 10 \log_{10} \left(\frac{P_t G_t G_r \lambda^2}{P_l (4\pi d)^2} \right) \tag{4}$$

Note that transmitting power, P_t, and antenna gains, G_t and G_r, are typically expressed in dBm. For brevity, Equation (4) expresses these variables in mW, and therefore, conversions between mW and dBm should be made using Equation (1), as appropriate.

The two-ray model considers the line-of-sight path assumed in the FSPL model, as well as a single path that is reflected off of the ground, based on the heights of the two antennas. The ground surface is characterized by a ground reflection coefficient, R. The full two-ray model is given in Equation (5) [11].

$$P_r = P_t \left(\frac{\lambda}{4\pi} \right)^2 \left| \frac{\sqrt{G_t G_r}}{\sqrt{d^2 + (h_t - h_r)^2}} + R \frac{\sqrt{G_t G_r} e^{-j \left(\frac{2\pi \left(\sqrt{d^2 + (h_t + h_r)^2} - \sqrt{d^2 + (h_t - h_r)^2} \right)}{\lambda} \right)}}{d^2 + (h_t + h_r)^2} \right|^2 \tag{5}$$

where h_t is the height of the transmitting antenna in m, h_r is the height of the receiving antenna in m, and R is the ground reflection coefficient (unitless); reference Equation (3) for the remaining variable names. A ground reflection coefficient of $R = -1$ represents perfect ground reflection, and $R = 0$ represents zero ground reflection [11–13].

It is common practice to approximate the two-ray path model given in Equation (5) using three piecewise approximation functions based on the distance from the transmitting antenna. This equation set is given in Equation (6) [11].

$$
P_r = \begin{cases}
\dfrac{P_t G_t G_r \lambda^2}{(4\pi)^2 \left(d^2 + h_t^2\right) P_l}, & if\ d < h_t,\ region\ 1 \\[3mm]
\dfrac{P_t G_t G_r \lambda^2}{(4\pi)^2 d^2 P_l}, & if\ h_t \leq d \leq d_c,\ region\ 2 \\[3mm]
\dfrac{P_t G_t G_r h_t^2 h_r^2}{d^4 P_l}, & if\ d > d_c,\ region\ 3
\end{cases}
\tag{6}
$$

In Equation (6), d_c is called the crossover distance, and is it defined in Equation (7).

$$
d_c = \frac{4 h_t h_r}{\lambda}
\tag{7}
$$

There are important intuitions to be drawn from the piecewise two-ray approximation model in Equation (6). Firstly, the extent of region one is defined by the height of the transmitting antenna, which is usually small relative to the total operating range of the wireless system. As such, region one is typically of little concern to researchers since its relevant range is very close to the transmitting antenna, on the order of meters. Secondly, the extent of region two is defined between region one and the crossover distance in Equation (7). In this region, the two-ray approximation model is identical to the FSPL model given in Equation (4), and power falls off at −20 dB/decade. Thirdly, region three consists of everything outside of the crossover distance. In this region, power falls off at −40 dB/decade.

In this work, the measured RSSI data is compared against the full, and approximate two-ray models, assuming that a generalized path loss variable, P_l, is introduced into the real data. We hypothesize that the transition from over-land to over-seawater will manifest as an increase to this path loss variable.

Employing both the approximate and full version of the two-ray model is beneficial because the approximate two-ray model is useful for identifying path loss, whereas the full two-ray model also captures the constructive-destructive interference effect of ground reflection, and varying antenna heights. Because the RSSI data was collected with equivalent antenna heights for all experiments, fitting the data to the full two-ray model should enable us to observe a change in ground reflectivity in the transition from over-land to over-seawater. In addition, the approximate two-ray model is more commonly cited in preceding literature (e.g. Ref. [1–3,5,14,15]), which is important to note when drawing comparisons between our work and previous literature.

2.2. Experiment Description

We selected the Sand Island Boat Launch Facility, located in Oahu, Hawaii, as the location for this study. This location is well suited for this experiment because it has a flat, approximately 0.5 km long entrance road with full line of sight. The over-land experiments were conducted on this entrance road, and the over-seawater experiments were conducted from the beach over the Sand Island channel. The two ground stations/origin points for each experiment were within 150 m of each other. The approximate over-land and over-seawater propagation paths are shown in Figure 1.

Data were collected on two different days: 27 July 2019, and 15 August 2019. The protected nature of the Sand Island channel demonstrated a Beaufort Sea state of one (rippling waters and light air) for both days of testing. The height of the transmitting and receiving antennas was set based on the height of the seawater surface at the time of testing. At this location, the tide can fluctuate between from 0 m to 0.6 m over a period of roughly 6 h. Data collection over seawater lasted no more than 2 h on either day, resulting in a maximum error of 0.2 m in height due to shifting tides.

Figure 1. Over-land data collection path on Sand Island access road (green), and over-seawater data collection path in the Sand Island channel (blue). Map data © 2018 Google. Reproduced with permission from Google Maps Geoguidelines, 2019 [16].

The Sand Island access road contains an occasional passing vehicle, and the Sand Island channel contains the occasional passing small boats or jet skis. During the experiment, if a passing car or vessel obstructed the line-of-sight between the transmitting antenna on the ground station and the receiving antenna on the mobile station, the data collection was paused, preventing those points from being recorded.

Two WLAN frequency bands were tested in this experiment, 2.4 GHz and 5 GHz. Note that the names "2.4 GHz" and "5 GHz" do not refer to the exact frequency employed, rather, they refer to the family of standardized IEEE 802.11 wireless channels, which specifies center frequency and bandwidth. The Airmax WLAN protocol used in this experiment adopts the IEEE 802.11 channel specification. Specific details about each system are listed in Table 1.

Table 1. Radio frequency (RF) System Specifications.

Specification	RF System	
	2.4 GHz WLAN	**5 GHz WLAN**
IEEE 802.11 channel	1	48
Center frequency	2412 MHz	5240 MHz
Channel width	20 MHz	20 MHz
Access point model	Ubiquiti BulletAC-IP67	Ubiquiti BulletAC-IP67
WLAN protocol	Ubiquiti Airmax	Ubiquiti Airmax
Transmitting power	18 dBm (63.1 mW)	16 dBm (39.8 mW)
Antenna model	Trendnet TEW-AO57	Trendnet TEW-AO57
Antenna type	Omnidirectional	Omnidirectional
Antenna gain	5 dBi	7 dBi
Antenna vertical beam width	30°	15°
Antenna polarization	Vertical	Vertical
Transmit EIRP	23 dBm	23 dBm

The settings and equipment were identical for the transmitting and receiving ends of each RF system. The selected access point model (Ubiquiti BulletAC-IP67) is equipped with both 2.4 GHz, and 5 GHz transmitters, allowing us to test both frequencies while eliminating any variability due to differences in the access point hardware.

The RSSI data was recorded directly from the Ubiquiti dashboard software. This software runs on the computer connected to the access point. Because this dashboard software does not feature a RSSI data logging feature, we developed a JavaScript program to "scrape" and timestamp the relevant RSSI data from the dashboard back end at 1s intervals. This program and its development can be found at the GitHub repository [17], or in the Supplementary Materials.

The separation distance between the two antennas was measured using a global navigation satellite system (GNSS) sensor mounted to the stationary transmitting antenna ground station, and a second GNSS sensor mounted to the receiving antenna mobile station. The Adafruit Ultimate GPS was the selected GNSS sensor; this was connected to a laptop computer was running the Robot Operating System (ROS) Melodic Morenia software. The ROS package "nmea_navsat_driver" [18] was used to communicate with the GNSS sensor. ROS features a message data logging service known as "bags" [19], which was used to log and timestamp the GNSS sensor data. This bag file was later converted to a comma separate value (.csv) file for post-processing. For details on reproducing the GNSS data collection process, see Ref. [17], or the Supplementary Materials.

During the experiment, the transmitting antenna was located at the stationary ground station, and was mounted to a leveled tripod with a height-adjustable boom between 2 and 5 m. The receiving antenna was set to a fixed height of 2 m, and rigidly connected to the mobile station, which represents a robot. The mobile station comprised of a 9'4" rigid tender, which was pulled on a trailer during the over-land experiments, and manually driven using an outboard transom motor during the over-seawater experiments. An image of the ground station, transmitting antenna, mobile station, and receiving antenna for the over-land experiments is shown in Figure 2. A pair of images of the ground station, transmitting antenna, mobile station, and receiving antenna for the over-seawater experiments is shown in Figure 3.

Figure 2. Over-land experimental setup. Ground station transmitting antenna (mounted on tripod, right), and trailered rigid tender serving as mobile station with receiving antenna (left).

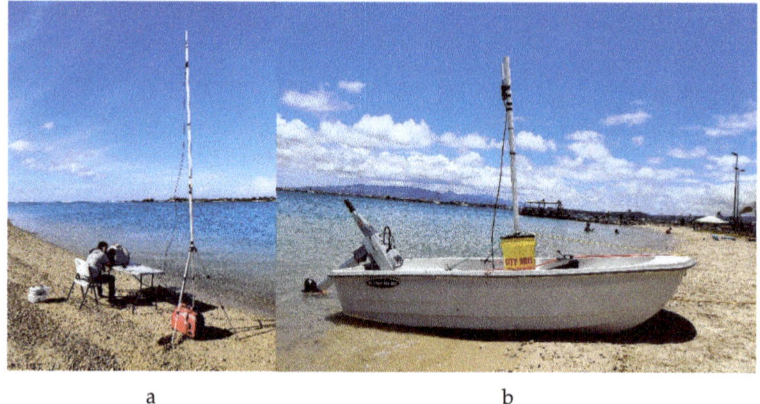

a   b

Figure 3. Over-seawater experimental setup. Ground station transmitting antenna mounted on tripod (**a**), and rigid tender serving as mobile station with receiving antenna (**b**).

3. Results

The collected RSSI measurements are plotted against distance in Figure 4. In all experiments, the receiving antenna height was set to $h_r = 2$ m. The transmitting antenna height was set to two different heights, $h_t = \{2,5\}$ m, and we evaluated two different WLAN frequencies, $f = \{2.412, 5.240\}$ GHz. All of the collected RSSI data is shown in Figure 4.

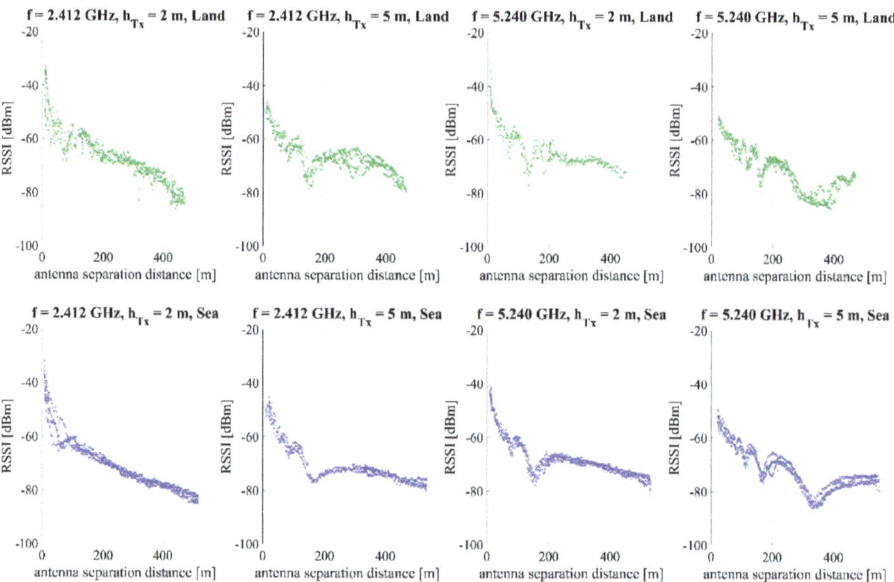

Figure 4. Compiled raw received signal strength indication (RSSI) data over land (green) and seawater (blue) plotted against distance for all eight experiments. RSSI versus distance measurements from all eight experiments (f = frequency, h_{Tx} = transmitting antenna height).

A statistical analysis of the binned mean, standard deviation, and confidence interval can be found in Appendix A.

The measured RSSI data is plotted against three theoretical path loss curves in Figure 5. These are:

- The full two-ray path loss model from Equation (5)
- The approximate region-two, two-ray path loss model ($h_t \leq d \leq d_c$). This is identical to the free space path loss model
- the approximate region-three, two-ray path loss model ($d > d_c$). The respective frequencies and transmitting antenna heights are labeled in their respective subplots.

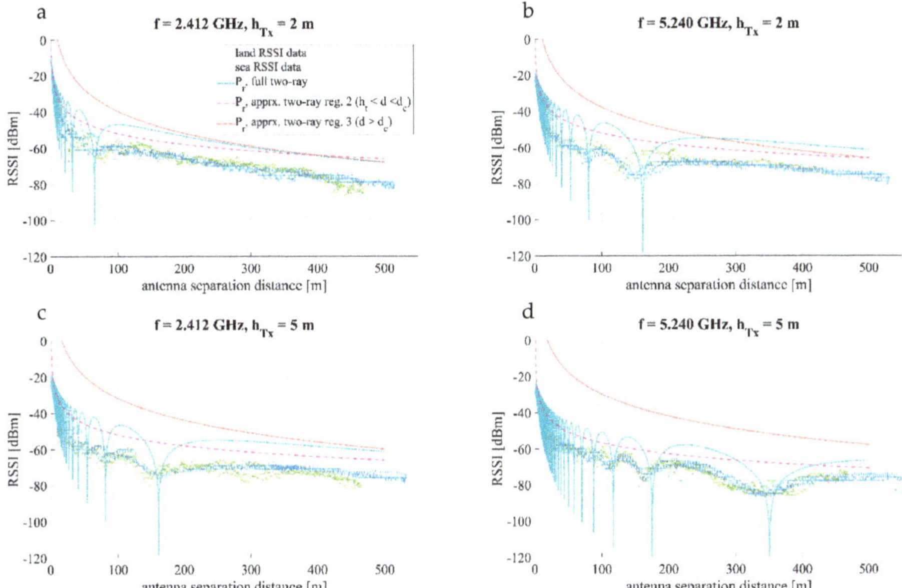

Figure 5. RSSI data over land (green), RSSI data over and seawater (blue), full two-ray model prediction (cyan), approximate region two, two-ray model prediction (magenta), and approximate region-three, two ray model (red), all plotted against distance. Subplot a plots the $f = 2.412$ GHz, $h_{Tx} = 2$ m data, subplot b plots the $f = 5.240$ GHz, $h_{Tx} = 2$ m data, subplot c plots the $f = 2.412$ GHz, $h_{Tx} = 5$ m data, and subplot d plots the $f = 5.240$ GHz, $h_{Tx} = 5$ m data. Model predictions assume RF system specifications given in Table 2, and perfect ground reflection coefficient ($R = -1$).

The theoretical path loss models plotted in Figure 5 demonstrate significant deviation from the measured RSSI data; this is expected since the theoretical models assume ideal power transmission, geometry, and ground reflectivity. In the case of the approximate two-ray models, there is an additional path loss variable, P_l, demonstrated by a constant power difference between the models and true data. In the case of the full two-ray model, there is a path loss variable, P_l, demonstrated by a constant power difference between the model and true data; a non-ideal ground reflection coefficient, R, demonstrated by the decreased amplitude of the constructive-destructive interference pattern; and small variations in the transmitting antenna height, h_{Tx}, demonstrated by the difference in the frequency of the constructive-destructive interference pattern. To characterize these coefficients, we fit the approximate two-ray model, and full two-ray model classes to the real data using a least squares minimization against these characteristic coefficients.

As discussed in Equation (6), the approximate two-ray model consists of three piecewise approximations, which are defined as a function of distance. The boundary between the first and second region is defined by the transmitting antenna height, h_t, which is negligible in this case due to the short antenna height. The boundary between the second and third region is defined by the crossover distance, d_c, defined by Equation (7). These crossover distances are listed in Table 2.

Table 2. Crossover Distance for Each Experiment.

Experiment	Crossover Distance, d_c [m]
$f = 2.412$ GHz, $h_t = 2$ m	404
$f = 2.412$ GHz, $h_t = 5$ m	1011
$f = 5.240$ GHz, $h_t = 2$ m	879
$f = 5.240$ GHz, $h_t = 5$ m	2196

In all of the experiments except for the 2 GHz, 2m experiment, the crossover distance is greater than the maximum distance for which measurements were taken. Therefore, the region-two, two-ray model in Equation (6) is the most appropriate approximation for this data. In the 2 GHz, 2m experiment, the approximate region-two, two-ray model was only fitted to data before the 404 m crossover distance.

In Figure 6, we fit the approximate region-two, two-ray model to all of the measured RSSI data. Because the only unknown variable in the approximate region-two, two-ray model is path loss, P_l, a single-variable least squares minimization against path loss, P_l, was sufficient to fit the theoretical model to the data.

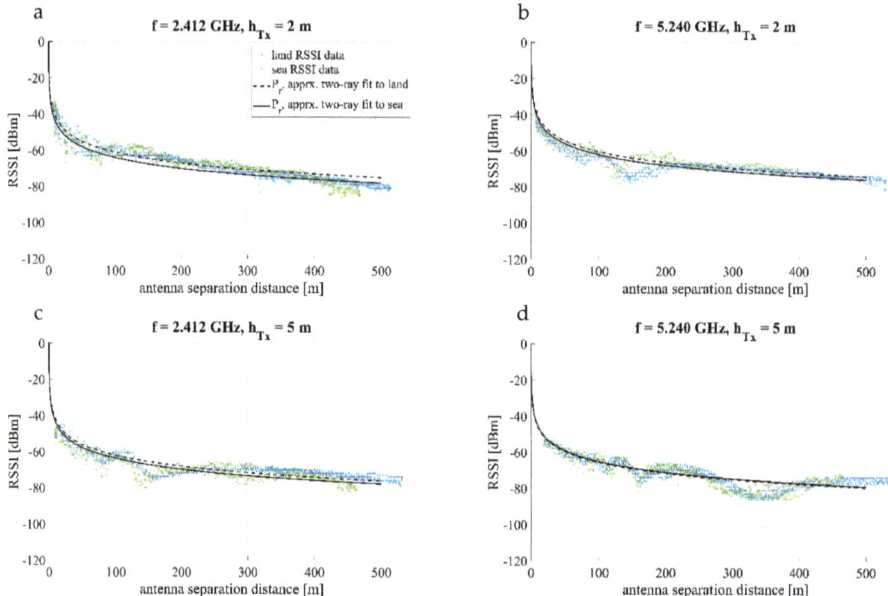

Figure 6. RSSI data plotted with unconstrainted single-variable least squares minimization against the two-ray, region two model, assuming a variable generalized path loss, P_l. Subplot a plots the $f = 2.412$ GHz, $h_{Tx} = 2$ m data, subplot b plots the $f = 5.240$ GHz, $h_{Tx} = 2$ m data, subplot c plots the $f = 2.412$ GHz, $h_{Tx} = 5$ m data, and subplot d plots the $f = 5.240$ GHz, $h_{Tx} = 5$ m data.

The fitted coefficients from the approximate region-two, two-ray model are compiled in Table 3.

Table 3. Approximate Region-Two, Two-Ray Fitted Path Loss Coefficients.

Experiment		P_l [dBm]	r-Squared Fit
$f = 2.412$ GHz	Land	−8.7	0.82
$h_t = 2$ m	Sea	−11.9	0.94
$f = 2.412$ GHz	Land	−10.0	0.55
$h_t = 5$ m	Sea	−11.7	0.87
$f = 5.240$ GHz	Land	−3.4	0.82
$h_t = 2$ m	Sea	−5.1	0.81
$f = 5.240$ GHz	Land	−8.8	0.74
$h_t = 5$ m	Sea	−8.3	0.75

The path loss difference when transitioning from over-land to over-seawater for each experiment is compiled in Table 4.

Table 4. Approximate Two-Ray Fitted Path Loss Delta Between Over-Land and Over-Seawater.

Experiment		$P_{l,land}-P_{l,sea}$ [dBm]
$f = 2.412$ GHz	$h_t = 2$ m	−3.0
$f = 2.412$ GHz	$h_t = 5$ m	−1.7
$f = 5.240$ GHz	$h_t = 2$ m	−1.7
$f = 5.240$ GHz	$h_t = 5$ m	+0.5

The approximate two-ray model path loss deltas compiled in Table 4 shows that the 2.4 GHz, 2 m experiment demonstrates a 3 dBm increase in path loss when transitioning from over-land to over-seawater; the 2.4 GHz, 5 m, and 5 GHz, 2 m experiments demonstrate a 1.7 dBm path loss when transitioning from over-land to over-seawater; and the 5 GHz, 5 m experiment demonstrates a 0.5 m *decrease* in path loss when transitioning from over-land to over-seawater. This indicates that a combination of higher antenna height and frequency contribute to reduced path loss when transitioning from over-land to over-seawater.

In Figure 7, we fit the full two-ray model to all of the measured RSSI data. In the full two-ray model, the path loss, P_l, ground reflection coefficient, R, and exact transmitting antenna height, h_{Tx}, are unknown and/or sensitive variables; therefore, a constrained multi-variable nonlinear least squares minimization against P_l, R, and h_{Tx}, was used to fit the theoretical model to the data. P_l was constrained to vary over the range $−25 < P_l < 0$ dBm, R over the range, $−1 < R < 0$, and h_{Tx} over the range $1.7 < h_{Tx} < 2.3$ m for the 2 m antenna experiments, and $4.7 < h_{Tx} < 5.3$ m for the 5 m antenna experiments. As a general rule, the path loss varies the height of the curve on the y-axis, the transmitter antenna height varies the frequency of the constructive-destructive interference pattern, and the reflection coefficient varies the amplitude of the constructive-destructive interference pattern.

The fitted coefficients from the full two-ray model are compiled in Table 5.

The path loss difference when transitioning from over-land to over-seawater for each experiment is compiled in Table 6.

The full two-ray model path loss deltas compiled in Table 6 demonstrate similar results to the approximate two-ray model path loss deltas given in Table 4, except for the 2.4 GHz, 2 m experiment, where the full two-ray model demonstrated a 2 dBm path loss delta, compared to the 3 dBm path loss delta of the approximate two-ray model. Inside the cutoff distance, d_c, where the RSSI decreases at a rate of roughly 20 dBm/decade, a path loss of 2 dBm results in an overall range decrease of 26%, and a path loss of 3 dBm is roughly equivalent to an overall range decrease of 41%. Because the 5 GHz 5 m experiment does not demonstrate this path loss, the combination of the higher 5 GHz frequency band in combination with 5 m transmitting antenna height results in a substantially smaller path loss over seawater.

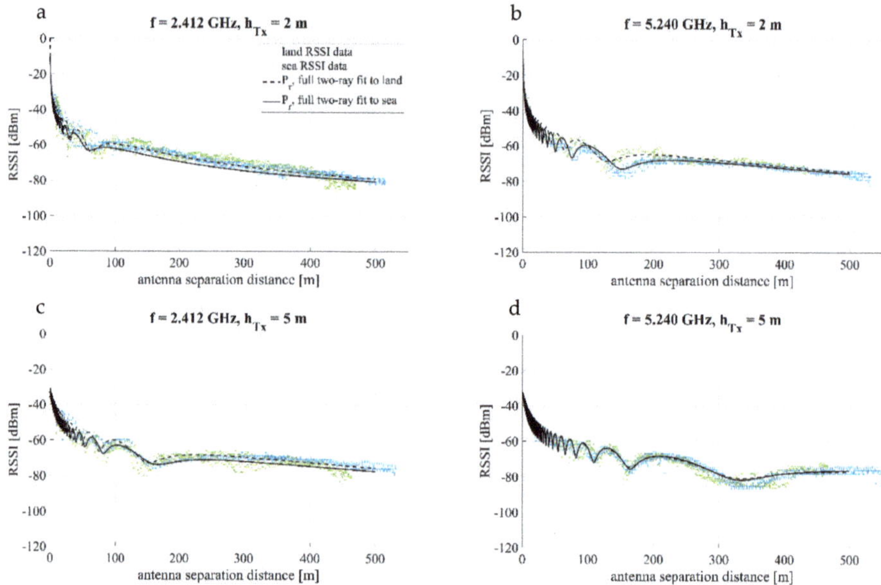

Figure 7. RSSI data plotted with constrained multi-variable nonlinear least squares minimization against the full two-ray model, assuming a variable generalized path loss, $-25 < P_l < 0$ dBm, ground reflectivity, $-1 < R < 0$, and transmitting antenna height, $1.7 < h_{Tx} < 2.3$ m for the 2 m antenna experiments and $4.7 < h_{Tx} < 5.3$ m for the 5 m antenna experiments. Subplot a plots the $f = 2.412$ GHz, $h_{Tx} = 2$ m data, subplot b plots the $f = 5.240$ GHz, $h_{Tx} = 2$ m data, subplot c plots the $f = 2.412$ GHz, $h_{Tx} = 5$ m data, and subplot d plots the $f = 5.240$ GHz, $h_{Tx} = 5$ m data.

Table 5. Full Two-Ray Fitted Path Loss, Transmitting Antenna Height, and Ground Reflectivity.

Experiment		P_l [dBm]	R []	h_t [m]	r-Squared Fit
$f = 2.412$ GHz	Land	−0.49	−0.49	1.9	0.86
$h_t = 2$ m	Sea	−0.33	−0.33	1.8	0.96
$f = 2.412$ GHz	Land	−0.48	−0.48	4.9	0.76
$h_t = 5$ m	Sea	−0.39	−0.39	5.1	0.94
$f = 5.240$ GHz	Land	−0.45	−0.45	1.9	0.89
$h_t = 2$ m	Sea	−0.50	−0.50	2.1	0.94
$f = 5.240$ GHz	Land	−0.45	−0.45	4.7	0.86
$h_t = 5$ m	Sea	−0.51	−0.51	4.7	0.91

Table 6. Full Two-Ray Fitted Path Loss Delta Between Over-Land and Over-Seawater.

Experiment	$P_{l,land} - P_{l,sea}$ [dBm]
$f = 2.412$ GHz $h_t = 2$ m	−2.0
$f = 2.412$ GHz $h_t = 5$ m	−1.8
$f = 5.240$ GHz $h_t = 2$ m	−2.0
$f = 5.240$ GHz $h_t = 5$ m	+0.1

Regarding ground reflection coefficient, *R*, the full two-ray model demonstrated a relatively constant $-0.49 \leq R \leq -0.45$ over the four different over-land experiments (2 GHz, 2 m; 2 GHz, 5 m; 5 GHz, 2 m; 5 GHz, 5 m); however, *R* changed significantly over the four different over-seawater experiments. The 2 GHz, 2 m experiment demonstrated $R = -0.33$; the 2 GHz, 5 m experiment demonstrated $R = -0.39$; the 5 GHz, 2 m experiment demonstrated $R = -0.50$; and the 5 GHz, 5 m experiment demonstrated $R = -0.51$. These results indicate that the R, as described in Equation (5) in [11], is constant when operating over land, but varies with frequency when operating over seawater. Because the ground reflection coefficient affects the amplitude of the constructive-destructive interference pattern inside the cutoff distance, a large $|R|$ can result in large changes in RSSI over relatively short distances; however, it does not have a significant effect on the overall "range" of the wireless system, especially when the antenna separation distance approaches and exceeds the cutoff distance.

4. Discussion

In this work, we studied the received signal strength indication (RSSI) of 2.4 GHz and 5 GHz wireless local area network (WLAN) systems over land and seawater, using transmitting antenna heights of $h_{Tx} = \{2, 5\}$ m, with a maximum antenna separation distance of 0.5 km. We fit these measurements to the approximate two-ray propagation model (also called the free space path loss model) and the full two-ray path loss model. Results showed that the 2.4 GHz system demonstrated an increase in path loss of 2 to 3 dBm when transitioning from over-land to over-seawater for both antenna heights; in practice, a 2 to 3 dBm path loss is equivalent to reduction of 25-40% in range for a WLAN. Interestingly, the 5 GHz system with a 5m transmitting antenna height did not demonstrate any significant change in path loss when transitioning from over-land to over-seawater. Therefore, if all other variables are equal, given the choice between the 2.4 GHz and 5 GHz frequency bands for near-shore, WLAN operations, the 5 GHz frequency band coupled with an antenna height of roughly 5 m will minimize the path loss penalty of operating over seawater, which also indicates future near-shore unmanned systems designers can expect accurate range predictions from 5 GHz, 5 m systems in practice. To the authors' knowledge, no other work prior has demonstrated this difference in path loss delta between 2.4 GHz and 5 GHz systems. Because only two frequencies were tested in this work, it is dangerous to extrapolate this assumption for all frequencies; however, previous work in [9] demonstrated a 10 dBm loss when transitioning from over-land to over-seawater for a 412 MHz system; this result weakly agrees with our experiment.

In addition, we also investigated the effect of the ground reflection coefficient, *R*, by fitting the full two-ray path model given in Equation (5), to the collected data. We found that *R* remained relatively constant $(-0.49 \leq R \leq -0.45)$ for all of the over-land experiments; however, *R* changed significantly over the four different over-seawater experiments. The 2 GHz, 2 m experiment demonstrated $R = -0.33$; the 2 GHz, 5 m experiment demonstrated $R = -0.39$; the 5 GHz, 2 m experiment demonstrated $R = -0.50$; and the 5 GHz, 5 m experiment demonstrated $R = -0.51$. These results indicate that the *R*, as described in Equation (5) in [11], is constant when operating over land, but varies with frequency when operating over seawater. *R* is generally considered to be a function of the antenna polarization direction, the angle of incidence with the ground, θ, and the relative complex permittivity of the land and/or seawater surface, η [11,13]. In particular the η of a medium depends slightly on frequency because of the Debye-Falkenhagen effect; however, this effect is generally considered to be negligible [20], and does not account for the significant change in *R* measured in this work. Previous work also suggests that the form of the constructive-destructive interference pattern is sensitive to sea-surface roughness, and the refraction of the propagating waves [21,22]; which may account for this discrepancy.

In future work, one should focus on models explaining why the 2.4 GHz frequency band demonstrates a greater path loss than the 5 GHz frequency band when transitioning from over-land to over-seawater. The approximate two-ray/free space and full two-ray models can be fitted to real data to describe this path loss; however, these models are currently insufficient to predict this phenomena. In addition, the ground reflection coefficient, R, also demonstrated a frequency dependence that is not predicted by current models. A similar experiment involving more frequencies should better describe the nature of this frequency dependence.

Supplementary Materials: The code and open-source software packages used to collect experimental data, the raw data, and the code used to generate the figures in this text are available online at https://github.com/riplaboratory/wireless_benchmarking.

Author Contributions: Contributions from each author in this work can be described by: conceptualization, B.Y. and A.Z.T.; methodology, A.W., K.J., and B.Y.; software, B.Y. and R.A.; validation, B.Y., A.Z.T., and A.W.; formal analysis, B.Y. and A.Z.T.; investigation, P.J.A., A.Z.T., A.W., D.G., B.Y., and R.A.; resources, A.W., K.J., and P.J.A.; data curation, B.Y., R.A., A.W., and P.J.A.; writing—original draft preparation, B.Y., A.W, K.J., and P.J.A.; writing—review and editing, B.Y., A.Z.T., A.W., P.J.A., K.J., D.G., and R.A.; visualization, B.Y.; supervision, A.Z.T.; project administration, B.Y. and A.W.; funding acquisition, A.Z.T.

Funding: This research was funded by the Office of Naval Research, grant number N00024-08-D-6323-TO28.

Acknowledgments: The authors would like to thank Serena Kobayashi, Kealoha Moody, Jordan Romanelli, and Daniel Truong for their assistance in collecting experimental data. Thank you also to Google Maps for making their map data available for publication in research papers under Ref. [16].

Appendix A

In this section, we analyze the mean, standard deviation, and the 95% confidence interval of the collected RSSI versus distance data. In this experiment, the distance data was collected using a global navigation satellite system (GNSS) sensor mounted to the transmitting antenna ground station, and another GNSS sensor mounted to the receiving antenna mobile station. The RSSI data was collected from the access point management software. Both the GNSS and RSSI data were collected at approximately 1 Hz; however, because both sources of information were collected separately, they are not time synchronized. The time synchronization of this data was performed in post-processing by linearly interpolating the distance data to the collected RSSI data. The mobile station traveled at roughly 1.5 m/s over the roughly 500 m data collection distance. This results in roughly 330 data points per data run. For each of the eight experiments (2 GHz/2 m/land, 2 GHz/2 m/sea, 2 GHz/5 m/land, 2 GHz/5 m/sea, 5 GHz/2 m/land, 5 GHz/2 m/sea, 5 GHz/5 m/land, and 5 GHz/5 m/sea) four data runs were taken, with the exception of the 5 GHz/2 m/land experiment, where one of the four data runs was removed due to clearly erroneous data. In total, this results in roughly 1320 data points per experiment. To take the mean, the RSSI data was separated into equally-spaced 2.5 m "bins" from 0 to 550 m. At a bin size of 2.5 m, this resulted in roughly 6 data points per bin. The mean, standard deviation, and 95% confidence interval was taken for the data in each bin. This information is plotted on Figure A1.

The mean standard deviation and 95% confidence interval for each experiment is given in Table A1.

Table A1. Mean Standard Deviation and Mean 95% Confidence Interval.

Experiment		$\bar{\sigma}$ [dBm]	$\overline{95\%\ CI}$
$f = 2.412\ \text{GHz}$ $h_t = 2\ \text{m}$	Land	2.00	1.68
	Sea	1.34	0.93
$f = 2.412\ \text{GHz}$ $h_t = 5\ \text{m}$	Land	2.07	1.82
	Sea	0.99	0.87
$f = 5.240\ \text{GHz}$ $h_t = 2\ \text{m}$	Land	1.44	1.64
	Sea	1.04	0.80
$f = 5.240\ \text{GHz}$ $h_t = 5\ \text{m}$	Land	1.57	1.42
	Sea	1.46	1.08

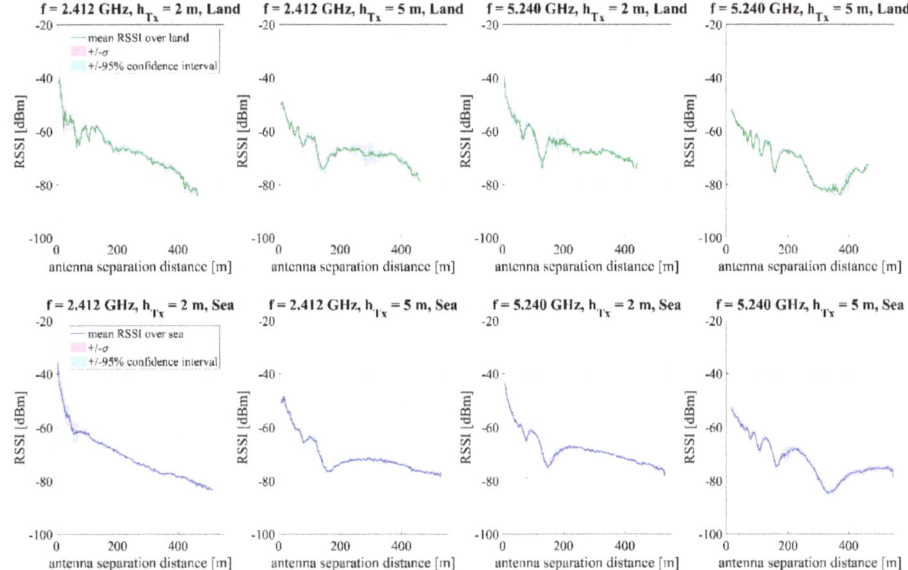

Figure A1. Mean (green for land data, blue for sea data), standard deviation (translucent magenta), and 95% confidence interval (translucent cyan) of raw RSSI versus distance data with a bin size of 2.5 m.

The standard deviation describes the "spread" of normally distributed data, and is given in Equation (A1).

$$\sigma = \sqrt{\frac{\Sigma(x - \overline{x})^2}{n}} \tag{A1}$$

where x is a data point, $\overline{x}$ is the mean of all data points, and n is the total number of data points. The confidence interval describes the "spread" of the data normalized by the number of data points taken; this interval gives you information about the precision/repeatability of the collected data points. It is given in Equation (A2).

$$CI = Z^* \cdot \frac{\sigma}{\sqrt{n}}, \; where \; Z^* = \begin{cases} 1.645 \; for \; \alpha = 90\% \\ 1.96 \; for \; \alpha = 95\% \\ 2.575 \; for \; \alpha = 99\% \end{cases} \tag{A2}$$

where α is the confidence level and Z^* is the confidence coefficient. In Figure A1, a confidence level of 95% was chosen; i.e., taking an additional data point is 95% likely to fall within the calculated confidence interval. Because the 95% confidence interval is generally smaller than the standard deviation in Figure A1 and Table A1, we are confident that sufficient data was collected to ensure that the calculated standard deviation is a good reflection of the true standard deviation of the data.

References

1. Balkees, S.; Sasidhar, K.; Rao, S. A survey based analysis of propagation models over the sea. In *Proceedings of the 2015 International Conference on Advances in Computing, Communications and Informatics (ICACCI)*; IEEE: Kochi, India, 2015; pp. 69–75.
2. Reyes-Guerrero, J.C.; Bruno, M.; Mariscal, L.A.; Medouri, A. Buoy-to-ship experimental measurements over sea at 5.8 GHz near urban environments. In *Proceedings of the Mediterranean Microwave Symposium (MMS), 2011 11th*; IEEE: Hammamet, Tunisia, 2011; pp. 320–324.

3.	Lee, J.H.; Choi, J.; Lee, W.H.; Choi, J.W.; Kim, S.C. Measurement and Analysis on Land-to-Ship Offshore Wireless Channel in 2.4 GHz. *IEEE Wirel. Commun. Lett.* **2017**, *6*, 222–225. [CrossRef]
4.	Reyes-Guerrero, J.C.; Mariscal, L.A. 5.8 GHz propagation of low-height wireless links in sea port scenario. *Electron. Lett.* **2014**, *50*, 710–712. [CrossRef]
5.	Yee Hui, L.E.E.; Dong, F.; Meng, Y.S. Near sea-surface mobile radiowave propagation at 5 GHz: Measurements and modeling. *Radioengineering* **2014**, *23*, 825.
6.	Choi, D.Y. Measurement of radio propagation path loss over the sea for wireless multimedia. In *Proceedings of the International Conference on Research in Networking*; Springer: Heidelberg, Germany, 2006; pp. 525–532.
7.	Le Roux, Y.-M.; Ménard, J.; Toquin, C.; Jolivet, J.-P.; Nicolas, F. Experimental measurements of maritime radio transmission channels. In *Proceedings of the 9th International Conference on Intelligent Transport Systems Telecommunications (ITST)*; IEEE: Lille, France, 2009.
8.	Joe, J.; Hazra, S.K.; Toh, S.H.; Tan, W.M.; Shankar, J.; Hoang, V.D.; Fujise, M. Path loss measurements in sea port for WiMAX. In *Proceedings of 2007 IEEE Wireless Communications and Networking Conference*; IEEE: Kowloon, China, 2007; pp. 1873–1878.
9.	Sim, C.Y.D. The propagation of VHF and UHF radio waves over sea paths. Ph.D. Thesis, University of Leicester, Leicester, UK, November 2002.
10.	Friis, H.T. A note on a simple transmission formula. *Proc. IRE* **1946**, *34*, 254–256. [CrossRef]
11.	Mathuranathan, V. (Ed.) *Wireless Communication Systems in MATLAB*; 2018; independently published; ISBN 9781720114352.
12.	Rappaport, T.S. *Wireless Communications: Principles and Practice*; Prentice Hall: Upper Saddle River, NJ, USA, 2001; ISBN 978-0130422323.
13.	Hubert, W.; Le Roux, Y.-M.; Ney, M.; Flamand, A. Impact of ship motions on maritime radio links. *Int. J. Antennas Propag.* **2012**, *2012*, 1–6. [CrossRef]
14.	Zhao, Y.; Ren, J.; Chi, X. Maritime mobile channel transmission model based on ITM. In *Proceedings of the 2nd International Symposium on Computer, Communication, Control and Automation*; Atlantis Press: Singapore, 2013.
15.	Ang, C.-W.; Wen, S. Signal strength sensitivity and its effects on routing in maritime wireless networks. In *Proceedings of the 2008 33rd IEEE Conference on Local Computer Networks (LCN)*; IEEE: Montreal, QC, Canada, 2008; pp. 192–199.
16.	Google Maps & Google Earth GeoGuidelines. Available online: https://www.google.com/permissions/geoguidelines/ (accessed on 20 June 2019).
17.	Riplaboratory Kanaloa Wireless Benchmarking. Available online: https://github.com/riplaboratory/wireless_benchmarking (accessed on 20 June 2019).
18.	Venator, E. Ros Developers ROS nmea_navsat_driver. Available online: http://wiki.ros.org/nmea_navsat_driver (accessed on 20 June 2019).
19.	ROS. ROS Bags. Available online: http://wiki.ros.org/Bags (accessed on 20 June 2019).
20.	Spiwak, R.R. Equations for Calculating the Dielectric Constant of Saline Water. *IEEE Trans. Microw. Theory Tech.* **1970**, *19*, 733–736.
21.	Timmins, I.J.; O'Young, S. Marine communications channel modeling using the finite-difference time domain method. *IEEE Trans. Veh. Technol.* **2009**, *58*, 2626–2637. [CrossRef]
22.	Benhmammouch, O.; Caouren, N.; Khenchaf, A. Influence of sea surface roughness on electromagnetic waves propagation in presence of evaporation duct. In *Proceedings of the Radar Conference-Surveillance for a Safer World*; IEEE: Bordeaux, France, 2009; Volume 1, pp. 1–6.

Journal of
Marine Science and Engineering

MDPI

Article

Modeling Analysis and Simulation of Viscous Hydrodynamic Model of Single-DOF Manipulator

Minglu Zhang, Xiaoyu Liu and Ying Tian *

School of Mechanical Engineering, Hebei University of Technology, Tianjin 300132, China
* Correspondence: flyserelo@126.com; Tel.: +86-136-8207-8144

Received: 15 July 2019; Accepted: 7 August 2019; Published: 9 August 2019

Abstract: Hydrodynamic modeling is the basis of the precise control research of underwater manipulators. Viscous hydrodynamics, an important part of the hydrodynamic model, directly affects the accuracy of the dynamic model and the control model of the manipulator. Considering the limited research on viscous hydrodynamics of underwater manipulators and the difficulty in measuring viscous hydrodynamic coefficients, the viscous hydrodynamic model in the form of Taylor expansion is analyzed and established. Through carrying out simulation calculations, curve fitting and regression analysis, positional derivatives, rotational derivatives, and coupling derivatives in the viscous hydrodynamic model, are determined. This model provides a crucial theoretical foundation and reference data for subsequent related research.

Keywords: viscous hydrodynamics; numerical calculation; underwater manipulator

1. Introduction

With the rapid development of the marine industry, the application of robots mounted on underwater vehicles has become increasingly widespread. Underwater manipulators are mainly employed in the acquisition and exploration, etc. Their underwater performance can be determined by their speed and accuracy. Underwater manipulators are usually subjected to large forces and moments during their operation. These forces and moments mainly include gravity, inertial hydrodynamics, and viscous hydrodynamics. There are more solutions to the determination of gravity and inertial hydrodynamics. Therefore, the viscous hydrodynamic model has become a research focus that needs to be broken through because of its large number of coefficients and the difficulty of measurement.

References [1–3] verified the accuracy of fluid simulation calculations using software, such as FLUENT, by comparing the simulated calculation values of hydrodynamics with the measured values of engineering methods. References [4–6] studied the drag and additional mass force of the manipulator underwater and obtained the inertial hydrodynamic model by the CFD (Computational Fluid Dynamics) simulation, but the viscous hydrodynamics were not considered as key targets in the study. References [7–10] numerically calculated the viscous hydrodynamic coefficients of submarine and ship models with complex shapes under the specified navigation conditions and further improved the handling performance during navigation. Reference [11] determined the viscous kinetic coefficients of the manipulator through aerodynamic experiments and carried out a sea trial.

At present, research on viscous hydrodynamics mostly focus on specific ship models and underwater vehicles, and less on manipulators loaded on them. Therefore, a numerical calculation method of the viscous hydrodynamic coefficient based on single-DOF (Degree of Freedom) manipulators is put forward. The method is based on ANSYS Fluent. UDF (user-defined function) and dynamic mesh which are used to simulate the rotational motion of the manipulator. The inlet and outlet conditions, as well as boundary conditions of the fluid domain, are changed. Also, the viscous hydrodynamic model could be obtained by curve fitting and regression analysis.

2. Mathematical Model and Theoretical Analysis

2.1. Mathematical Modeling of Viscous Hydrodynamics

Since its surrounding flow field is changed during the operation of the underwater manipulator; the arm is subjected to the reaction of the water body caused by force generated by the water body. The hydrodynamics of the moving manipulator in water include two parts: Inertial hydrodynamics caused by acceleration and viscous hydrodynamics caused by friction. Inertial hydrodynamics is generally determined by empirical formulas, while coefficients of viscous hydrodynamics are complex and difficult to measure.

This paper is intended to use a single-DOF manipulator model mounted on a fixed base. The arm is a homogeneous lightweight rod with a circular cross-section, 50 mm in diameter (d), 500 mm in arm length (l), at length to diameter ratio of 10, belonging to the slender pole. The establishment of its coordinate system is shown in Figure 1.

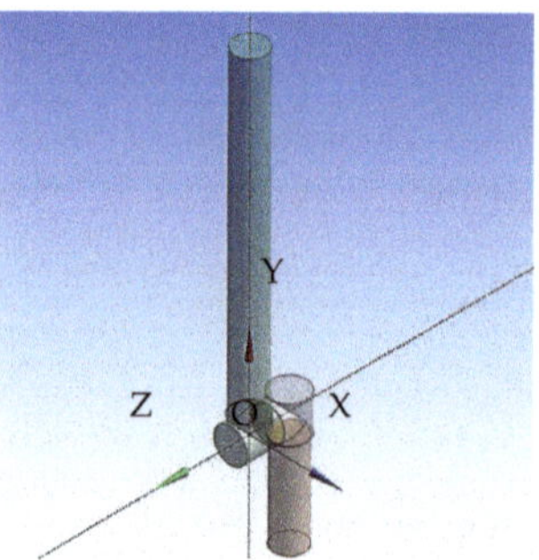

Figure 1. Manipulator model and coordinate system.

When the manipulator performs the constant motion, the influence of acceleration and angular acceleration on the motion variable could be ignored based on the "slow-motion" assumption. So the mere considerations for viscous hydrodynamics are the effects of the velocity and angular velocity in the motion variable. It can be represented as: $U = [u\ v\ w\ p\ q\ r]^T$, where u stands for the velocity in the OX direction, v for the velocity in the OY direction, w for the velocity in the OZ direction, p for the angular velocity of the rotation around the OX axis, q for the angular velocity of the rotation around the OY axis, and r for the angular velocity of the rotation around the OZ axis.

The viscous hydrodynamics can be expressed as a multivariate function $F = f(u, v, w, p, q, r)$, and the six-dimensional component of viscous hydrodynamics $F = [X\ Y\ Z\ K\ M\ N]^T$ could also be displayed in the above functional form, the direction of which is shown in Figure 2.

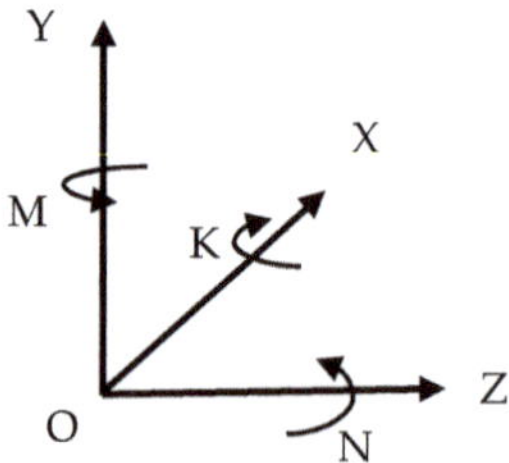

Figure 2. Six-dimensional component of viscous hydrodynamic F.

The viscous hydrodynamic F performs Taylor expansion in the form of Equation (1):

$$\begin{aligned} f(U) = \ & f(U_k) + (U - U_k)^T \nabla f(U_k) \\ & + \tfrac{1}{2!}(U - U_k)^T H(U_k)(U - U_k) + n \end{aligned} \tag{1}$$

where $U_k = \begin{bmatrix} u_0 & v_0 & w_0 & p_0 & q_0 & r_0 \end{bmatrix}^T$, U_k is the initial constant.

$$H(U_k) = \begin{bmatrix} \dfrac{\partial^2 f(U_k)}{\partial u^2} & \dfrac{\partial^2 f(U_k)}{\partial u \partial v} & \cdots & \dfrac{\partial^2 f(U_k)}{\partial u \partial r} \\ \dfrac{\partial^2 f(U_k)}{\partial u \partial v} & \dfrac{\partial^2 f(U_k)}{\partial v^2} & \cdots & \dfrac{\partial^2 f(U_k)}{\partial v \partial r} \\ \vdots & \vdots & \ddots & \vdots \\ \dfrac{\partial^2 f(U_k)}{\partial u \partial r} & \dfrac{\partial^2 f(U_k)}{\partial v \partial r} & \cdots & \dfrac{\partial^2 f(U_k)}{\partial r^2} \end{bmatrix}$$

Considering the left-right and front-back symmetry and the motion limit of the manipulator, several coefficients in the second-order expansions are zero, and the remaining others are non-negligible hydrodynamic coefficients, as in Equation (2):

$$\begin{cases} X = X_{uu}u^2 + X_{vv}v^2 + X_{ww}w^2 + X_{rr}r^2 + X_{vr}vr \\ Y = Y_v v + Y_r r + Y_{v|v|}v|v| + Y_{r|r|}r|r| + Y_{v|r|}v|r| \\ Z = Z_w w + Z_{w|w|}w|w| + Z_{vv}v^2 + Z_{rr}r^2 + Z_{vr}vr \\ K = K_v v + K_{v|v|}v + K_r r + K_{r|r|}r|r| \\ M = M_w w + M_{w|w|}w|w| + M_{vv}v^2 + M_{rr}r^2 + M_{vr}vr \\ N = N_v v + N_r r + N_{v|v|}v|v| + N_{r|r|}r|r| + N_{v|r|}v|r| \end{cases} \tag{2}$$

Viscous hydrodynamics of the manipulator could be calculated in FLUENT. The regression analysis of the calculated viscous hydrodynamics and corresponding velocities could be performed to obtain unknown coefficients in Equation (2) in MATLAB. The first derivative coefficients associated only with the linear velocity (u, v, w) are the position derivatives, and the first derivative coefficients related to the angular velocity (p, q, r) are the rotational derivatives. The coefficients caused by joint changes of two or more parameters are the coupling derivatives.

2.2. Control Equation

To analyze hydrodynamics of underwater manipulators, it is generally assumed that the fluid is isothermal and incompressible, also as a constant flow that magnitude and direction do not change with time. The continuity equation (Equation (3)) and the Navier–Stokes equation (Equation (4)) serve as the two basic equations necessary to solve the flow problem of viscous fluid. These equations are generally described in the form of partial differentials:

$$\nabla \cdot \vec{u} = 0 \tag{3}$$

$$\frac{\partial \vec{u}}{\partial t} + \left(\vec{u} \cdot \nabla \right)\vec{u} = \vec{f} - \frac{1}{\rho}\nabla p + \nu \nabla^2 \vec{u} \tag{4}$$

The form of expression of the time-averaged continuity equation does not change. Instead, the tensor of the Reynolds stress is added to the formula after the N–S equation time-averaged, which leads to the closure problem of the original equation. The Reynolds stress is about 10^{-2} Pa, which is indicative of the turbulent flow and cannot be directly ignored. Therefore, it is necessary to introduce a proper turbulence model to make a modeling description of the increased Reynolds stress.

For the incompressible isothermal turbulent water environment, the basic equations of turbulence include the DNS equations (direct numerical simulation), LES equations (large eddy simulation), and RANS equations (Renault time average). The two equations (DNS equations and LES equations) are of limited use due to the requirements of a large number of computational grids. At present, most engineering calculations adopt the RANS equation to solve closed equations formed by introducing the turbulence model and, thus, obtain the time-average value of the turbulent elements. To solve the viscous hydrodynamics under steady-state, the appropriate turbulence model is the key to the numerical simulations in this paper.

3. Calculation Method Analysis

3.1. Calculation Domain Establishment

Generally speaking, the larger the size of the calculation domain is, the closer it is to the real working condition. The downside is that it will increase the amount of calculation and prolong the calculation period. If the calculation domain is too small, the boundary conditions and calculation results are difficult to match the real working conditions. Therefore, it is very important to choose the size of the calculation domain reasonably.

Based on previous experience and multiple numerical practices [6,12,13], this paper establishes the domain as the computational domain, shown in Figure 3b. The specific sizes are as follows:

Front boundary: 1.5 H
Back boundary: 2 H
Side boundary: 1.5 H

where, H stands for the sum of the height of the arm and the base.

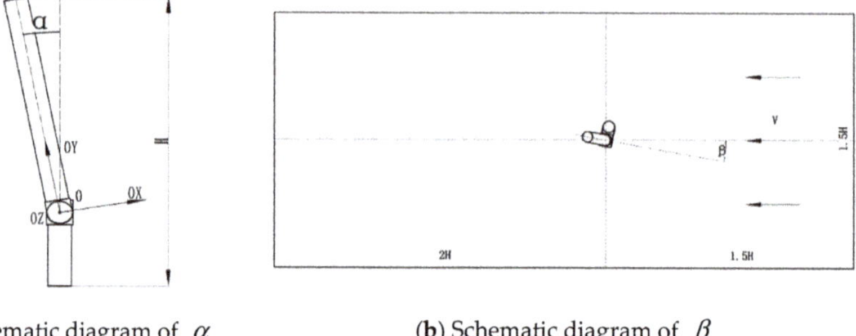

(**a**) Schematic diagram of α (**b**) Schematic diagram of β

Figure 3. Schematic diagram of the calculation domain and α/β.

The motion parameters of the simulation are shown in Figure 3, α stands for the rotation angle around the OZ axis, β for the angle of the arm to the flow direction, V for the inlet flow velocity of the domain, and the linear velocity components of the manipulator are as shown in Equation (5):

$$\begin{cases} u = V\cos\beta \cos\alpha \\ v = V\cos\beta \sin\alpha \\ w = V\sin\beta \end{cases} \tag{5}$$

3.2. Meshing

The calculation of CFD requires well-distributed grids. Usually, grids are classified into structured and unstructured grids. The unstructured grid means that internal points in the mesh area do not have the same adjacent unit, with no regular topology, not arranged in layers, and the distribution of mesh nodes is arbitrary. Therefore, it is more flexible than structured grid. Unstructured grids can be optimized by using certain criteria in the process of their generation. Ultimately, they can be displayed as high-quality meshes, which are suitable for complex geometry, easy to control grid size, and node density. Moreover, the adoption of random data structures is conducive to mesh adaptation. It is difficult to model the structured mesh due to the shape of the manipulator, instead, the unstructured mesh is easy to combine with the dynamic mesh technology. Thus, the unstructured mesh is adopted in the research.

We compare basic mesh to dense mesh for the sake of grid independence verification. Basic mesh in the vicinity of the manipulator is shown in Figure 4a and dense mesh is shown in Figure 4b. The number of elements and nodes are shown in Table 1.

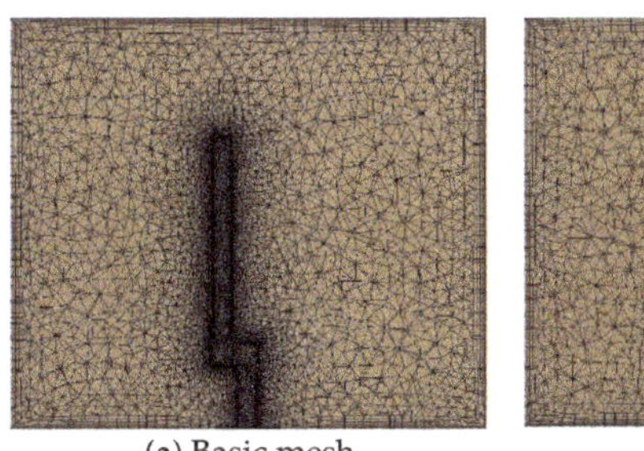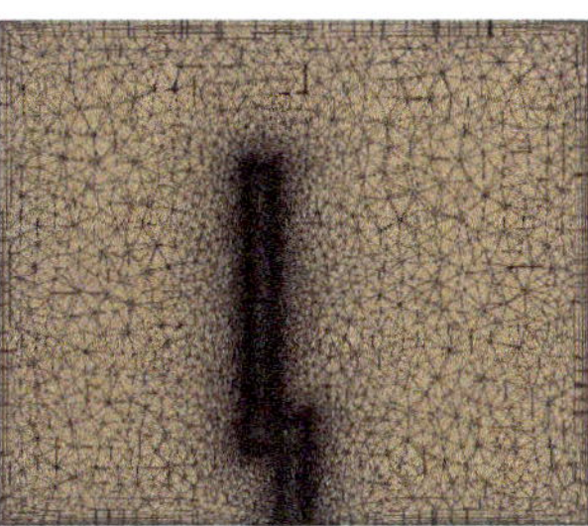

(a) Basic mesh   (b) Dense mesh

Figure 4. Meshing of manipulator section.

Table 1. Parameters of basic and dense mesh.

	Maximum Layers	Growth Rate	Elements	Nodes
Basic Mesh	5	1.2	636,457	114,559
Dense Mesh	5	1.2	1,429,231	401,609

The rotation process of the manipulator is realized by UDF and dynamic mesh technology in Fluent. Dynamic meshing is performed by using the spring smoothing model, which approximates connecting lines between the grid nodes as springs, and the position of nodes after smoothing is obtained by calculating force equilibrium equations between them. In the calculation process, meshes with a large aberration rate or huge change in size are grouped together to re-divide the partial meshes.

3.3. The Solution of Position Derivatives, Rotation Derivatives, and Coupling Derivatives

Solving the unknown coefficients in Equation (2) is the key to the research task, where Y_v, N_v, Z_w, and M_w are positional derivatives, Y_r and N_r are rotational derivatives, and the remaining unknown coefficients are coupled derivatives.

The positional derivatives could be obtained by simulating the low-speed wind tunnel test, and the rotational derivatives are obtained by measuring the viscous hydrodynamic of the model at different rotational angular velocities. The number of coupled derivatives is large, and the viscous hydrodynamic subjected to the rotational manipulator is measured when β is not 0. After extensive experiments, the viscous hydrodynamic coefficients were obtained by least-square regression analysis.

The calculation domain inlet is set as the velocity boundary, the outlet as the free-flow condition, and the calculation domain wall as the non-slip fixed wall.

According to the research of the turbulence model in the reference [13], the standard k–ω model has many advantages, such as good numerical stability, accurate solution of pressure gradient, low Reynolds number influence, compressibility effect, and shear flow diffusion. It has better adaptability in calculating the flow problem of the boundary layer separation. It is one of the most widely used turbulence models for the viscous hydrodynamic solution. Therefore, the standard k–ω model is used as the turbulence model in this paper. The equation of the kinetic energy k and the turbulent frequency

ω are as shown in Equation (6). The determined model parameters are shown in Table 2, according to the reference [13]:

$$\begin{cases} \dfrac{\partial \rho k}{\partial t} + \nabla \cdot \left(\rho \vec{U} k\right) = & \nabla \cdot \left[\left(\mu + \dfrac{\mu_t}{\sigma_k}\right)\nabla k\right] \\ & + P_k - \beta' \rho k \omega \\ \dfrac{\partial \rho \omega}{\partial t} + \nabla \cdot \left(\rho \vec{U} \omega\right) = & \nabla \cdot \left[\left(\mu + \dfrac{\mu_t}{\sigma_\omega}\right)\nabla \omega\right] \\ & + \alpha_t \dfrac{\omega}{k} P_k - \beta_t \rho \omega^2 \end{cases} \tag{6}$$

Table 2. Standard k–ω model parameters.

σ_k	σ_ω	α_t	β_t	β'
2.0	2.0	5/9	0.075	0.09

4. Result Analysis

Comparing the viscous hydrodynamics of the manipulator measured by adopting the basic and dense mesh, respectively, we conclude that the differences are in the range from 1.7% to 4.3%. The differences are sufficiently low and negligible.

Computations are executed in a 64-bit processor consist of CPU (Intel Core i5-8400 @ 2.80 GHz) and 8 GB accessible memory, and take 37 h with at least 40 iterations per time step.

Using the size of calculation domain set in Section 3.1, the flow velocity of the calculation domain is kept constant, the angle β between the manipulator and the incoming flow direction is adjusted, the viscous hydrodynamics at different angles are calculated, and then the viscous hydrodynamic position derivative is obtained by linear analysis. The position derivative calculation contents are shown in Table 3.

Table 3. Solving cases of position derivatives.

α	0°, ±2°, ±5°, ±7°, ±10°
β	0°, ±2°, ±4°, ±6°, ±8°, ±10°
r	0 rad/s
V	1 m/s

Since the different degrees of α and β cause the manipulator to have different linear velocities in the OX, OY, and OZ directions, viscous hydrodynamics of the arm is measured in the horizontal plane XOZ and the vertical plane XOY, respectively. The values of the vertical force Y and the pitching moment N measured at the different linear velocity v in the OY direction are shown in Table 4.

Table 4. Measurements of vertical force Y and pitching moment N at different linear velocity v.

v (m/s)	Y (N)	N (N·m)
−0.1714	−0.0373	0.0677
−0.1200	−0.0235	0.0673
−0.0860	−0.0135	0.0670
−0.0340	0.0003	0.0660
0	0.0104	0.0658
0.0340	0.0190	0.0652
0.0860	0.0331	0.0649
0.1200	0.0420	0.0640
0.1714	0.0545	0.0628

The position derivatives Yv and Nv are the first derivative coefficients of the linear velocity v, so the least squares curve fitting of the vertical force Y and the pitching moment N in viscous hydrodynamics to the linear velocity v is performed, as shown in Figures 5 and 6. The derivative value at the median 0 points of the line velocity v is taken as the position derivative.

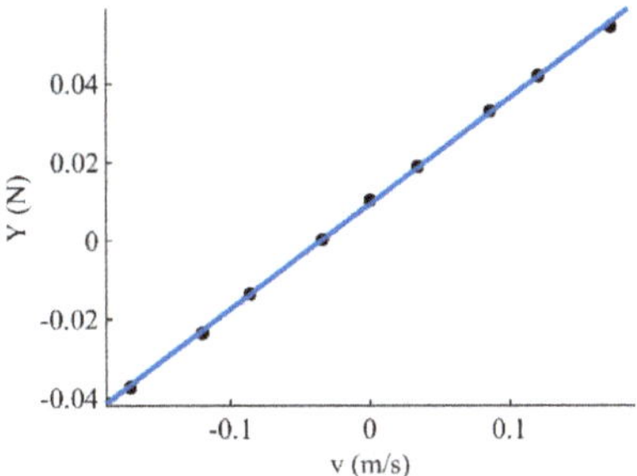

Figure 5. Fitted curve of vertical force Y and line velocity v.

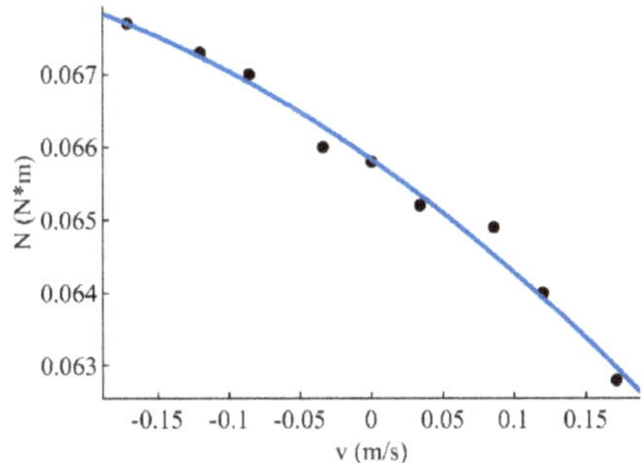

Figure 6. Fitted curve of pitching moment N and linear velocity v.

Same as above, the values of the lateral force Z and the yaw moment M measured at the different linear velocity w in the OZ direction are shown in Table 5.

Table 5. Measurement of the lateral force Z and the yaw moment M at the different linear velocity w.

w (m/s)	Z (N)	M ($e^{-5} \cdot$N·m)
−0.1740	−0.0364	−4.8199
−0.1390	−0.0280	−2.5920
−0.1050	−0.0184	1.5570
−0.0700	−0.0100	2.0860
−0.0350	−0.0025	2.1628
0	0.0066	4.1640
0.0350	0.0165	5.4699
0.0700	0.0250	8.9650
0.1050	0.0328	9.3461
0.1390	0.0430	12.3140
0.1740	0.0526	16.7242

The position derivatives Z_w and M_w are the first derivative coefficients of the linear velocity w, so the least squares curve fitting of the viscous hydrodynamic lateral force Z and the yawing moment M to the linear velocity w is performed, as shown in Figures 7 and 8. The derivative at 0 point of the median line velocity w is taken as the position derivative.

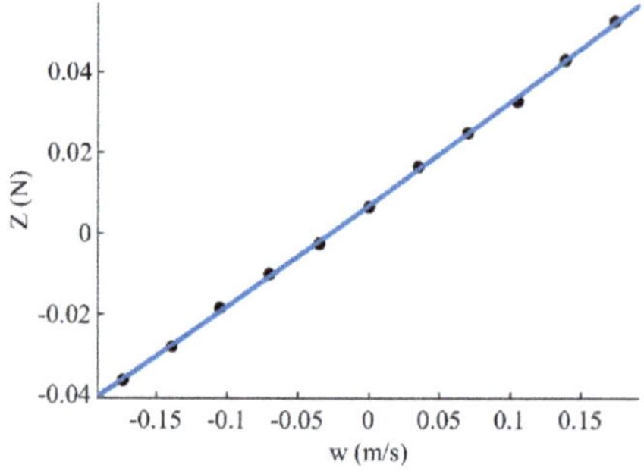

Figure 7. Fitted curve of lateral force Z and linear velocity w.

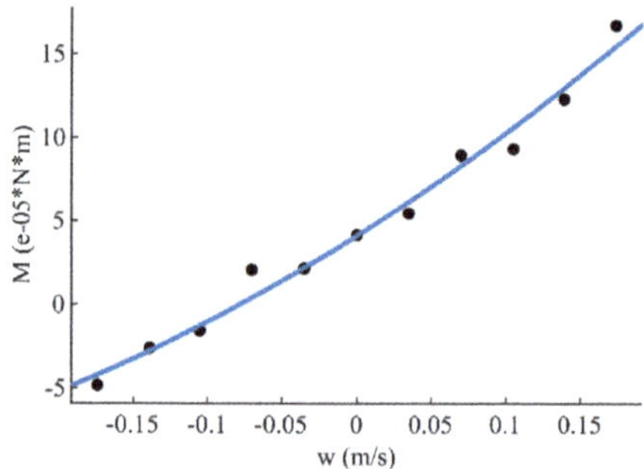

Figure 8. Fitted curve of yaw moment M and line speed w.

The positional derivatives could be obtained as shown in Table 6 below:

Table 6. Position derivatives of the manipulator in viscous hydrodynamics.

Y_v	N_v	Z_w	M_w
1.4352	−0.1472	1.3520	0.0117

Since the manipulator studied in this paper is a single DOF, only the viscous hydrodynamics are measured when it rotates around the OZ axis. UDF is applied to adjust the rotational velocity of the arm. The measurement conditions of the rotation derivatives are shown in Table 7.

Table 7. Solving cases of rotational derivatives.

α	−10°–10°
β	0°
r	0.5, 0.75, 1, 1.25, 1.5, 1.75, 2.0 (rad/s)
V	0 m/s

This paper calculates the hydrodynamics of the rotating underwater manipulator when the flow is stationary. The arm rotates in the XOY plane. The angular velocity r is adjusted according to Table 7, and the instantaneous viscous hydrodynamics of the arm during the rotation from −10° to 10° around the OZ axis are measured. The measured vertical force Y and pitching moment N are shown in Table 8.

Table 8. Measurements of vertical force Y and pitching moment N at different angular velocities r.

r (rad/s)	Y (N)	N (N·m)
0.5	0.0033	0.0444
0.75	0.0113	0.0731
1.0	0.0182	0.1023
1.25	0.0306	0.1462
1.5	0.0399	0.1750
1.75	0.0492	0.2181
2.0	0.0592	0.2613

The rotational derivatives Y_r and N_r are the first derivative coefficients of the angular velocity r, similar to the solution method of the position derivatives. The least squares curve fitting of the vertical force Y and the pitching moment N in viscous hydrodynamics to the angular velocity r is performed, as shown in Figures 9 and 10. The derivative at the median angular velocity r of 1.25 rad/s is taken as the rotation derivative.

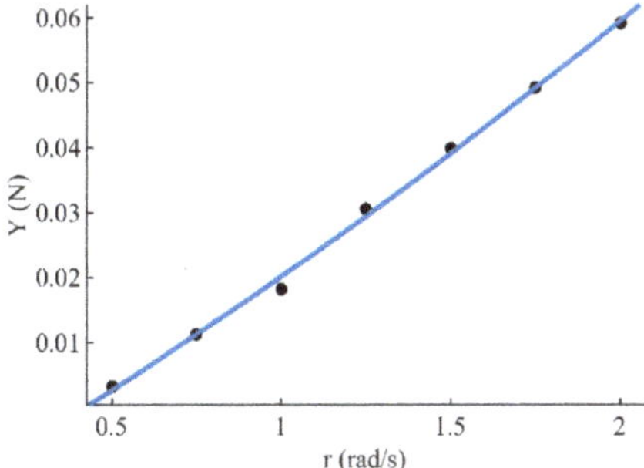

Figure 9. Fitted curve of vertical force Y and angular velocity r.

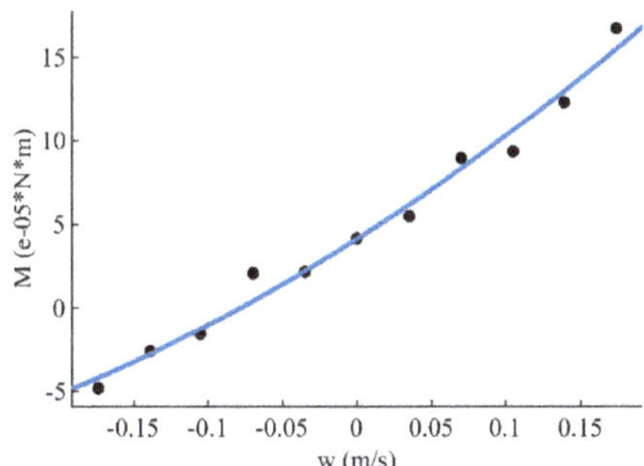

Figure 10. Fitted curve of pitching moment N and angular velocity r.

The rotational derivatives could be obtained as shown in Table 9 below:

Table 9. Rotation derivatives of the manipulator in viscous hydrodynamics.

Y_r	N_r
0.4043	3.0869

The coupled derivative is the second-order viscous hydrodynamic coefficient of the arm under complex motion conditions. Through the transient iterative calculation of the motion of the manipulator at different α and β angles and various angular velocities, the viscous hydrodynamics of each motion state are measured, and many coupling derivatives are obtained by least-squares regression analysis. The coupling derivative solution conditions are shown in Table 10 below.

Table 10. Solving cases of coupling derivatives.

α	$0°$, $\pm2°$, $\pm5°$, $\pm7°$, $\pm10°$
β	$0°$, $\pm2°$, $\pm4°$, $\pm6°$, $\pm8°$, $\pm10°$
r	0.5 rad/s, 1 rad/s
V	1 m/s

The coupled derivatives are obtained by regression analysis, and the obtained coupled derivatives are processed without dimensioning. The values of the parameters are as shown in Table 11.

Table 11. Coupling derivatives of the manipulator in viscous hydrodynamics.

X_{uu}	-0.6960	X_{vv}	-3.8896
X_{ww}	-74.4904	X_{rr}	44.8128
X_{vr}	-275.0688	$Y_{v\lvert v\rvert}$	1.172
$Y_{r\lvert r\rvert}$	-6.2688	$Y_{v\lvert r\rvert}$	31.6064
$Z_{w\lvert w\rvert}$	13.0336	Z_{vv}	2.1088
Z_{rr}	-0.0512	Z_{vr}	-0.1264
K_v	-0.0011	$K_{v\lvert v\rvert}$	-0.0160
K_r	0.1515	$K_{r\lvert r\rvert}$	0.0256
$M_{w\lvert w\rvert}$	-1.0816	M_{vv}	0.0144
M_{rr}	0.0384	M_{vr}	-0.2144
$N_{v\lvert v\rvert}$	32.9184	$N_{r\lvert r\rvert}$	-25.0688
$N_{v\lvert r\rvert}$	130.4128		

5. Conclusions

The hydrodynamics of underwater manipulators during operation are complex and difficult to predict. It is analyzed that components of the hydrodynamics include inertial hydrodynamics caused by acceleration and viscous hydrodynamics caused by friction. This paper takes viscous hydrodynamics as the research target.

According to the research on dynamics of AUV (Autonomous Underwater Vehicle), ROV(Remote Operated Vehicle) and ships in other references, the viscous hydrodynamic model of the manipulator is analyzed in the form of Taylor expansion, and the viscous hydrodynamics are measured by using a single-DOF manipulator to simulate the underwater motion, and the viscous hydrodynamic coefficients among the model are calculated via regression analysis. An accurate viscous hydrodynamic model is obtained to predict viscous hydrodynamics of manipulators during operation at any attitude.

This model is the basis for the feedforward control and is helpful to study control stability of underwater manipulators. We believe that although the simulations in this paper were performed for single-DOF manipulators, the modeling method may be extended for the manipulator with multi-DOF and more complicated shapes as is in the case of real underwater manipulators.

Author Contributions: M.Z. Provided supervision and funding support for the project. X.L. conducted formal analysis, designed all the experiments, and subsequently drafted the manuscript. Y.T. conceived the original ideas, provided supervision to the project and reviewed the writing.

Funding: This research was supported by the National Key Research and Development Program of China (Grant No. 2018YFB1309401), and the Hebei Provincial Higher Education Science and Technology Research Project (Grant No. QN2018090).

Conflicts of Interest: The authors declare no conflict of interest.

J. Mar. Sci. Eng. **2019**, *7*, 261

References

1. Stern, F.; Agdrup, K.; Kim, S.Y.; Hochbaum, A.C.; Rhee, K.P.; Quadvlieg, F.; Perdon, P.; Hino, T.; Broglia, R.; Gorski, J. Experience from SIMMAN 2008—The First Workshop on Verification and Validation of Ship Maneuvering Simulation Methods. *J. Ship Res.* **2011**, *55*, 135–147.
2. Larsson, L. *Numerical Ship Hydrodynamics*; Springer: Dordrecht, The Netherlands, 2013.
3. Irwin, R.P.; Chauvet, C. *Quantifying Hydrodynamic Coefficients of Complex Structures*; IEEE: New York, NY, USA, 2007; pp. 1341–1345.
4. Mclain, T.W.; Rock, S.M. Development and Experimental Validation of an Underwater Manipulator Hydrodynamic Model. *Int. J. Robot. Res.* **1998**, *17*, 748–759. [CrossRef]
5. Pazmiño, R.S.; Cena, C.E.G.; Arocha, C.A.; Santonja, R.A. Experiences and results from designing and developing a 6 DoF underwater parallel robot. *Robot. Auton. Syst.* **2011**, *59*, 101–112. [CrossRef]
6. Kołodziejczyk, W. Preliminary Study of Hydrodynamic Load on an Underwater Robotic Manipulator. *J. Autom. Mob. Robot. Intell. Syst.* **2015**, *9*, 11–17. [CrossRef]
7. Arabshahi, A.; Beddhu, M.; Briley, W.R.; Chen, J.P.; Gaither, A.; Gaither, K.; Janus, J.M.; Jiang, M.; Marcum, D.; Mcginley, J. A Perspective on Naval Hydrodynamic Flow Simulations. In Proceedings of the Symposium on Naval Hydrodynamics, Washington, DC, USA, 9–14 August 1998.
8. Ohmori, T.; Fujino, M.; Miyata, H. A study on flow field around full ship forms in maneuvering motion. *J. Mar. Sci. Technol.* **1998**, *3*, 22–29. [CrossRef]
9. Tahara, Y.; Longo, J.; Stern, F. Comparison of CFD and EFD for the Series 60 $C_B = 0.6$ in steady drift motion. *J. Mar. Sci. Technol.* **2002**, *7*, 17–30. [CrossRef]
10. Ueno, M. Hydrodynamic derivatives and motion response of a submersible surface ship in unbounded water. *Ocean Eng.* **2010**, *37*, 879–890. [CrossRef]
11. Filaretov, V.F.; Konoplin, A.J.; Getman, A.V. Experimental Determination of the Viscous Friction Coefficients for Calculation of the Force Impacts on the Moving Links of the Underwater Manipulators. *Mekhatronika Avtom. Upr.* **2015**, *16*, 738–743. [CrossRef]
12. Kolodziejczyk, W. The method of determination of transient hydrodynamic coefficients for a single DOF underwater manipulator. *Ocean Eng.* **2018**, *153*, 122–131. [CrossRef]
13. Zhiqiang, H.U.; Ruiwen, Y.I.; Lin, Y.; Haitao, G.U.; Wang, C. Numerical calculation methods for hydrodynamics of unmanned underwater vehicles based on body-fixed coordinate frames. *Chin. Sci. Bull.* **2013**, *58*, 55.

Article

Second Path Planning for Unmanned Surface Vehicle Considering the Constraint of Motion Performance

Jiajia Fan [1], Ye Li [1,*], Yulei Liao [1,*], Wen Jiang [1], Leifeng Wang [1], Qi Jia [1] and Haowei Wu [2]

[1]   Science and Technology on Underwater Vehicle Laboratory, Harbin Engineering University,
     Harbin 150001, China; fanjiajia@hrbeu.edu.cn (J.F.); 15171493813@163.com (W.J.);
     wangleifeng1992@foxmail.com (L.W.); yeondy@hotmail.com (Q.J.)
[2]   Beijing Institute of Astronautically Systems Engineering, Beijing 100076, China; walk2newyork@163.com
*   Correspondence: liyeheu103@163.com (Y.L.); liaoyulei@hrbeu.edu.cn (Y.L.); Tel.: +86-186-8685-1807 (Y.L.);
     +86-186-4562-3860 (Y.L.)

Received: 11 March 2019; Accepted: 11 April 2019; Published: 17 April 2019

Abstract: When utilizing the traditional path planning method for unmanned surface vehicles (USVs), 'planning-failure' is a common phenomenon caused by the inflection points of large curvatures in the planned path, which exceed the performances of USVs. This paper presents a second path planning method (SPP), which is an initial planning path optimization method based on the geometric relationship of the three-point path. First, to describe the motion performance of a USV in conjunction with the limited test data, a method of integral nonlinear least squares identification is proposed to rapidly obtain the motion constraint of the USV merely by employing a zig zag test. It is different from maneuverability identification, which is performed in combination with various tests. Second, the curvature of the planned path is limited according to the motion performance of the USV based on the traditional path planning, and SPP is proposed to make the maximum curvature radius of the optimized path smaller than the rotation curvature radius of the USV. Finally, based on the 'Dolphin 1' prototype USV, comparative simulation experiments were carried out. In the experiment, the path directly obtained by the initial path planning and the path optimized by the SPP method were considered as the tracking target path. The artificial potential field method was used as an example for the initial path planning. The experimental results demonstrate that the tracking accuracy of the USV significantly improved after the path optimization using the SPP method.

Keywords: unmanned surface vehicle; maneuverability identification; second path planning; motion constraints

---

1. Introduction

Unmanned surface vehicle (USV), which has attracted significant research attention in recent years, is used for dangerous and inhospitable missions [1,2]. At present, a USV mainly consists of a motion control system [3], sensor system, and communication system.

Path planning [4,5], as the core of USV research, represents the intelligence level of the USV to a certain extent. The path planning method can be used to determine an optimal path from the starting point to the end point, which mainly includes the A* algorithm [6–8], genetic algorithm [9], artificial potential field method [10], ant colony algorithm [11], and particle swarm algorithm [12,13]. However, all the abovementioned methods of USV path planning present several drawbacks. The optimal set of path planning is solved based on the shortest distance standard, mostly on the premise that the USV is regarded as a mass point, while ignoring its maneuvering and turning performances. This results in the inability of the underactuated USV to track path nodes with large curvatures beyond the limitations of its own motion performance. In practical engineering applications, planning failures, such as the inaccessibility to the target, the inability to complete the obstacle avoidance task, or detouring, may occur during the trajectory tracking of the USV, thus increasing the difficulty of the tasks.

At present, there are several path planning smoothing methods. For the path planning of a mobile robot, based on the A* algorithm [6,14], all nodes in the planning path are traversed in the grid environment. When there is no obstacle on the connecting line of a node before and after, the intermediate node of the extended line is removed to reduce the number of path turns. However, this method does not consider the maneuverability of the USV in optimization, only the reduction of the number of turns to achieve the optimization effect. For the path planning of an automatic underwater vehicle (AUV), based on the genetic algorithm [15,16], the B spline interpolation method was introduced into the objective function for path smoothing. However, when a few path nodes exceed the turning ability of the AUV, irrespective of the fit between the nodes, the curve formed is unreachable. For the path planning of the USV, based on the statistical method of regression [17], the path smoothing method is dependent on a large amount of actual navigation data, and it is based on an improved A* algorithm (FAA*) [18], to achieve path smoothing. This method limits the initial heading angle of the USV and removes the acute angle section in the planned path. For the path planning of a high-speed numerical control machine, the hyperboloid sheet method is used to solve the path smoothing problem of the continuous path, combined with a straight line and arc [19,20]. However, this method is used to perform curve fitting on planned path points, and the slice of the high-speed cutting machine is more flexible and demonstrates good maneuverability.

Although the abovementioned methods are beneficial to the smooth processing of path planning to a certain extent, in the application of USVs, the methods demonstrated poor adaptability, and the results were accidental. In this paper, using the zig zag test, a rapid and one-time identification and analysis method for the motion performance of the USV is proposed. A second path planning (SPP) method is then proposed to consider the constraints of the USV performance. This method can realize the re-optimization based on the existing path planning method to obtain the optimal path within the range of the motion ability of the USV, and to effectively prevent the occurrence of planning failure. Finally, using the improved artificial potential field path planning method [21] as an example, a second optimization strategy was introduced and applied to the 'Dolphin 1' mini-type USV. Moreover, a simulation comparison test was carried out to verify the feasibility and effectiveness of the method.

2. Analyzing the Motion Performance of USV

When the traditional path planning method is combined with the trajectory tracking guidance algorithm of the USV, the planning failure phenomena occur, such as detouring and collision, given that this method considers the USV as a mass point, which is an ideal state. Therefore, in the process of the path planning of the USV, the motion ability of the USV, especially the minimum radius of the turning circle, should be considered.

2.1. Modeling the USV Maneuverability

To describe the motion of the USV, two right-handed rectangular coordinate systems were adopted [22]. The first is the inertial frame, o_0-$x_0y_0z_0$, and the second is the body-coordinate system, o-xyz (Figure 1). The inertial frame was fixed on the earth. The plane, o_0-x_0y_0, was on the undisturbed free surface, and the axis, x_0, was directed forward to the initial straight course of the USV with the axis, z_0, directed downward. The body-coordinate system was fixed on the vehicle, and the origin was located in the middle of it. The plane, o-xy, was on the undisturbed free surface, the axis, x, was in the direction of the bow, the axis, y, was in the direction of the starboard, and the axis, z, was directed downward.

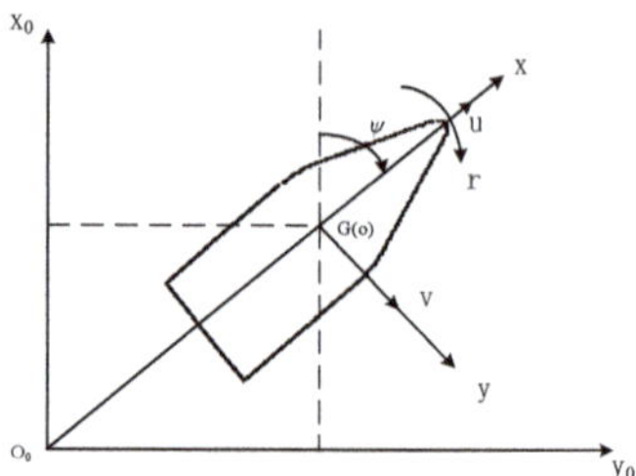

Figure 1. USV coordinate system.

The USV motion equation of six degrees of freedom can be expressed as follows:

$$\begin{cases} m(\dot{u} + qw - rv) = X \\ m(\dot{v} + ru - pw) = Y \\ m(\dot{w} + pv - qu) = Z \\ I_x\dot{p} + (I_z - I_y)qr = L \\ I_y\dot{q} + (I_x - I_z)rp = M \\ I_z\dot{r} + (I_y - I_x)pq = N \end{cases} \tag{1}$$

where the first three equations are the expressions of the motion theorem of the mass center in the body coordinate system, and the last three equations are the Euler dynamics equations of the rotation of a rigid body around a fixed point.

Only considering the horizontal motion, $p = q = w = 0$, the linear Equation (2) of the vehicular heading response can be obtained through the transformation of Equation (1):

$$T_1 T_2 \ddot{r} + (T_1 + T_2)\dot{r} + r = K\delta + KT_\delta\delta \tag{2}$$

In 1957, Nomoto [23] suggested that the first order linear differential equation can be approximately substituted for the second order equation, in the case wherein steering is infrequent, as shown in Equation (3):

$$T\dot{r} + r = K\delta \tag{3}$$

The first order nonlinear Equation (4) is obtained by adding the nonlinear term to the motion of the large rudder angle, low speed, and the study of small ships:

$$T\dot{r} + r + ar^3 = K\delta \tag{4}$$

The mini-type USV evaluated in this study uses two electric propellers as the power device, with a maximum voltage of 12 V. Different voltage values correspond to different speed values. In combination with the model of thrust and velocity in [24], and with reference to the rudder and heading model, the corresponding relationship between the speed and voltage of the USV was established. The speed model of the USV can be expressed as follows:

$$u = k_1 n^2 + k_2 n + k_3 \tag{5}$$

where u is the longitudinal speed of the USV, n is the propeller voltage, and k_1, k_2, k_3 represent the system parameters. The transverse speed can be neglected due to the low speed of the USV evaluated in this study.

In summary, the maneuverability model of the USV can be expressed as follows:

$$\begin{cases} \dot{x} = u\cos\psi - v\sin\psi \\ \dot{y} = u\sin\psi + v\cos\psi \\ \dot{\psi} = r \\ \dot{r} = \frac{1}{T}(K\delta - r - \alpha r^3) \\ u = k_1 n^2 + k_2 n + k_3 \end{cases} \tag{6}$$

2.2. Integral Nonlinear Least Squares Method

The least squares method [25] is one of the most commonly used motion parameter identification methods. However, this method was typically used to identify the nonlinear motion equation by a combination of experiments, i.e., the zig zag test and the turning circle test. In this paper, an integral nonlinear least squares method is proposed. The nonlinear parameters of the motion equation are transformed and processed. Using a set of zig zag test data, the nonlinear motion equation can be rapidly identified. Based on the identification results, the course control parameters can be predicted, and the motion ability can be analyzed.

Based on the 'Dolphin 1' prototype USV, the integral nonlinear least square method was adopted to identify the parameters of the first-order nonlinear control model using the zig zag test data. According to the identified parameters, the corresponding zig zag test manipulation simulation model was established using MATLAB, and the reliability of the method was confirmed by the error between the simulation and actual case. The experimental data were derived from the zig zag test in the Songhua River in Harbin, Heilongjiang, China. Figure 2 presents the layout of the 'Dolphin 1' in the Songhua River, and the test data are shown in Figure 3. The 'Dolphin 1' is a mini unmanned catamaran with a single floating body length of 2.0 m, diameter of 0.25 m, and a two floating body spacing of 1.1 m. It is propelled by two propellers with a rear-mounted rudder plate, and the maximum speed is 1.2 m/s.

Figure 2. The 'Dolphin 1' developed by Harbin Engineering University.

At 10.0 volts, the 'Dolphin 1' carried out the zig zag test control experiment at a speed of 1.08 m/s. The variation curve of the heading and rudder angle with respect to time is shown in Figure 3.

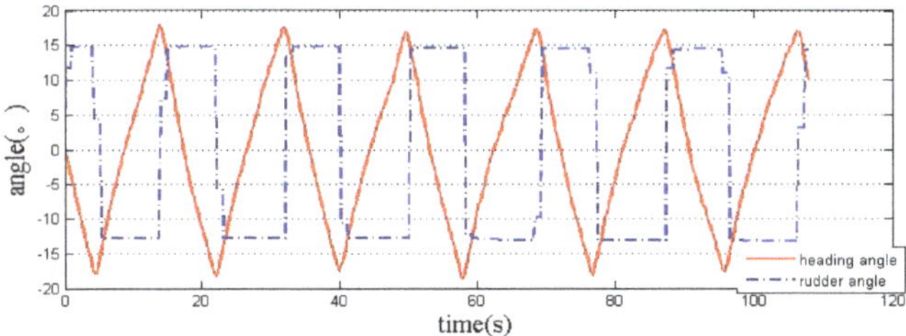

Figure 3. The zig zag test.

According to the test, the first order nonlinear K-T Equation (6) was identified using integral nonlinear least squares. Referring to Figure 4, given that the vehicle has no real-time measurement record of angular acceleration, the integral of both sides of Equation (6) was carried out based on the time region, [*a*,*b*]. By means of the integral, the angular acceleration was eliminated, and the heading angle data was introduced to identify the mini-type USV maneuvering model:

$$T \int_a^b r dt + \int_a^b r dt + \alpha \int_a^b r^3 dt = K \int_a^b \delta_m dt \tag{7}$$

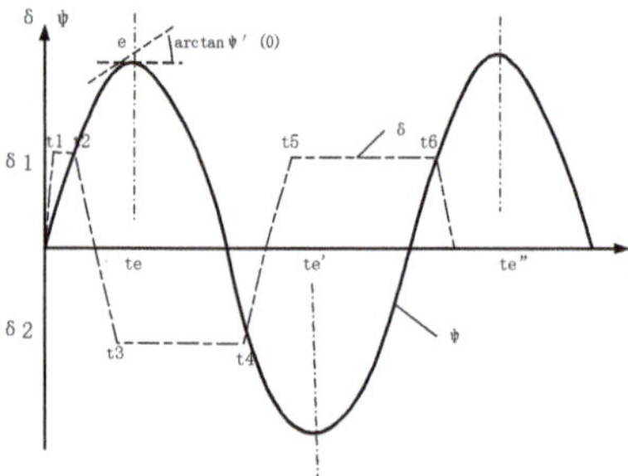

Figure 4. K and T from the zig zag test.

According to the actual field test data, the operating tempo of the program of the USV control system was 0.15 s. Hence, the N equal interval was adopted, and the equal spacing value was 0.15. For the *i*-th interval, the linear terms were integrated and then discrete:

$$\begin{cases} a_i = \int_a^i \delta_m(t)dt \\ b_i = -[\dot{\varphi}(i) - \dot{\varphi}(a)] = -[r(i) - r(a)] \\ \theta_i = \varphi(i) - \varphi(a) \end{cases} \tag{8}$$

For the nonlinear term, $c_i = \int_a^i r^3 dt$, given that r is a function of t, the integral of $r^3(t)$ with respect to the time, *t*, is complex and difficult to solve. When the experimental data is discrete, the Newton-Cotes quadrature equation is adopted to calculate the product, as shown in Equation (9).

Definition 1. [26] *Integrand, $f(x) \in [a, b]$, given a set of nodes, $a \le x_0 < x_1 < \ldots < x_n \le b$, and given the value of the function, $f(x)$, on these nodes, the Lagrangian interpolation polynomial, $L_n(x)$, is applied; thus:*

$$\begin{aligned} I_n(f) = \int_a^b f(x)dx &\cong \int_a^b L_n(x)dx = \int_a^b [\sum_{k=0}^n l_k(x)f(x_h)]dx \\ &= \sum_{k=0}^n f(x_k) \int_a^b l_k(x)dx \\ &= \sum_{k=0}^n A_k f(x_k) \end{aligned} \tag{9}$$

where $A_k = \int_a^b l_k(x)dx$. Equation (9) is referred to as the interpolation formula. The remainder of the interpolation formula is shown in Equation (10):

$$E(f) = \int_a^b [f(x) - L_n(x)]dx = \int_a^b R_n(x)dx = \int_a^b \frac{f^{(n+1)}(\zeta)}{(n+1)!}\omega_{n+1}(x)dx \tag{10}$$

where $\zeta \in (x_0, x_n)$ is dependent on x. $\omega_{n+1}(x) = (x - x_0)(x - x_1)\ldots(x - x_n)$.

The quadrature coefficient, A_k, is independent of $f(x)$, and it is related to the equidistant nodes, x_k, and integral interval, $[a, b]$. The integral formula of the Newton-Cotes (11) can be obtained by transforming A_k:

$$I_n(f) = (b - a)\sum_{k=0}^{n} C_k^{(n)} f(x_k) \tag{11}$$

$$C_k^{(n)} = \frac{(-1)^{n-k}}{nk!(n-k)!}\int_0^n t(t-1)\cdots(t-k+1)(t-k-1)\cdots(t-n)dt \tag{12}$$

where $C_k^{(n)}$ is referred to as the cotes coefficient. The coefficient of cotes is shown in Table 1.

Table 1. Coefficient.

n	$C_k^{(n)}$				
1	$\frac{1}{2}$	$\frac{1}{2}$	/	/	/
2	$\frac{1}{6}$	$\frac{4}{6}$	$\frac{1}{6}$	/	/
3	$\frac{1}{8}$	$\frac{3}{8}$	$\frac{3}{8}$	$\frac{1}{8}$	/
4	$\frac{7}{90}$	$\frac{32}{90}$	$\frac{12}{90}$	$\frac{32}{90}$	$\frac{7}{90}$

Based on the actual test data, $h = 0.15$, the method takes $n = 1.0$, divides the interval, $[a, i]$, into m equal parts, counts $c = 1.0$, and uses a composite integral formula to reduce the remaining terms.

According to Equation (10), the remainder of the composite newton-cotes interpolation can be obtained:

$$E_f = \sum_{k=0}^{m-1}\left[-\frac{h^3}{12}f''(\zeta_k)\right] \approx -\frac{h^2}{12}[f(i) - f(a)] \tag{13}$$

When $n = 1.0$, the composite function, $g(x) = r^3(x)$, can be considered as $g(r) = r^3$. Moreover, by considering the concavity and convexity of the function, the right side of the equation is divided into two parts in the process of the composite Newton-Cotes quadrature, i.e., the exact solution and the approximate solution. In addition, the gain coefficient is introduced to reduce the interpolation remainder, and c_i is shown in Equation (14):

$$c_i = \int_a^i r^3(t)dt = \alpha \cdot \frac{1}{2} \cdot (r^3(i) - r^3(a)) \cdot \Delta t + r^3(a) \cdot \Delta t \tag{14}$$

where $\alpha = 0.5$.

Utilizing the least square method, the left and right sides of Equation (15) are respectively considered as functions, and the square of their difference can be minimized to obtain the values of K, T, α:

$$J = \sum_{i=1}^{N}(Ka_i + Tb_i + \alpha c_i - \theta_i)^2 \tag{15}$$

$$\begin{cases} \frac{\partial J}{\partial K} = 0 & 2\sum \left(Ka_i^2 + Tb_ia_i + \alpha c_ia_i - \theta_ia_i \right) = 0 \\ \frac{\partial J}{\partial T} = 0 & 2\sum \left(Ka_ib_i + Tb_i^2 + \alpha c_ib_i - \theta_ib_i \right) = 0 \\ \frac{\partial J}{\partial \alpha} = 0 & 2\sum \left(Ka_ic_i + Tb_ic_i + \alpha c_i^2 - \theta_ic_i \right) = 0 \end{cases} \tag{16}$$

The same method can also be used to solve the parameters of the velocity model.

2.3. Identification of Maneuverability and Analysis of Motion Ability

2.3.1. Identification of USV Maneuverability by Field Test Data

Given that the data from the zig zag test conducted in the Songhua River had wild values, the outliers were primarily deleted by data fitting. The curve fitting heading angle was a 7th order Fourier function [27], and the curve fitting heading angular velocity was the piecewise cubic polynomial function. The fitting curves of the heading angle and heading angular velocity are shown in Figure 5a–d.

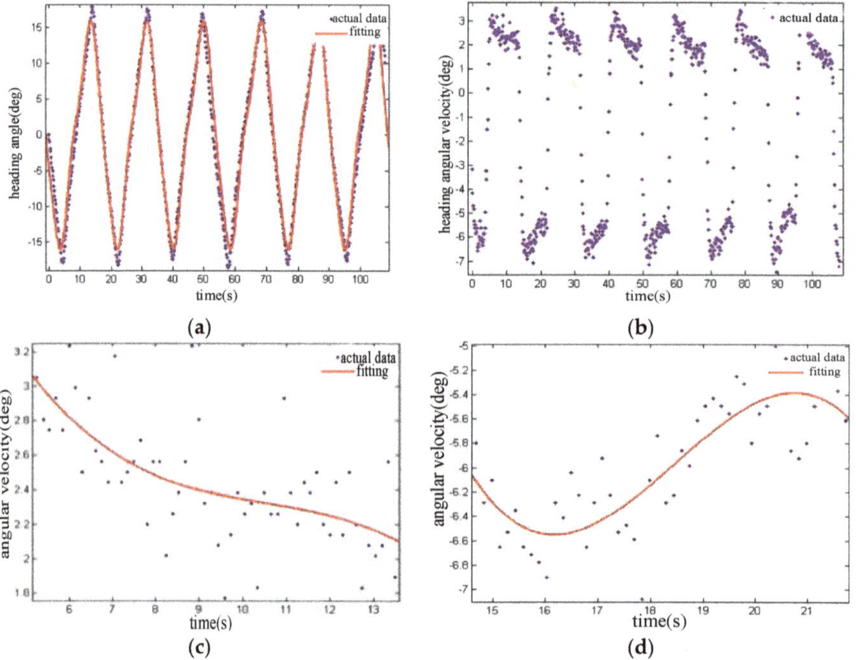

Figure 5. (**a**) Description of heading angle fitting curve; (**b**) the distribution of heading angular velocity; (**c**) the first part of the heading angular velocity's fitting curve; and (**d**) the second part of the heading angular velocity's fitting curve.

According to the fitting curve, the wild values (data points with large deviation) were eliminated.
$$\begin{cases} K = 0.286642 \\ T = 0.410205 \\ \alpha = 0.008477 \end{cases}$$
, as obtained from Equation (16). A zig zag test manipulation experimental simulation model was established in MATLAB. Based on the solved values of K, T, α, the predicted heading angle obtained by the simulation under the same rudder angle input was compared with the actual test data. The results are presented in Figures 6 and 7.

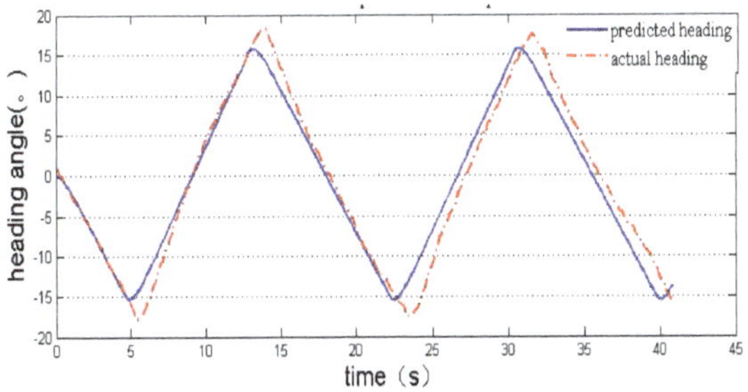

Figure 6. Comparison between the actual and predicted Z-type manipulation.

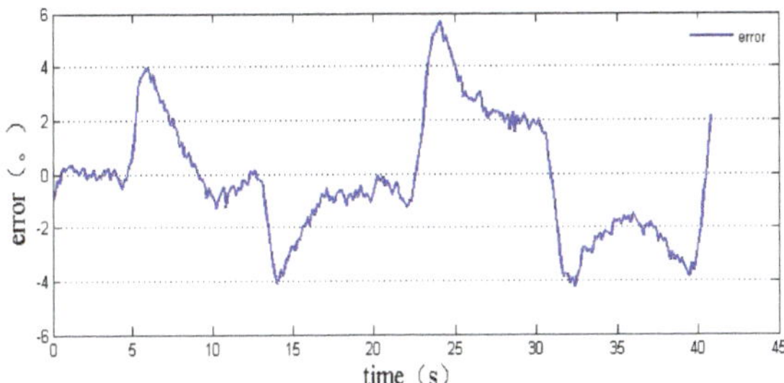

Figure 7. Description of the Z-type error curve.

According to Figure 7, at the same input rudder angle, the heading angle error was in the $\pm 5°$ interval. Considering the communication delay, inertia of the USV, accuracy of the rudder angle feedback, and that of the heading angle sensor, the fitting curve was consistent with the actual data.

Based on the velocity and voltage data, the least squares identification method was used to obtain $k_1 = -0.006$, $k_2 = 0.159$, and $k_3 = 0.004$. The comparison results are presented in Figures 8 and 9.

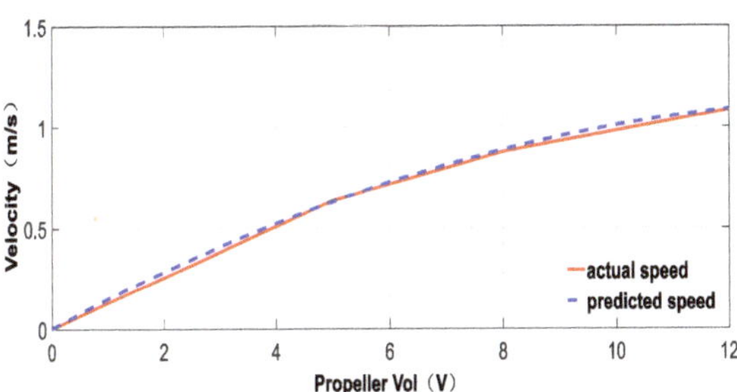

Figure 8. Comparison between the actual and predicted speed.

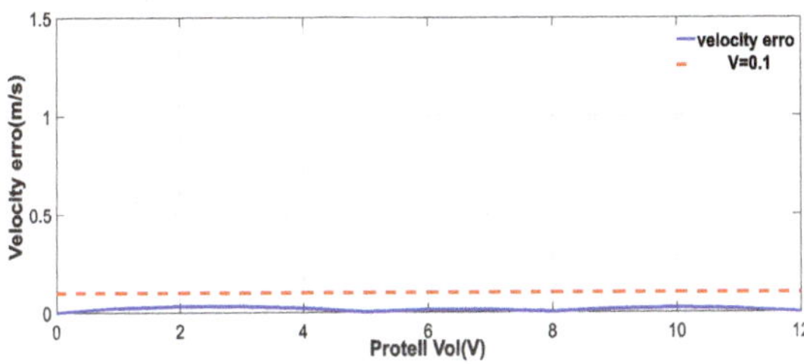

Figure 9. Description of the speed error curve.

2.3.2. Simulation and Analysis of USV Motion

When the USV is sailing in a straight line, the vehicle deviates from the original route and makes a turning movement at a certain rudder angle. The radius of the circle formed by the trajectory of the USV center of gravity is referred to as the radius of steady rotation. K is the constant rotation angular velocity due to the unit rudder angle, which is directly proportional to the radius of rotation, and inversely proportional to the curvature of steady rotation. Therefore, the maximum rotational curvature can be considered as an evaluation index of the USV motion ability.

Using Equation (16), the dynamic model of the 'Dolphin 1' USV was solved, and the turning circle simulation model was established based on MATALB software. Moreover, a proportional–integral–derivative (PID) control method was adopted to establish the control law [28], as shown in Equation (17):

$$u = K_p e + K_i \int edt + K_d \frac{de}{dt} \tag{17}$$

where u is the output of the controller, and e is the deviation between the given expected value and the actual output value. The simulated steady rotation trajectory is presented in Figure 10. The initial position coordinate of the USV was (0,0), the rudder angle was constant at 30.0°, and the iterative step length and control cycle were 0.15 s.

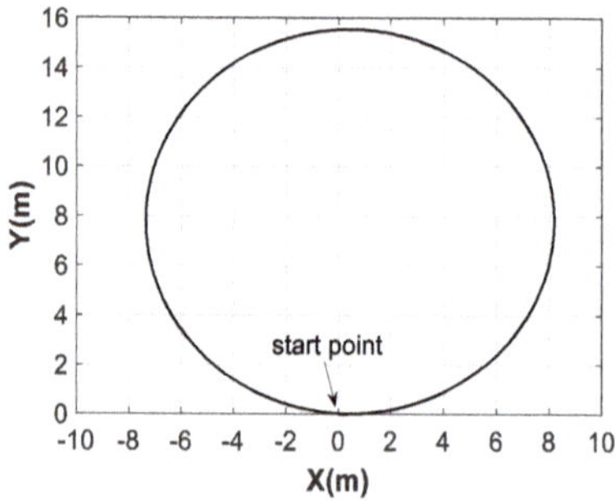

Figure 10. Description of the turning circle simulation test.

To confirm the reliability of the motion analysis, the field steady turning test data of 'Dolphin 1' were used for comparison. The data were collected and recorded at a rudder angle with a fixed value of 30.0°. The trajectory of the USV is shown in Figure 11.

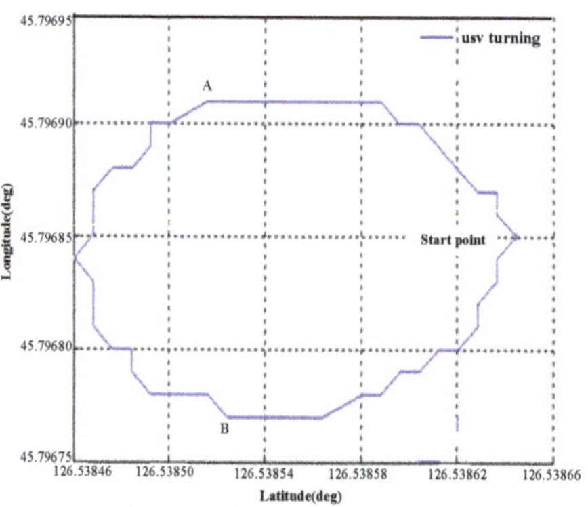

Figure 11. Description of 'Dolphin 1' field turning test.

In Figure 11, the distance between two course path points at A(126.53852,45.79677) and B(126.53851,45.79691) were calculated. Using Equation (18), the longitude and latitude coordinates can be converted from angular units to metric units:

$$R_{tur} = \frac{\pi}{360} \sqrt{(A_{lon} - B_{lon})^2 + (A_{lat} - B_{lat})^2} R_{globle} \tag{18}$$

where $R_{globle} = 6371.393$ km is the radius of the earth, and the radius, R_{tur}, of 'Dolphin 1' is 7.8 m. By analyzing and comparing Figures 10 and 11, the two turning trajectories were found to be identical.

The integral nonlinear least square method was employed to analyze and determine the performance index of the USV, which was found to demonstrate a high efficiency, convenience, and high accuracy.

3. SPP of USV

In this paper, a second path planning method (SPP) is presented, which is not restricted by the USV motion ability. The improved artificial potential field method proposed by Huang Xinghua [21] was selected as the first path planning method in this study. On the premise of the existing planning path, the second path optimization was carried out to calculate an optimal path constrained by the motion performance of the vehicle. This method was used to solve the problem and deficiency of the traditional USV path planning.

3.1. SPP Model

The SPP of the USV refers to the optimal path constrained by the performance of the USV, which was selected according to the existing optimal path and the three-point geometric relationship. This method preserves the preferred criteria of the traditional path planning.

The SPP model can be described as follows.

Assuming that x, y are the coordinates in the inertial coordinate system, and $p_i = [x_i, y_i]^T$ is defined as the coordinate position of a path point, the set of n path points derived from the traditional path planning can be expressed as $p = [p_1^T \ldots \ldots p_n^T]^T$. The check point at time k can be expressed as $p_k = [x_k, y_k]^T$, and ω is the second planning task variable. The task function is established by combining the geometric constraint relationship of the path, and the corresponding task function can be expressed as shown in Equation (19):

$$\omega = f(p_{k-1}, p_k, p_{k+1}) \tag{19}$$

$$\begin{cases} (\omega - \omega_d) < 0, & \begin{cases} p_{k-1} = p_k \\ p_k = p_{k+1} \end{cases} \\ (\omega - \omega_d) \geq 0, & \begin{cases} p_k = p_{k+1} \\ p_{k+1} = p_{k+2} \end{cases} \end{cases} \tag{20}$$

where ω_d is the constraint index variable of the motion performance of the vehicle. When $(\omega - \omega_d) < 0$, the current check point is maintained and the next path point is discussed. When $(\omega - \omega_d) \geq 0$, the current checkpoint is discarded and the next point of the optimal set of paths is tested until the checkpoint is the endpoint of the path. The schematic diagram of SPP is shown in Figure 12.

The second path planning is applicable to the traditional USV path planning method, which can be used to solve a series of discrete path point sets.

Using the improved artificial potential field method proposed by Huang Xinghua [21] as an example, as shown below, the path planning of the USV involved the implementation of the secondary optimization subject to its motion constraint.

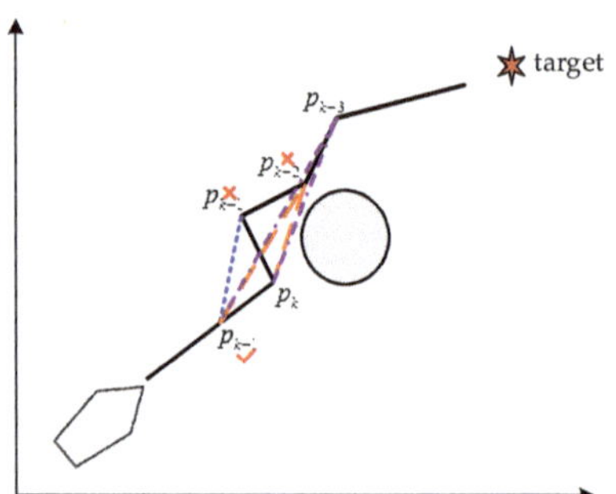

Figure 12. The schematic diagram of the second path planning.

3.2. An Improved Path Planning for the Artificial Potential Field Method

The artificial potential field method (APF), initially proposed by Khatib, was applied to the obstacle avoidance motion planning of a robot manipulator, which had as its objective the realization of the real-time obstacle avoidance of a mechanical arm. The essence of the APF is to define an abstract potential field artificially for the operating space of the robot, which is the superposition of the gravitational field of the target position and the repulsive force field of the obstacle in the environment. The negative gradient of the force field is the virtual force that acts on the robot, and the resultant force of gravity and repulsion represent the accelerating force of the robot.

Although the path planned by the artificial potential field method is efficient, smooth, and safe, there are several drawbacks related to this method [29–34], i.e., (1) the local minimum problem; (2) the inability to navigate through closely-situated obstacles; and (3) oscillations near the obstacle. The main objective of the improved artificial potential field method [21] proposed by Huang Xinghua is the solution of these problems to establish a new potential field function. Hence, during the approach of the target point by the robot, the repulsion field is close to zero and the position of the robot target point is the minimum point of the entire situation field; thus, the target can be reached. The repulsion force potential field function of the improved artificial potential field method can be expressed as shown in Equation (21):

$$U_{rep}(X) = \begin{cases} \frac{1}{2}k_{rep}(\frac{1}{X-X_o} - \frac{1}{\rho_0})^2(X-X_g)^n & (X-X_o) \le \rho_0 \\ 0 & (X-X_o) \ge \rho_0 \end{cases} \tag{21}$$

where $0 \le n < 1$.

The negative gradient of the potential field function is also considered as the corrected repulsive force of the repulsion field, as follows:

$$F_{rep}(X) = -\nabla U_{rep}(X) = \begin{cases} F_{rep1} + F_{rep2} & (X-X_o) \le r_o \\ 0 & (X-X_o) > r_o \end{cases} \tag{22}$$

where:

$$F_{rep1} = k_{rep}(\frac{1}{X-X_o} - \frac{1}{\rho_o})\frac{1}{(X-X_o)^2}\frac{\partial(X-X_o)}{\partial X}(X-X_g)^n \tag{23}$$

$$F_{rep2} = -\frac{1}{2}k_{rep}(\frac{1}{X-X_o} - \frac{1}{\rho_o})^2\frac{\partial(X-X_g)^n}{\partial X} \tag{24}$$

where U_{rep} is the repulsive field of the obstacle, ρ is the distance between the robot and the obstacle, ρ_0 is the maximum influencing range of the obstacle, k_{rep} is the constant of the repulsive field, n is positive, X represents the coordinates of the robot, X_g represents the coordinates of the target point, and ∇ represents the sign of the gradient.

The gravitational potential field function can be expressed as follows:

$$U_{att}(X) = \frac{1}{2}k(X - X_g)^2 \tag{25}$$

where U_{att} is the gravitational field generated by the target point, and k is the gain constant. Gravity can be expressed as shown in Equation (26):

$$F_{att}(X) = -\nabla U_{att}(X) = k(X_g - X) \tag{26}$$

3.3. SPP of the Improved Artificial Potential Field Method

Combined with the improved artificial potential field method, the theoretical block diagram of the second path planning constrained by the motion of the USV is shown in Figure 13.

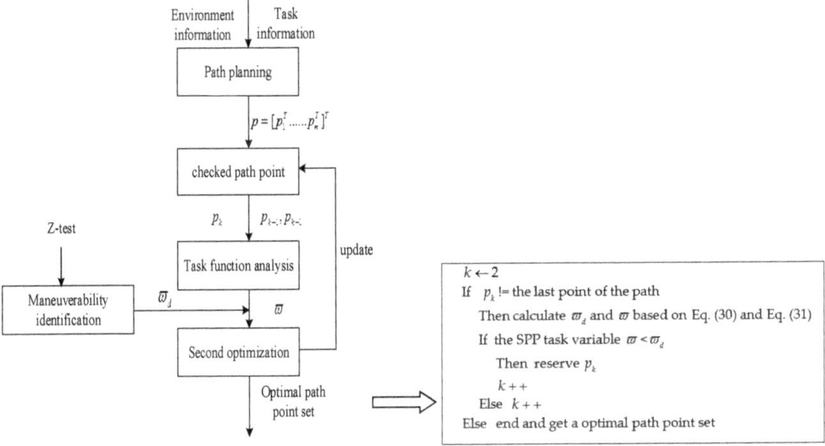

Figure 13. The theoretical block diagram of SPP.

When the vehicle is performing the navigation task, the starting position, target position, and obstacle position are determined according to the mission information and sea chart information. According to the objective function of the artificial potential field method, a set of path points are obtained. The starting position is reserved, and the second planning task function is analyzed from the second point, as shown in Equation (27):

$$D_k = \|p_k - p_{k-1}\| \quad D_{k+1} = \|p_{k+1} - p_k\| \quad D_{k-1} = \|p_{k-1} - p_{k+1}\| \tag{27}$$

$$\widetilde{D} = \sum_{i=0}^{2} D_{k-1+i} \tag{28}$$

$$J_k = \frac{1}{2}[\widetilde{D}(\widetilde{D} - 2D_k)(\widetilde{D} - 2D_{k-1})(\widetilde{D} - 2D_{k+1})]^{\frac{1}{2}} \tag{29}$$

where p_k represents the location coordinates of the path point that correspond to the current check moment, k; p_{k+1}, p_{k-1} are the path points of the current test point forward and backward once, respectively; D_i is the distance of the two path points; $\widetilde{D}$ is the distance between the path points; and J_k is the size of the region across the three path points. The second planning task variable can be expressed as follows:

$$\varpi = f(p_{k-1}, p_k, p_{k+1}) = \frac{4J_k}{D_k D_{k-1} D_{k+1}} \tag{30}$$

$$\omega_d = \frac{1}{r} \tag{31}$$

where ω_d is the constraint index variable of the USV motion performance, and its value can be determined by the rotational curvature of the vehicle. Substituting Equations (30) and (31) into Equation (19), the second optimization can be evaluated.

4. Simulation Test of Trajectory Tracking of USV by SPP Method

Considering the 'Dolphin 1' mini-type USV as the test object and using the improved artificial potential field method proposed by Huang Xinghua [21] as an example, the path planning optimization method presented in this paper was simulated and verified. The simulation platform was developed using MATLAB software, the PID control method was employed to control the course and speed, and the trajectory tracking method was implemented in conjunction with the PP method.

4.1. Comparative Experiment of Small Planning Step

The initial position of the USV was (10 m, 5 m), the target position was (1000 m, 800 m), the length of the vehicle was $l = 2.0$ m, and the planning step, t_p, was $t_p = 2.0$ m. The locations of the obstacles were (200 m, 150 m) and (500 m, 400 m), respectively. The radius of the obstacles, r, corresponded to 10.0 m and 30.0 m, and the influencing radius was defined as $r\_in = r + \lceil t_p/l \rceil * l$, where $\lceil\ \rceil$ is the integer sign considered upward.

The results of the comparison between the initial path planning using the improved artificial potential field method and the second path planning under the consideration of the motion performance of the USV are shown in Figure 14a. Figure 14b presents the local magnification of the path planning near the first obstacle. In Figure 14a, the solid line represents the initial planning path, and the dotted line represents the optimized path by the secondary planning. The figure illustrates that the initial planning path near the obstacle produces zig-zag oscillations, and the intensity of the path oscillation is related to the impact of the obstacle on the connectivity of the starting point and the end point, in addition to the size of the obstacle. As shown in Figure 14b, given that the SPP method is optimized based on the initial path; it retains the fitness function standard of the initial planning and reduces the oscillation of the initial planning path through optimization.

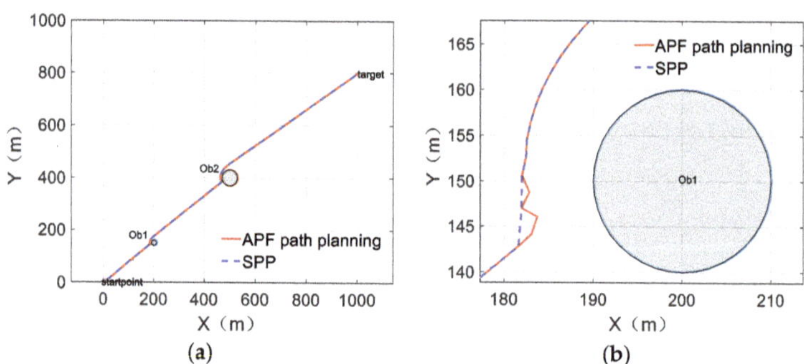

Figure 14. (a) Description of the path planning comparison diagram; and (b) local enlargement of the first obstacle area for path planning.

On the Simulink simulation platform, a USV motion model was developed. The iterative step length and control cycle were 0.15 s. The parameters of the course PID controller were set as $p = 1.0, i = 0.001, d = 1.0$, the parameters of the speed controller were $p = 2.0, i = 0.001, d = 0.001$, and the trajectory tracking adopted the pure tracking (PP) method [35–37]. In the simulation experiment, when the time exceeded 50.0 s, the target point was replaced, although the vehicle had not reached the distance threshold of the target in the current tracking stage. The simulation results of the USV trajectory tracking of the two methods are respectively shown in Figure 15a,c. The tracking deviation of the two methods is shown in Figure 15b,d.

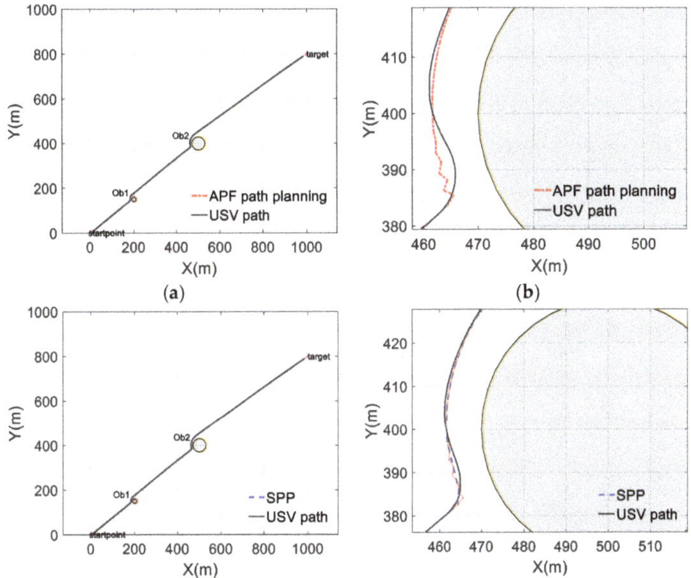

Figure 15. (**a**) Description of USV trajectory by APF; (**b**) local enlargement of the second obstacle area for the USV trajectory by APF; (**c**) description of the USV trajectory by SPP; and (**d**) local enlargement of the second obstacle area for the USV trajectory by SPP.

As can be seen from Figure 15a,c, under the premise of a small planning step length, $t_p = 2.0$ m, the vehicle can track the optimized path after the initial planning using an artificial potential field method and secondary planning method, respectively, and successfully complete the obstacle avoidance task. The comparison and analysis of Figure 15b,d illustrate that although both methods can be used to achieve obstacle avoidance and complete the tracking tasks by SPP, the actual course path of the vehicle was more consistent with the planned path than by APF. This is because secondary programming is an optimization standard of the maneuverability of the USV, which leads to a higher tracking accuracy.

4.2. Comparative Experiment of Large Planning Step

With the same simulation environment and control parameters as in the abovementioned experiment, the USV sailed from the same position to the same target, and the planning step, t_p, was set greater than the length of the vehicle by a factor of 5 ($t_p = 10.0$ m). The results of the comparison between the two methods with respect to the USV trajectory tracking are respectively shown in Figure 16a,b.

As can be seen from Figure 16a, the USV generated roundabout routes when it tracked the planned path points during its approach of the first and second obstacles, which resulted in a failure to complete the obstacle avoidance task (given that the obstacle was a virtual object in the simulation, the task was not interrupted due to collision, and the experiment was continued). Figure 17 presents the tracking deviation of the USV for the duration of the trajectory tracking task. A maximum path deviation of 34 m was generated near the first obstacle, which gradually decreased to a value of approximately 4.0 m, and then remained stable. It should be noted that the distance between the navigation position of the vehicle and the tracking target was reduced to 4.0 m in the simulation; thus, the task was considered as completed. In the vicinity of the second obstacle, the maximum deviation was approximately 77.0 m, and due to its course deviation, the tracking process always sailed in an approximate 3/4 minimum turning circle in the following trajectory tracking process. This resulted in the oscillation of the tracking deviation within the range of 8 m until the target point was reached.

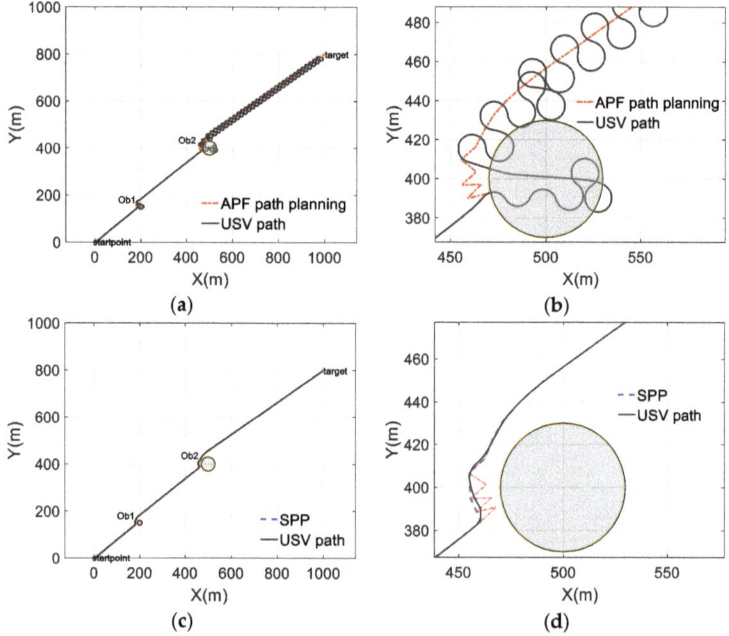

Figure 16. (**a**) Description of the USV trajectory by APF; (**b**) local enlargement of the second obstacle area for the USV trajectory by APF; (**c**) description of the USV trajectory by SPP; and (**d**) local enlargement of the second obstacle area for the USV trajectory by SPP.

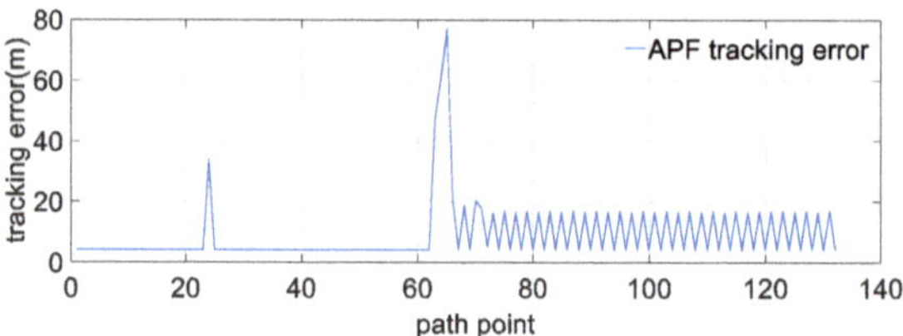

Figure 17. Description of the USV along tracking error by APF.

According to the analysis results in Figures 16b and 18, during the large planning step, the USV tracked the planned path optimized by the secondary planning, and then successfully completed the obstacle avoidance task. The tracking deviation, namely, the along track error, was less than 4 m; thus, the vehicle was considered to reach the target point. As can be seen from Figure 18, the obstacle has no effect on the tracking accuracy, and the tracking deviation was approximately 4 m. Figures 17 and 18 illustrate the decrease in the number of track points after optimization, which reduced the possibility of the occurrence of detour road events. The experimental results indicate that the SPP method improves the tracking accuracy of the USV and reduces the length of the detour route, thus reducing the energy consumption and preventing collisions.

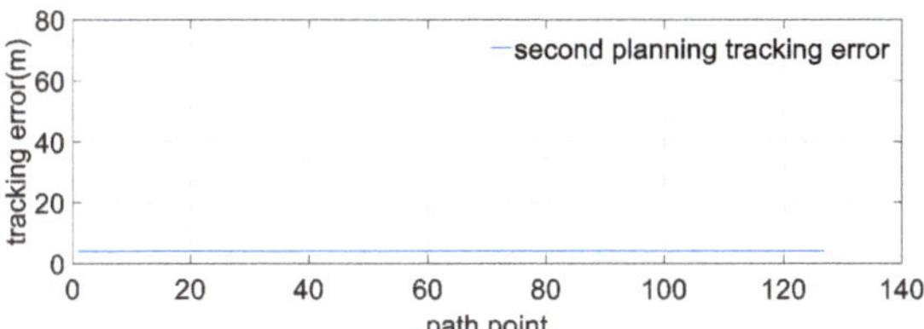

Figure 18. Description of the USV along tracking error by SPP.

4.3. Comparative Experiment of Multi-Obstacle Path Planning

According to the abovementioned experiment, in the case of two obstacles in the small planning step, the USV successfully tracked the path planned by APF and SPP, and completed the obstacle avoidance task. With an identical simulation environment and control parameters to those of the abovementioned experiment, the vehicle navigated from the same starting position to the same target point, thus avoiding six obstacles in the course, which had position coordinates of (60 m, 50 m), (200 m, 150 m), (400 m, 300 m), (500 m, 400 m), (600 m, 400 m), and (720 m, 570 m). In addition, the obstacle radius corresponded to 5.0 m, 10.0 m, 30.0 m, 50.0 m, 30.0 m, and 20.0 m, respectively. The results of the comparison between the initial path planning utilizing the improved artificial potential field method and the second path planning under the consideration of the motion performance of the USV are shown in Figure 19a,b. The trajectory tracking deviations for the two methods are shown in Figures 20 and 21.

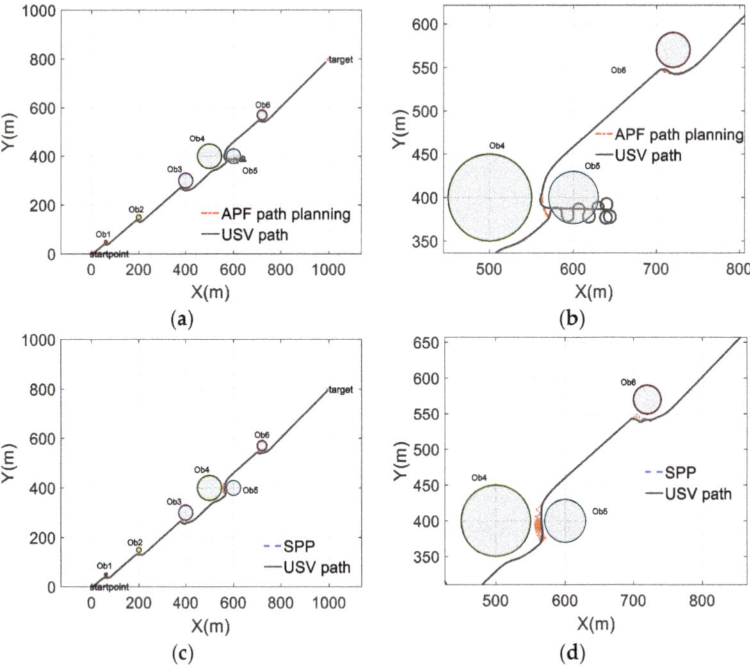

Figure 19. (**a**) Description of the USV trajectory by APF; (**b**) local enlargement for the USV trajectory by APF; (**c**) description of the USV trajectory by SPP; and (**d**) local enlargement for the USV trajectory by SPP.

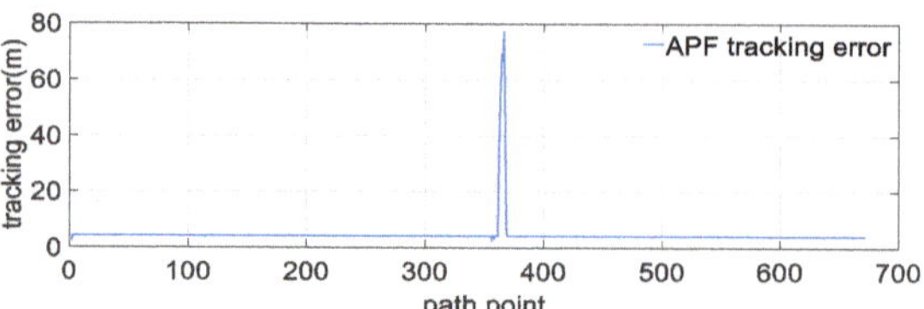

Figure 20. Description of the USV along tracking error by APF.

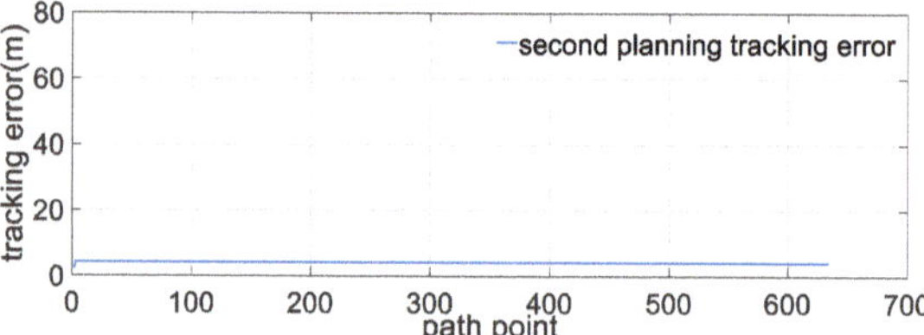

Figure 21. Description of the USV along tracking error by SPP.

As can be seen from Figures 19a and 20, using the APF method, the USV successfully completed the obstacle avoidance task of the first three obstacles, and the trajectory tracking deviation was constant at approximately 4 m. However, when sailing near the fourth and fifth obstacles, due to the limitation of the motion performance of the vehicle, the planned path exceeded its minimum turning radius, which resulted in a collision with the fifth obstacle. Moreover, the trajectory tracking deviation was approximately 78 m, and the obstacle avoidance task was not completed.

Figures 19b and 21 reveal that on the basis of the SPP method, the USV successfully completed the obstacle avoidance task of all six obstacles, and the track tracking deviation was constant at approximately 4.0 m. The optimized path of the second path planning method can therefore effectively ensure the completion efficiency of the obstacle avoidance task and improve the precision of the trajectory tracking control.

5. Discussion

(1) By the analysis of the path planning theory and the USV control model, the traditional path planning method was found to lead to the 'planning failure' phenomenon when applied to the trajectory tracking field of the USV path planning.

(2) In this study, an integral nonlinear least squares method was developed. In the case of limited test data, a nonlinear motion model of USV was rapidly identified by merely conducting a type of maneuvering experiment, which can effectively predict the motion performance indexes, such as the rotatory curvature of the vehicle.

(3) The SPP method was presented under the consideration of the USV motion performance, which reduces the influence of the motion performance of the vehicle during the trajectory tracking and helps lower the risk of failing to complete the obstacle avoidance task using the traditional path planning method. The proposed SPP method can effectively prevent the issue of an untraceable USV and improve the USV tracking accuracy.

6. Conclusions

This paper presented an integral nonlinear least squares method to rapidly and effectively obtain the motion constraint of USVs, and an SPP method was proposed under the consideration of the USV motion performance.

In future work, the path optimization method in a bounded environment, especially an uneven boundary, should be considered. In addition, a field test of the method should be carried out.

Author Contributions: Conceptualization, Y.L. (Ye Li) and J.F.; methodology, Y.L. (Yulei Liao) and J.F.; software, J.F. and W.J.; validation, Y.L. (Ye Li), J.F. and L.W.; formal analysis, H.W.; investigation, W.J. and H.W.; data curation, J.F. and W.J.; writing—Original draft preparation, J.F.; writing—review and editing, J.F. and Q.J.; supervision, Y.L. (Ye Li) and Y.L. (Yulei Liao); funding acquisition, Y.L. (Ye Li) and Y.L. (Yulei Liao) Authorship has been limited to those who have contributed substantially to the work reported.

Funding: This research was funded by the National Key R&D Program of China under Grant 2017YFC0305700, in part by the National Natural Science Foundation of China under Grant 51779052, U1806228, and 51879057, in part by the Natural Science Foundation of Heilongjiang Province of China under Grant QC2016062, in part by the Research Fund from Science and Technology on Underwater Vehicle Laboratory under Grant 614221503091701, in part by the Fundamental Research Funds for the Central Universities under Grant HEUCFP201741, and HEUCFG201810, and in part by the Qingdao National Laboratory for Marine Science and Technology under Grant QNLM2016ORP0406.

Conflicts of Interest: The authors declare no conflict of interest.

References

1. Liao, Y.L.; Zhang, M.J.; Wan, L.; Li, Y. Trajectory tracking control for underactuated unmanned surface vehicles with dynamic uncertainties. *J. Cent. South Univ. Technol.* **2016**, *23*, 370–378. [CrossRef]

2. Li, Y.; Wang, L.F.; Liao, Y.L.; Jiang, Q.Q.; Pan, K.W. Heading MFA control for unmanned surface vehicle with angular velocity guidance. *Appl. Ocean Res.* **2018**, *80*, 57–65. [CrossRef]

3. Liu, Y.; Bucknall, R. A survey of formation control and motion planning of multiple unmanned vehicles. *Robotica* **2018**, *36*, 1019–1047. [CrossRef]

4. Singh, Y.; Sharma, S.; Sutton, R.; Hatton, D. Path planning of an autonomous surface vehicle based on artificial potential fields in a real-time marine environment. In Proceedings of the 16th International Conference on Computer and IT Applications in the Maritime Industries, Cardiff, UK, 15–17 May 2017.

5. Singh, Y.; Sharma, S.; Sutton, R.; Hatton, D.; Khan, A. Feasibility study of a constrained Dijkstra approach for optimal path planning of an unmanned surface vehicle in a dynamic maritime environment. In Proceedings of the 2018 IEEE International Conference on Autonomous Robot Systems and Competitions (ICARSC), Torres Vedras, Portugal, 25–27 April 2018; pp. 117–122.

6. Fu, B.; Chen, L.; Zhou, Y.T.; Zheng, D. An improved A algorithm for the industrial robot path planning with high success rate and short length. *Robot. Auton. Syst.* **2018**, *106*, 26–37. [CrossRef]

7. Singh, Y.; Sharma, S.; Sutton, R.; Hatton, D.; Khan, A. A constrained A* approach towards optimal path planning for an unmanned surface vehicle in a maritime environment containing dynamic obstacles and ocean currents. *Ocean Eng.* **2018**, *169*, 187–201. [CrossRef]

8. Zhu, L.; Fan, J.Z.; Zhao, J.; Wu, X.G.; Liu, G. Global path planning and local obstacle avoidance of searching robot in mine disasters based on grid method. *J. Cent. South Univ.* **2011**, *11*, 3421–3428.

9. Qu, Y.H.; Zhang, Y.T.; Zhang, Y.M. A global path planning algorithm for fixed-wing UAVs. *J. Intell. Robot. Syst.* **2018**, *91*, 691–707. [CrossRef]

10. Sfeir, J.; Saad, M.; Saliah, H. An improved artificial potential field approach to real –time mobile robot path planning in an unknown environment. In Proceedings of the IEEE International Symposium on Robotic and Sensors Environments, Montreal, QC, Canada, 17–18 September 2011; pp. 208–213.

11. Pei, Z.B.; Chen, X.B. Improved ant colony algorithm and its application in obstacle avoidance for robot. *CAAI Trans. Intell. Syst.* **2015**, *1*, 90–96.

12. Wang, J.; Wu, X.X.; Guo, B.L. Robot path planning using improved particle swarm optimization. *Comput. Eng. Appl.* **2012**, *48*, 240–244.

13. Yan, Z.P.; Li, J.Y.; Wu, Y.; Zhang, G.S. A real-time path planning algorithm for AUV in unknown underwater environment based on combining PSO and waypoint guidance. *Sensors* **2019**, *19*, 20. [CrossRef] [PubMed]

14. Wang, H.W.; Ma, Y.; Xie, Y.; Guo, M. Mobile robot optimal path planning based on smoothing A* algorithm. *J. Tongji Univ. (Nat. Sci.)* **2010**, *38*, 1647–1650.

15. Conte, G.; Capua, G.P.; Scaradozzi, D. Designing the NGC system of a small ASV for tracking underwater targets. *Robot. Auton. Syst.* **2016**, *76*, 46–57. [CrossRef]

16. Doostie, S.; Hoshiar, A.K.; Nazarahari, M. Optimal path planning of multiple nanoparticles in continuous environment using a novel Adaptive Genetic Algorithm. *Precis. Eng.* **2018**, *53*, 65–78. [CrossRef]

J. Mar. Sci. Eng. **2019**, *7*, 104

17. Wu, C.; Wu, Q.; Ma, F.; Wang, S.W. A situation awareness approach for USV based on Artificial Potential Fields. In Proceedings of the International Conference on Transportation Information and Safety 2017, Banff, AB, Canada, 8–10 August 2017; pp. 232–235.

18. Yang, J.M.; Tseng, C.M.; Tseng, P.S. Path planning on satellite images for unmanned surface vehicles. *Int. J. Nav. Archit. Ocean Eng.* **2015**, *7*, 87–99. [CrossRef]

19. Abbas, S.; Abbas, K.; Troy, R.; Saeid, N. Smooth path planning using biclothoid fillets for high speed CNC machines. *Int. J. Mach. Tools Manuf.* **2018**, *132*, 36–49.

20. Brezak, M.; Petrovic, I. Path smooth using clothoids for differential drive mobile robots. *IFAC Proc. Vol.* **2011**, *44*, 1133–1138. [CrossRef]

21. Shi, W.F.; Huang, X.H.; Zhou, W. Planning of mobile robot based on improved artificial potential. *J. Comput. Appl.* **2010**, *30*, 2021–2023.

22. Liao, Y.L.; Zhang, M.J.; Wan, L. Serret-Frenet frame based on path following control for underactuated unmanned surface vehicles with dynamic uncertainties. *J. Cent. South Univ. Technol.* **2015**, *22*, 214–223. [CrossRef]

23. Nomoto, K.; Taguchi, K.; Honda, K.; Hirano, S. On the steering qualities of ships. *Int. Shipbuild. Prog.* **1957**, *4*, 354–370. [CrossRef]

24. Christian, R.S.; Craig, A.W. Modeling, identification, and control of an unmanned surface vehicle. *J. Field Robot.* **2013**, *30*, 371–398.

25. Nikola, M.; Zoran, V. Fast in-field identification of unmanned marine vehicles. *J. Field Robot.* **2011**, *28*, 101–120.

26. Podisuk, M.; Chundang, U.; Sanprasert, W. Single step formulas and multi-step formulas of the integration method for solving the initial value problem of ordinary differential equation. *Appl. Math. Comput.* **2007**, *190*, 1438–1444. [CrossRef]

27. Li, H.L.; Zhang, A.W.; Meng, X.G.; Hu, S.X. A remote sensing image sub-pixel registration method based on curve fitting and improved Fourier-Mellin transform. *J. Chin. Comput. Syst.* **2015**, *36*, 2763–2768.

28. Liao, Y.L.; Li, Y.M.; Wang, L.F.; Li, Y.; Jiang, Q.Q. The heading control method and experiments for an unmanned wave glider. *J. Cent. South Univ. Technol.* **2017**, *24*, 1–9. [CrossRef]

29. Pradhan, S.K.; Parhi, D.H.; Panda, A.K. Potential field method to navigate several mobile robots. *Appl. Intell.* **2006**, *25*, 321–333. [CrossRef]

30. Liu, Z.X.; Yang, L.X.; Wang, J.G. Soccer robot path planning based on evolutionary artificial field. *Adv. Mater. Res.* **2012**, *562*, 955–958. [CrossRef]

31. Koren, Y.; Borenstein, J. Potential field methods and their inherent limitations for mobile robot navigation. In Proceedings of the IEEE International Conference on Robotics and Automation, Sacramento, CA, USA, 9–11 April 1991; Volume 2, pp. 1398–1404.

32. Liang, K.; Li, Z.Y.; Chen, D.Y. Improved artificial potential field for unknown narrow environments. In Proceedings of the IEEE International Conference on ROBIO, Shenyang, China, 22–26 August 2004; pp. 688–692.

33. Geva, Y.; Shapiro, A. A combined potential function and graph search approach for free gait generation of quadruped robots. In Proceedings of the IEEE International Conference on Robotics and Automation River Centre, Saint Paul, MN, USA, 14–18 May 2012; pp. 5371–5376.

34. Qi, N.N.; Ma, B.J.; Liu, X.E.; Zhang, Z.X.; Ren, D.C. A modified Artificial Potential Field Algorithm for Mobile Robot Path Planning. In Proceedings of the 7th World Congress on Intelligent Control and Automation, Chongqing, China, 25–27 June 2008; Volume 7, pp. 2603–2607.

35. Liao, Y.L.; Wang, L.F.; Li, Y.M.; Li, Y.; Jiang, Q.Q. The intelligent control system and experiments for an unmanned wave glider. *PLoS ONE* **2016**, *11*, 1–24. [CrossRef]

36. Liao, Y.L.; Jia, Z.H.; Zhang, W.B.; Jia, Q.; Li, Y. Layered berthing method and experiment of unmanned surface vehicle based on multiple constraints analysis. *Appl. Ocean Res.* **2019**, *86*, 47–60. [CrossRef]

37. Liao, Y.L.; Jiang, Q.Q.; Du, T.P.; Jiang, W. Redefined output model-free adaptive control method and unmanned surface vehicle heading control. *IEEE J. Ocean. Eng.* **2019**. [CrossRef]

Journal of
*Marine Science
and Engineering*

MDPI

Article

UUV Simulation Modeling and its Control Method: Simulation and Experimental Studies

Ji-Hong Li *, Min-Gyu Kim, Hyungjoo Kang, Mun-Jik Lee and Gun Rae Cho

Marine Robotics R&D Division, Korea Institute of Robot and Convergence, Jigok-Ro 39, Nam-Gu, Pohang 37666, Korea; zxdwa0817@kiro.re.kr (M.-G.K.); hjkang@kiro.re.kr (H.K.); mcklee@kiro.re.kr (M.-J.L.); sandman@kiro.re.kr (G.R.C.)
* Correspondence: jhli5@kiro.re.kr; Tel.: +82-54-279-0467

Received: 27 February 2019; Accepted: 25 March 2019; Published: 28 March 2019

Abstract: This paper presents the development of an unmanned underwater vehicle (UUV) platform, especially the derivation of the vehicle's simulation model and its control method to overcome strong sea current. The platform is designed to have a flattened ellipsoidal exterior so as to minimize the hydrodynamic damping on the horizontal plane. Four horizontal thrusters with the identical specifications are symmetrically mounted on the horizontal plane, and each of them has the same thrust dynamics in both forward and reverse directions. In addition, there are three vertical thrusters used to handle the vehicle's roll, pitch and heave motions. Control strategy proposed in this paper to overcome strong current is that: maximizing the vectored horizontal thrust force against the sea current without or with the least of the vehicle's rotation on the horizontal plane. For the vehicle model, due to it being symmetric in all of three axes, the vehicle dynamics can be simplified and all of hydrodynamic coefficients are calculated through both of theoretical and empirically-derived formulas. Numerical simulations and experimental studies in both of the water tank and the circulating water channel are carried out to demonstrate the vehicle's capability of overcoming strong current.

Keywords: unmanned underwater vehicle (UUV); marine systems; vehicle dynamics; simulation model; overcome strong sea current

---

1. Introduction

In the sea around the Korean peninsula, especially in the West Sea also known as the Yellow Sea, strong sea currents usually pose tough technical challenges in the salvage operations as well as in the various scientific activities. In both cases of ROKS Cheonan sinking on 26 March 2010 [1] and sinking of MV Sewol on 16 April 2014 [2], one of the most difficult problems during the initial response for salvage was that there was not any underwater vehicle able to be deployed in the strong sea current environment [3] so as to collect the swift on-site disaster scene information. Recently, how to overcome strong sea current has become a hot topic in the UUV community, especially in Korea. In [4], a ROV called Crabster CR200 was introduced. The vehicle has about 188 kg of negative buoyancy in the water. So in the case of strong current which can be up to 3 knots, the vehicle can land on the sea floor and using its 6 legs can resist the current. In [5], the authors presented a two-body vehicle, where the lower body can land on the sea floor and overcome strong current using a sort of anchor system. In both cases, the whole vehicle or part of the body should be land on the sea floor to resist the sea current. This kind of mechanism might constrain the vehicle's precise underwater inspection capability. Other attempts to counter strong turbulence using the high velocity water itself include [6].

This paper presents the development of a UUV platform and its motion control technology for the purpose of overcoming strong sea current. The vehicle has the flattened ellipsoidal exterior to minimize the hydrodynamic damping in the water, as seen in Figure 1. Four horizontal thrusters with the identical specification are mounted symmetrically on the horizontal plane. Each thruster has the same thrust dynamics in both forward and reverse directions. This kind of horizontal thrust mechanism can guarantee the uniform distribution of vectored thrust forces in all horizontal directions. On the other hand, this is also beneficial for easily stabilizing the vehicle's horizontal motion in the dynamic sea current environment. Three vertical thrusters, as seen in Figure 1, are used to stabilize the vehicle's roll, pitch, and heave motions.

Figure 1. Developed unmanned underwater vehicle (UUV) platform with the flattened ellipsoidal exterior.

The control strategy for this vehicle to overcome strong current is to maximize the vectored horizontal thrust force along certain direction, usually against the sea current, while keeping its heading or with the least of heading rotation on the horizontal plane. Three vertical thrusters, as mentioned before, are used to stabilize the roll, pitch and heave motion. General PD controllers [7,8] are designed independently for each of the horizontal and vertical thrusters groups.

Some of the simulation and experimental studies are carried out to demonstrate the performance of the platform design and its motion control technologies. In the simulation, the hydrodynamic coefficients are calculated through both of the theoretical and empirically-derived formulas [9–11]. Furthermore, in the controller design, four horizontal thrusters are modeled under the consideration of the fact that the maximum thrust force will be reduced in compliance with the increasing of fluid speed flow through the thruster [12,13]. In addition, through circulating water channel test, it is observed that the vehicle can get forward motion while keeping its heading in the strong current environment where the current is up to 2.5 knots.

The remainder of this paper is organized as follows. Section 2 describes the vehicle's kinematic and hydrodynamic model, and the vehicle motion sensor's lever arm effects are considered in Section 3. Controllers for both of horizon keeping and maximum forward speed on the horizontal plane are presented in Section 4. Furthermore, Section 5 shows the simulation result, while experimental studies are presented in Section 6. A brief conclusion and some future works are discussed in Section 7.

2. Vehicle Modeling

2.1. Kinematics and Dynamics

Usually, the kinematics and dynamics of underwater vehicles can be expressed as follows [9],

$$\dot{\eta} = C_b^n \nu, \tag{1}$$

$$M_{RB}\dot{\nu} + C_{RB}\nu = \sum F_{ext}, \tag{2}$$

where $\eta = [x, y, z, \phi, \theta, \psi]^T$ is the position and attitude vector defined in the navigation frame (NED-frame), and $\nu = [u, v, w, p, q, r]^T$ is the linear and angular velocity vector defined in the vehicle's body-fixed frame; and C_b^n denotes the coordinate transformation matrix from the body-fixed frame to the navigation frame, and can be expressed as

$$C_b^n = \begin{bmatrix} J_1 & 0^{3\times3} \\ 0^{3\times3} & J_2 \end{bmatrix} \tag{3}$$

where

$$J_1 = \begin{bmatrix} c\psi c\theta & -s\psi c\phi + c\psi s\theta s\phi & s\psi s\phi + c\psi s\theta c\phi \\ s\psi c\theta & c\psi c\phi + s\psi s\theta s\phi & -c\psi s\phi + s\psi s\theta c\phi \\ -s\theta & c\theta s\phi & c\theta c\phi \end{bmatrix}, \quad J_2 = \begin{bmatrix} 1 & s\phi t\theta & c\phi t\theta \\ 0 & c\phi & -s\phi \\ 0 & s\phi/c\theta & c\phi/c\theta \end{bmatrix},$$

with $s(\cdot) = sin(\cdot)$, $c(\cdot) = cos(\cdot)$, $t(\cdot) = tan(\cdot)$.

For the vehicle developed as seen in Figure 1, its body-fixed frame is centered at the vehicle's buoyancy center as in [11] and the weight center is located at $(x_g, y_g, z_g) = (0, 0, 4.51 \text{ mm})$ after being neutrally ballasted. Therefore, the rigid-body inertia matrix M_{RB} and Coriolis and centripetal matrix C_{RB} can be simplified as follows

$$M_{RB} = \begin{bmatrix} m & 0 & 0 & 0 & mz_g & 0 \\ 0 & m & 0 & -mz_g & 0 & 0 \\ 0 & 0 & m & 0 & 0 & 0 \\ 0 & -mz_g & 0 & I_{xx} & 0 & 0 \\ mz_g & 0 & 0 & 0 & I_{yy} & 0 \\ 0 & 0 & 0 & 0 & 0 & I_{zz} \end{bmatrix}, \tag{4}$$

$$C_{RB} = \begin{bmatrix} 0 & 0 & 0 & mz_g r & mw & -mv \\ 0 & 0 & 0 & -mw & mz_g r & mu \\ 0 & 0 & 0 & -m(z_g p - v) & -m(z_g q + u) & 0 \\ -mz_g r & mw & m(z_g p - v) & 0 & I_{zz} r & -I_{yy} q \\ 0 & -mz_g r & mz_g q & -I_{zz} r & 0 & I_{xx} p \\ 0 & 0 & 0 & I_{yy} q & -I_{xx} p & 0 \end{bmatrix}, \tag{5}$$

where $m = 58.94$ kg is the rigid body mass, and I_{xx}, I_{yy}, and I_{zz} denote the inertia moments each along the X, Y, and Z axes. Calculated inertia moments are as shown in Table 1.

Table 1. Inertia Moments.

Parameters	Value	Unit
I_{xx}	$3.33e + 000$	kg·m^2
I_{yy}	$3.33e + 000$	kg·m^2
I_{zz}	$7.45e + 000$	kg·m^2

For the vehicle dynamics as in (2), the sum of external forces and moments can be expressed as follows as in [9,11]

$$\sum F_{ext} = F_{hydrostatics} + F_{drag} + F_{added\_mass} + F_{control}, \tag{6}$$

where $F_{hydrostatics} = [0, 0, 0, -z_g W c\theta s\phi, -z_g W s\theta, 0]^T$ with W the rigid body weight.

2.2. Hydrodynamic Damping Term F_{drag}

The vehicle is symmetric along all three of X, Y, and Z axes, see Figure 1. Therefore, the movement-induced moments $A_{b|b|}$ with $A = \{K, M, N\}$ and $b = \{u, v, w\}$, and the rotation-induced forces $B_{a|a|}$ with $B = \{X, Y, Z\}$ and $a = \{p, q, r\}$ are all negligible. In addition, to simplify the vehicle's model and therefore to avoid complicated mathematical calculations, the following assumptions similar to [11] are made in the vehicle modeling, where [14] describes deterministic artificial intelligence methods to deal with coupled disturbances. Preliminary research in overcome linear coupling is contained in [15], while methods to counter angular coupling is described in [16], following an established lineage of continuing research from [17–23].

- Linear and angular coupled damping terms are ignorable.
- Any damping terms greater than second-order are negligible.

Consequently, F_{drag} can be simplified as

$$F_{drag} = [X_{u|u|}u|u|, Y_{v|v|}v|v|, Z_{w|w|}w|w|, K_{p|p|}p|p|, M_{q|q|}q|q|, N_{r|r|}r|r|]^T. \tag{7}$$

Corresponding coefficients are calculated using the following formulas

$$X_{u|u|} = Y_{v|v|} = -0.5\rho c'_{dc}\pi aR - 4(0.5\rho s_D c_{dD}), \tag{8}$$

$$Z_{w|w|} = -0.5\rho c_{dc}\pi R^2 - 4(0.5\rho s_F c_{dF}), \tag{9}$$

$$K_{p|p|} = M_{q|q|} = 2\left[-0.5\rho c_{dc}\int_0^R D(r)r^3 dr - 2x_{TF}^3(0.5\rho s_F c_{dF})\right], \tag{10}$$

$$N_{r|r|} = 2\left[-0.5\rho c'_{dc}\int_0^R D'(r)r^3 dr - 2x_{TD}^3(0.5\rho s_D c_{dD})\right], \tag{11}$$

where $a = 0.128$ m, $R = 0.435$ m, $x_{TF} = x_{TD} = 0.543$ m, $s_F = 0.163$ m^2, $s_D = 0.023$ m^2, and the drag coefficients are derived through empirical graph as Figure 2.4 in (p. 19, [1]), and selected as $c_{dc} = 1.12, c'_{dc} = 0.3, c_{dF} = 0.63, c_{dD} = 0.9$.

Calculated coefficients are shown in Table 2.

Table 2. Hydrodynamic Damping Coefficients.

Parameters	Value	Unit
$X_{u|u|}$	$-6.74e+001$	kg/m
$Y_{v|v|}$	$-6.74e+001$	kg/m
$Z_{w|w|}$	$-5.38e+002$	kg/m
$K_{p|p|}$	$-3.73e+001$	kg·m^2/rad^2
$M_{q|q|}$	$-3.73e+001$	kg·m^2/rad^2
$N_{r|r|}$	$-6.92e+001$	kg·m^2/rad^2

2.3. Added Mass Term F_{added_mass}

As mentioned before, the vehicle is symmetric in all three axes. Therefore, only the diagonal terms in the vehicle's added mass matrix are considered in this paper. The coefficients are estimated through empirical graph as Figure 4.8 in Newman [10] (p. 147), and calculated as shown in Table 3.

Table 3. Added Mass Coefficients.

Parameters	Value	Unit
$X_{\dot{u}}$	$-1.75e+001$	kg
$Y_{\dot{v}}$	$-1.75e+001$	kg
$Z_{\dot{w}}$	$-2.09e+002$	kg
$K_{\dot{p}}$	$-5.03e+000$	kg·m^2/rad
$M_{\dot{q}}$	$-5.03e+000$	kg·m^2/rad
$N_{\dot{r}}$	$-1.17e+000$	kg·m^2/rad

3. Compensation of Motion Sensors Lever Arm Effect

For the majority of underwater vehicles, the motion sensors such as Doppler Velocity Log (DVL) and Attitude and Heading Reference System (AHRD) cannot be arranged on the same center point or center line. Instead, they are usually separated from each other and also from the center point of the body-fixed frame. Therefore, raw measurements of these motion sensors should be suitably compensated in order to acquire the accurate vehicle motion information, which is further used for the vehicle's navigation and motion control [24,25].

For the vehicle seen in Figure 1, DVL and AHRS arrangements are as shown in Figure 2a, where DVL is mounted at the center of the body-fixed frame and AHRS is 0.218 m away from the center line.

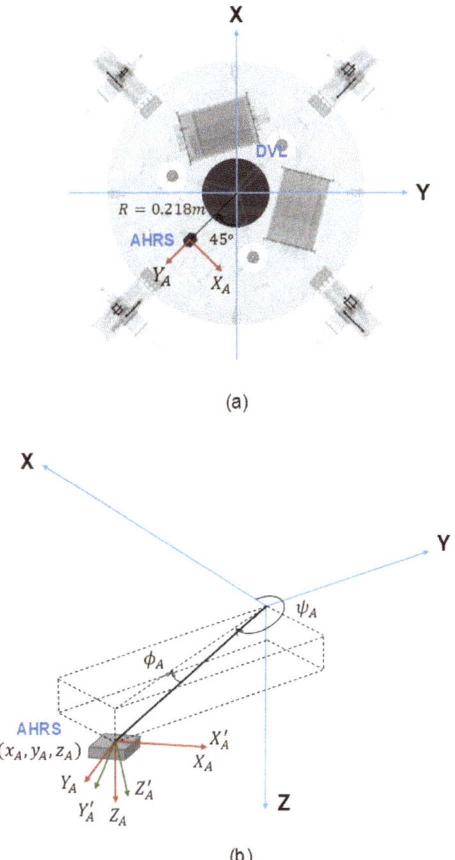

Figure 2. Motion sensor arrangement and lever arm coordinates for the developed vehicle: (**a**) is the motion sensor arrangement; and (**b**) shows the AHRS lever arm coordinates.

3.1. Angular Rate Correction

As seen in Figure 2, AHRS body frame $X_A Y_A Z_A$ is rotated from the vehicle's body-fixed frame XYZ at $\phi_A = 0°$, $\theta_A = 0°$, $\psi_A = 135°$. For this reason, the angular rate $\omega = [p, q, r]^T$ in the XYZ frame can be calculated through the following coordinate transformation

$$\begin{bmatrix} p \\ q \\ r \end{bmatrix} = \begin{bmatrix} \cos\psi_A & -\sin\psi_A & 0 \\ \sin\psi_A & \cos\psi_A & 0 \\ 0 & 0 & 1 \end{bmatrix} \begin{bmatrix} p_r \\ q_r \\ r_r \end{bmatrix}, \tag{12}$$

where $\omega_r = [p_r, q_r, r_r]^T$ is the angular velocity measured by AHRS.

Remark 1. *$X_A Y_A Z_A$ and XYZ are two of fixed frames in the vehicle's rigid body. In this case, it is necessary to mention that the coordinates transformations of all vectors including angular rate should obey the linear velocity transformation Equation (2.16) in [9], not the angular velocity transformation of Equation (2.26) in [9].*

3.2. Acceleration Correction

In the vehicle's body-fixed frame XYZ, AHRS center point coordinate is (x_A, y_A, z_A). In addition, lever arm effect usually causes additional acceleration at the AHRS center point in the case of vehicle's rotation. The velocity V_A at the AHRS center point in the frame $X'_A Y'_A Z'_A$ (see Figure 2b) and the velocity V_B in the vehicle's body-fixed frame XYZ has the following relationship [24,26]

$$V_A = V_B + \omega \times R, \tag{13}$$

where $R = [x_A, y_A, z_A]^T$.

Differentiating (13), have

$$\dot{V}_A = \dot{V}_B + \dot{\omega} \times R + \omega \times V_B + \omega \times (\omega \times R). \tag{14}$$

From Figure 2b, it is easy to see that the frame $X'_A Y'_A Z'_A$ is tilted from AHRS body frame $X_A Y_A Z_A$ at $\phi'_A = atan(z_A / \sqrt{x_A^2 + y_A^2})$. So, it has

$$\dot{V}_A = \begin{bmatrix} 1 & 0 & 0 \\ 0 & \cos\phi'_A & \sin\phi'_A \\ 0 & -\sin\phi'_A & \cos\phi'_A \end{bmatrix} \dot{V}_{AHRS} = C^A_{AHRS} \dot{V}_{AHRS}, \tag{15}$$

where $\dot{V}_{AHRS}$ is the acceleration measurement by AHRS.

Consequently, the acceleration vector V_B in the vehicle's body-fixed frame can be estimated from AHRS measurement through the following equation

$$\dot{V}_B = C^A_{AHRS} \dot{V}_{AHRS} - \dot{\omega} \times R - \omega \times V_B - \omega \times (\omega \times R), \tag{16}$$

where $\dot{\omega}$ is the angular acceleration and should be properly calculated. One option is that it can be estimated through a sort of low-pass filter as follows

$$\dot{\omega}(k+1) = (1-\lambda)\dot{\omega}(k) + \lambda\dot{\omega}_m(k+1), \quad \dot{\omega}(0) = \dot{\omega}_m(0), \tag{17}$$

where $\dot{\omega}_m(k+1) = [\omega(k+1) - \omega(k)]/\Delta T$ with ΔT sampling time, and $0 < \lambda \leq 1$ is a design parameter.

Remark 2. *Vehicle's angular rate ω and acceleration $\dot{V}_B$ in the body-fixed frame can be calculated through each of (12) and (16). Furthermore, the vehicle's attitude is calculated by $(\phi, \theta, \psi) = (\phi_m, \theta_m, \psi_m - \psi_A)$, where*

(ϕ_m, θ_m, ψ_m) is the measurement of AHRS and $\psi_m - \psi_A$ is defined in the domain $[0, 2\pi)$. Since the DVL is located at the center point, its measurement is directly used as V_B.

4. Control Design

As mentioned before, the control strategy proposed in this paper to overcome strong current is to maximize the vectored horizontal thrust force along the direction against the current, while keeping the vehicle's heading or with the minimum heading rotation. To do so, it needs to spread the vectored horizontal thrust force on the horizontal plane as uniformly as possible.

4.1. Maximum Horizontal Vectored Thrust Force

As shown in Figure 3, four horizontal thrusters are mounted symmetrically around the ellipsoidal shape of platform. Each of the thrusters has the same forward and reverse thrust dynamics.

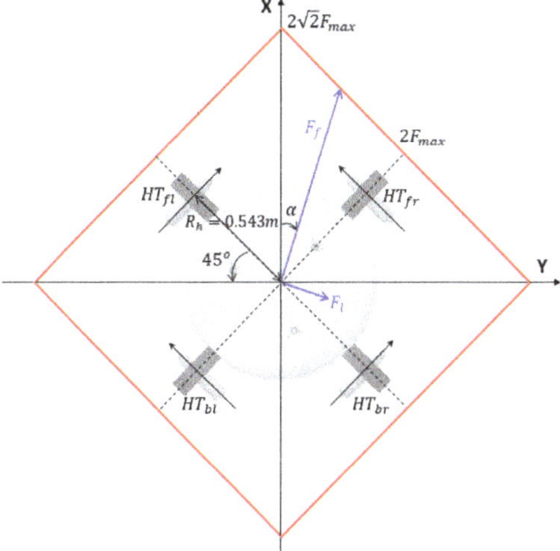

Figure 3. Horizontal thrusters arrangement and maximum vectored horizontal thrust force.

Definition 1. *Given the direction α as seen in Figure 3, the maximum horizontal vectored thrust force means the maximum of F_f caused by four symmetrically mounted horizontal thrusters with the following properties*

P1.　*The orthogonal component $F_l = 0$.*
P2.　*Horizontal rotation moment is zero.*

Remark 3. *In the case of maximum horizontal vectored thrust force, it has $F_{HT_{fl}} = F_{HT_{br}}$ and $F_{HT_{fr}} = F_{HT_{bl}}$ where F_{HT_a} denotes the thrust force for thruster HT_a with $a \in \{fl, fr, br, bl\}$. For example, consider the case $\alpha = 45°$. To maximize F_f, the thrusters HT_{fl} and HT_{br} have to take the maximum of forward thrusts. Due to the fact that these four thrusters have the same thrust dynamics, it has $F_{HT_{fl}} = F_{HT_{br}} = F_{max}$ with F_{max} the maximum thrust force provided by one thruster. On the other hand, in order to satisfy the P1 and P2 in Definition 1, the remainder thrusters have to set as $F_{HT_{fr}} = F_{HT_{bl}} = 0$.*

Lemma 1. *Given four of the symmetrically mounted horizontal thrusters as shown in Figure* 3, *the maximum horizontal vectored thrust force can be calculated as follows*

$$
F_f =
\begin{cases}
\dfrac{2F_{max}}{cos(\alpha - 45°)}, & if\ \ 0° \leq \alpha < 90° \\[2mm]
-\dfrac{2F_{max}}{cos(\alpha + 45°)}, & if\ \ 90° \leq \alpha < 180° \\[2mm]
-\dfrac{2F_{max}}{cos(\alpha - 45°)}, & if\ \ 180° \leq \alpha < 270° \\[2mm]
\dfrac{2F_{max}}{cos(\alpha + 45°)}, & if\ \ 270° \leq \alpha < 360°
\end{cases}
\tag{18}
$$

Proof of Lemma 1. From Figure 3, it is easy to get

$$
F_f = F_A cos(\alpha - 45°) + F_B cos(\alpha + 45°),
\tag{19}
$$
$$
F_l = F_A sin(\alpha - 45°) + F_B sin(\alpha + 45°),
\tag{20}
$$

where $F_A = F_{HT_{fl}} + F_{HT_{br}}$ and $F_B = F_{HT_{fr}} + F_{HT_{bl}}$.

In order for $F_l = 0$ with (20), it has

$$
\frac{\dot{F_A}}{F_B} = -cot(\alpha - 45°).
\tag{21}
$$

Therefore, F_A and F_B cannot be taken the maximum value at the same time if $|cot(\alpha - 45°)| \neq 1$. In the case of $\alpha \in \{[0°,\ 90°) \cup [180°,\ 270°)\}$, it has $|cot(\alpha - 45°)| \geq 1$. Substituting $F_B = -F_A \cdot tg(\alpha - 45°)$ into (19), it can be get

$$
F_f = F_A \left[cos(\alpha - 45°) - cos(\alpha + 45°)\frac{sin(\alpha - 45°)}{cos(\alpha - 45°)} \right] = \frac{F_A}{cos(\alpha - 45°)}.
\tag{22}
$$

In the case $\alpha \in [0°,\ 90°)$, it has $F_A = 2F_{max}$ and if $\alpha \in [180°,\ 270°)$ then $F_A = -2F_{max}$. Therefore, (22) can be rewritten as

$$
F_f =
\begin{cases}
\dfrac{2F_{max}}{cos(\alpha - 45°)} & if\ \ 0° \leq \alpha < 90°, \\[2mm]
-\dfrac{2F_{max}}{cos(\alpha - 45°)} & if\ \ 180° \leq \alpha < 270°
\end{cases}
\tag{23}
$$

Similarly, in the case of $\alpha \in \{[90°,\ 180°) \cup [270°,\ 360°)\}$, substituting $F_A = -F_B cot(\alpha - 45°)$ into (19), have

$$
F_f = F_B \left[cos(\alpha + 45°) - \frac{cos^2(\alpha - 45°)}{sin(\alpha - 45°)} \right] = -\frac{F_B}{sin(\alpha - 45°)}.
\tag{24}
$$

If $\alpha \in [90°,\ 180°)$, then $F_B = -2F_{max}$, and if $\alpha \in [270°,\ 360°)$, then $F_B = 2F_{max}$. Consequently, (24) can be rewritten as

$$
F_f =
\begin{cases}
-\dfrac{2F_{max}}{sin(\alpha - 45°)} & if\ \ 90° \leq \alpha < 180°, \\[2mm]
\dfrac{2F_{max}}{sin(\alpha - 45°)} & if\ \ 270° \leq \alpha < 360°
\end{cases}
\tag{25}
$$

Combining (23) and (25) can conclude the *Proof.* □

Remark 4. *Calculated maximum horizontal vectored thrust force field is the red-colored square shown in Figure* 3. *The maximum vectored thrust force is* $2\sqrt{2}F_{max}$ *and minimum value is* $2F_{max}$. *It is easy to see that to align the maximum vectored thrust force with the opposite direction of arbitrarily given sea current, the vehicle's maximum rotation angle is less than* $45°$.

4.2. Maximum Forward Speed Controller with Heading-Keeping

One of the important target specifications of this project is the vehicle's maximum forward speed while keeping its heading angle. Proposed control law is as follows:

$$
\begin{aligned}
&if\ \delta\psi \in \{[0°,\ 45°)\cup[315°,\ 360°)\}\\
&\qquad F_{HT_{fl}} = F_{HT_{fr}} = F_{max}\\
&\qquad if\ dt \geq 0\\
&\qquad\qquad F_{HT_{br}} = F_{max} - dt;\ F_{HT_{bl}} = F_{max}\\
&\qquad else\\
&\qquad\qquad F_{HT_{br}} = F_{max};\ F_{HT_{bl}} = F_{max} + dt\\
&\qquad end\\
&else\ if\ \delta\psi \in [45°,\ 135°)\\
&\qquad F_{HT_{fr}} = -F_{max};\ F_{HT_{br}} = F_{max}\\
&\qquad if\ dt \geq 0\\
&\qquad\qquad F_{HT_{bl}} = -F_{max} + dt;\ F_{HT_{fl}} = F_{max}\\
&\qquad else\\
&\qquad\qquad F_{HT_{bl}} = -F_{max};\ F_{HT_{fl}} = F_{max} + dt\\
&\qquad end\\
&else\ if\ \delta\psi \in [135°,\ 225°)\\
&\qquad F_{HT_{br}} = F_{HT_{bl}} = -F_{max}\\
&\qquad if\ dt \geq 0\\
&\qquad\qquad F_{HT_{fl}} = -F_{max} + dt;\ F_{HT_{fr}} = -F_{max}\\
&\qquad else\\
&\qquad\qquad F_{HT_{fl}} = -F_{max};\ F_{HT_{fr}} = -F_{max} - dt\\
&\qquad end\\
&else\\
&\qquad F_{HT_{bl}} = F_{max};\ F_{HT_{fl}} = -F_{max}\\
&\qquad if\ dt \geq 0\\
&\qquad\qquad F_{HT_{fr}} = F_{max} - dt;\ F_{HT_{br}} = -F_{max}\\
&\qquad else\\
&\qquad\qquad F_{HT_{fr}} = F_{max};\ F_{HT_{br}} = -F_{max} - dt\\
&\qquad end\\
&end
\end{aligned}
$$

where $\delta\psi = \psi_r - \psi$ with ψ_r the reference heading and $dt = K_{hp}\delta\psi + K_{hd}r$ with K_{hp} and K_{hd} control gain parameters.

4.3. Horizon Keeping Controller

Three vertical thrusters are mounted as seen in Figure 4.

4.3.1. Roll Motion Control

The roll motion control component for VT_1 is designed as

$$VT_{r1} = K_{vp}(\phi_r - \phi) + K_{vd}p, \tag{26}$$

where ϕ_r is the reference roll angle, K_{vp} and K_{vd} are two gain parameters.

In the case of roll motion control, the remainder of two control components VT_{r2} and VT_{r3} are designed through simultaneously satisfying the following two conditions

C1. $\quad VT_{r1} \cdot cos15° = -VT_{r2} \cdot cos45° - VT_{r3} \cdot sin15°.$

C2. $VT_{r1} \cdot sin15° + VT_{r2} \cdot cos45° = VT_{r3} \cdot cos15°.$

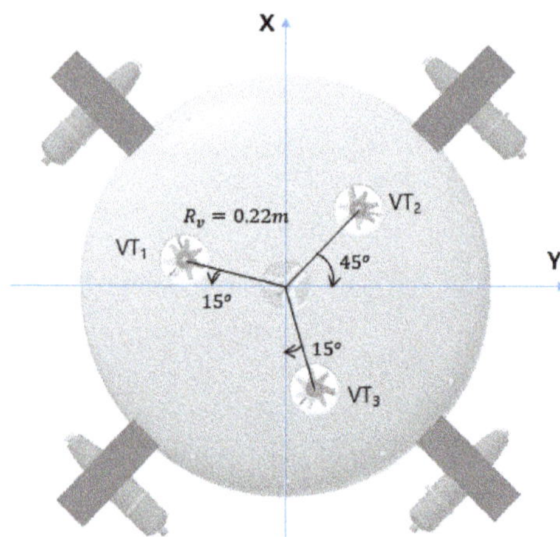

Figure 4. Vertical thrusters arrangement.

Remark 5. *Here, C1 is the condition to balance the right and left torques about the X-axis, and C2 is to neutralize the pitch torque during the roll motion control.*

Combining C1 and C2, it is easy to get

$$VT_{r2} = -1.1547 \cdot VT_{r1}, \tag{27}$$

$$VT_{r3} = -0.5774 \cdot VT_{r1}. \tag{28}$$

4.3.2. Pitch Motion Control

The design procedure is similar to the roll motion control. First, the pitch motion control component for VT_3 is chosen as

$$VT_{p3} = K_{vp}(\theta_r - \theta) + K_{vd}q, \tag{29}$$

where θ_r is the reference pitch angle.

Control law for other two vertical thrusters is to simultaneously satisfying the following two conditions

C3. $-VT_{p3} \cdot cos15° = VT_{p2} \cdot cos45° + VT_{p1} \cdot sin15°.$
C4. $VT_{p1} \cdot cos15° = VT_{p2} \cdot cos45° + VT_{p3} \cdot sin15°.$

Remark 6. *Here C3 is to balance the pitch torque about the Y-axis, and C4 is to neutralize the roll torque during the pitch motion control.*

Consequently, have

$$VT_{p1} = -0.5774 \cdot VT_{p3}, \tag{30}$$

$$VT_{p2} = -1.1547 \cdot VT_{p3}. \tag{31}$$

Remark 7. *In both of roll and pitch motion controls, any of three vertical thrusters can be selected and designed its thrust force using (26) or (29), and other two thrusters are designed through simultaneously satisfying C1 and C2, or C3 and C4. In the case of horizon keeping, ϕ_r and θ_r can be simply set as zero values.*

4.3.3. Depth Control

The following simple PD controller is designed for depth control

$$VT_{di} = K_{vp}^d(z_r - z) + K_{vd}^d w, \quad i = 1,2,3, \tag{32}$$

where VT_{di} is the depth control component for the thruster VT_i, z_r is the reference depth, and K_{vp}^d and K_{vd}^d are gain parameters.

4.4. Thruster Models

4.4.1. Vertical Thruster Model

For each vertical thruster, its thrust versus input relationship is as follows [27]

$$F_{VT}(x) = \begin{cases} 74.5N \cdot sat(x/5), & if\ x \geq 0 \\ -31.4N \cdot sat(x/5), & if\ x < 0 \end{cases} \tag{33}$$

where $sat(\cdot)$ is the saturation function.

Consequently, the thrust force provided by each vertical thruster can be calculated as

$$F_{VT_i} = F_{VT}(VT_{ri} + VT_{pi} + VT_{di}), \quad i = 1,2,3. \tag{34}$$

Remark 8. *In most of the practical cases, the vehicle takes low speed motion in the vertical direction. For this reason, in the case of vertical thrusters, it is not considered the effect where the thruster's maximum thrust force decreases in compliance with the increase of the fluid speed flow through the thruster [12].*

4.4.2. Horizontal Thruster Model

In the case of horizontal thrusters, the fact that the thruster's maximum thrust force decreases in compliance with the increase of fluid speed flow through the thruster has to be considered.

According to the thruster's specifications [13], the approximated relationship between the thruster maximum thrust force versus fluid flow speed (red dotted line in Figure 5) can be obtained as follows

$$F_{max}(U) = g \cdot (0.811U^2 - 6.3903U + 24.9384), \tag{35}$$

where F_{max} denotes the maximum thrust force with the unit N, $g = 9.8066$ m/s^2 is the standard gravity, and U is the fluid speed flow through the thruster.

According to Figure 3, fluid speed for each of the four horizontal thrusters can be approximated using DVL measurement as follows

$$\begin{bmatrix} U_{HT_{fl}} \\ U_{HT_{fr}} \\ U_{HT_{br}} \\ U_{HT_{bl}} \end{bmatrix} = \frac{\sqrt{2}}{2} \begin{bmatrix} 1 & 1 \\ 1 & -1 \\ 1 & 1 \\ 1 & -1 \end{bmatrix} \begin{bmatrix} u \\ v \end{bmatrix}. \tag{36}$$

As mentioned before, the horizontal thruster has the same forward and reverse dynamics which can be expressed as follows

$$F_{HT}(x) = F_{max}(U) \cdot sat(x/5). \tag{37}$$

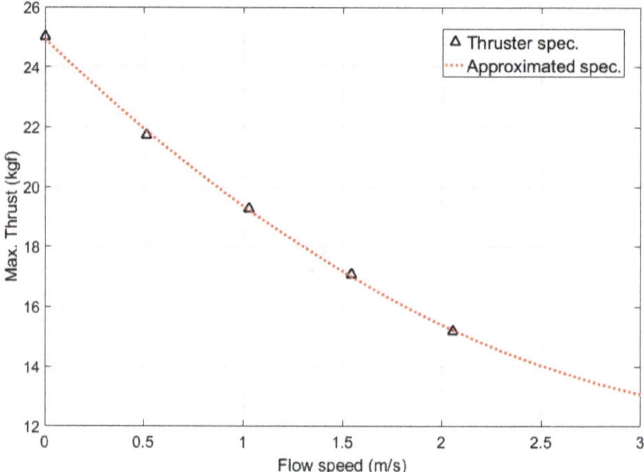

Figure 5. Horizontal thruster's specification: maximum thrust force vs. flow speed.

4.5. Calculation of $F_{control}$

Consequently, the control force $F_{control}$ in (6) can be calculated as follows

$$
F_{control} = \begin{bmatrix} \frac{\sqrt{2}}{2}\left(F_{HT_{fl}} + F_{HT_{fr}} + F_{HT_{br}} + F_{HT_{bl}}\right) \\ \frac{\sqrt{2}}{2}\left(F_{HT_{fl}} - F_{HT_{fr}} + F_{HT_{br}} - F_{HT_{bl}}\right) \\ -F_{VT_1} - F_{VT_2} - F_{VT_3} \\ F_{VT_1}R_v cos15° - F_{VT_2}R_v cos45° - F_{VT_3}R_v sin15° \\ F_{VT_1}R_v sin15° + F_{VT_2}R_v cos45° + F_{VT_3}R_v cos15° \\ R_h\left(F_{HT_{fl}} - F_{HT_{fr}} - F_{HT_{br}} + F_{HT_{bl}}\right) \end{bmatrix}. \tag{38}
$$

5. Simulation Studies

First, the vehicle's platform stability on the horizontal plane is observed. The initial condition is set as $\eta = [1.54, 0, 0, 10°, 20°, 0, 0, 0, 0, 0, 0, 0]^T$, and simulation result is shown in Figure 6, from which it can be seen that the vehicle possesses suitable self-stability. Obviously, this kind of stability is caused by the fact that the vehicle's gravity center is designed to be lower than the buoyancy center.

Then, it carries out the maximum forward speed simulation combined with the horizon and heading keeping tests. In the simulation, control gains are set as $K_{vp} = 5$, $K_{vd} = -10, K_{vp}^d = 15$, $K_{vd}^d = -45, K_{hp} = 2.5, K_{hd} = -3$, and other parameters are taken as $\phi_r = \theta_r = 0$, $\psi_r = 30°$, $z_r = 1.5$ m, and sampling time is $\Delta T = 0.1$ s. Figure 7 shows the control force $F_{control}$ calculated using (38), and Figure 8 presents the corresponding vehicle motion information. From Figure 8, it can be seen that the vehicle's maximum forward speed is about 2.56 m/s. This speed is lower than 3.2 m/s which is the simulation result in [28] where the effect of the relationship between the maximum thrust force and the fluid speed flow through the thruster was not considered. However, 2.56 m/s is still larger than the experimental result of 2.1 m/s, and this will be further discussed in the next section.

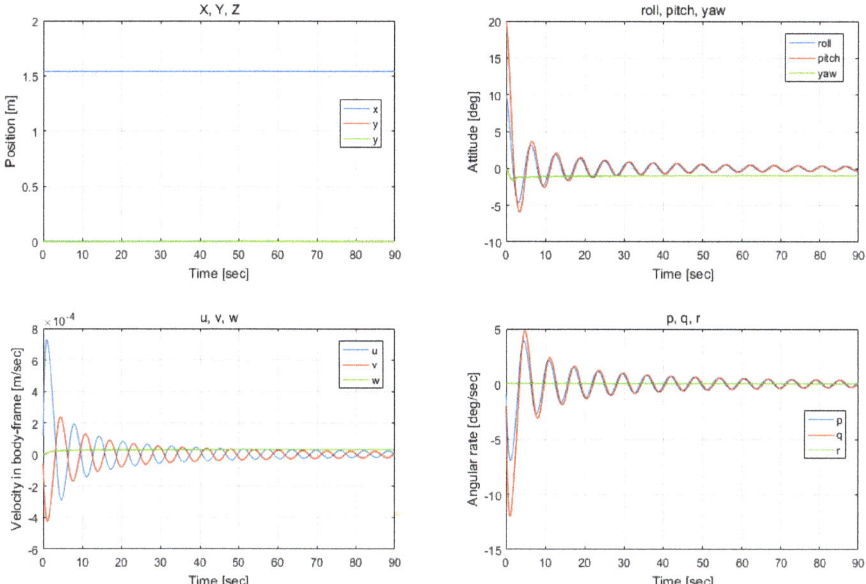

Figure 6. Simulation result of self-stabilization capability.

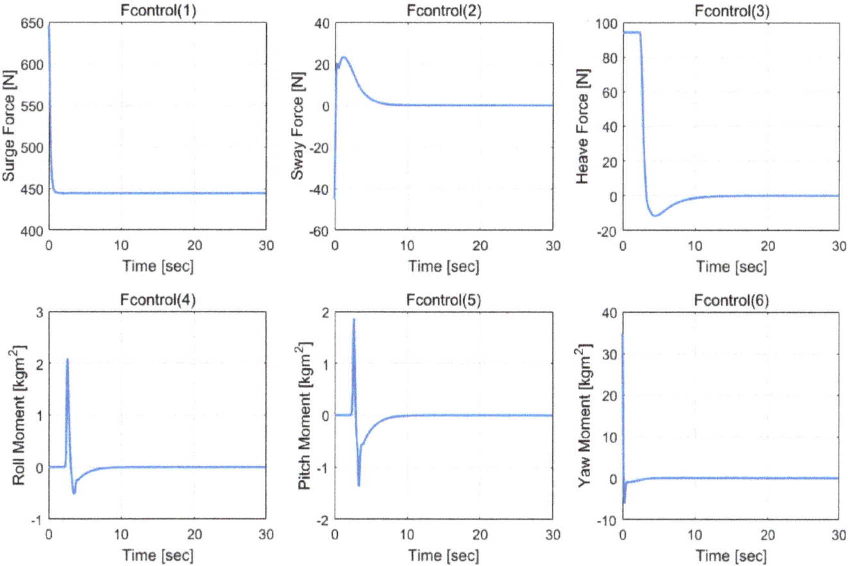

Figure 7. Calculated control forces in the maximum speed simulation.

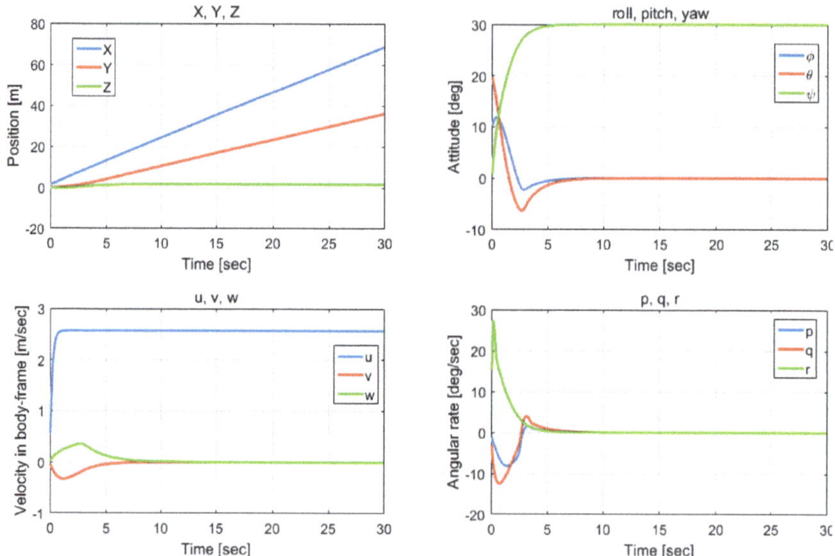

Figure 8. Vehicle's motion information in the maximum speed simulation.

6. Experimental Studies

Experimental tests are carried out in the water tank and circulating water channel both in the Underwater Construction Robotics R&D Center (UCRC) in the Korea Institute of Ocean Science and Technology (KIOST) [29].

Figure 9 shows the engineering basin where the maximum forward speed test is taken. The experimental result is shown in Figure 10, from which it can be seen that the vehicle's maximum forward speed is about 2.1 m/s. This speed is lower than the simulation result of 2.56 m/s. This might be caused by several reasons. One is that there are quite a number of coupled and complicated hydrodynamic terms that are not included in the vehicle's model in the simulation. The second is that the drag component caused by the tether cable is not considered in the simulation. Indeed, this drag term might be significant in the case of the small size of underwater vehicles. From this point of view, whether the vehicle can be operated in AUV mode might be an important issue for the vehicle to overcome strong current. Figure 11 shows the control inputs for four of horizontal thrusters in this maximum forward speed test.

In addition, the test is taken where the vehicle is installed in the circulating water channel as in Figure 12, and it investigates if the vehicle can take the forward speed motion while keeping its heading in the strong current environment. In the test, the current speed is adjusted from 1.0 knots to 2.0 knots and further up to 2.5 knots. From the experimental result shown in Figure 13, it can be concluded that the vehicle can take forward motion even in the case where the current increased up to 2.5 knots.

Figure 9. Maximum forward speed test in the engineering basin.

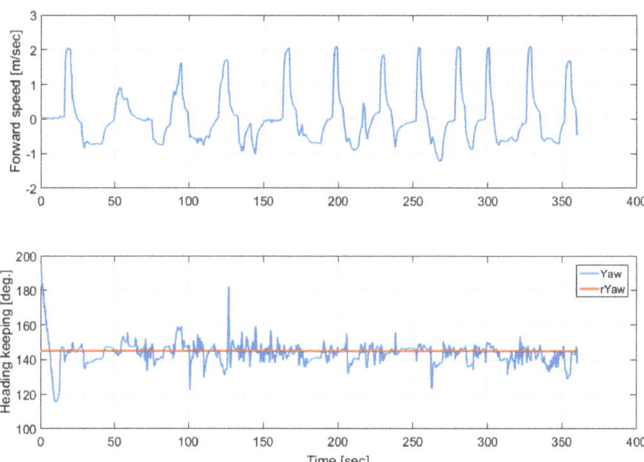

Figure 10. Experimental result of maximum forward speed with heading keeping.

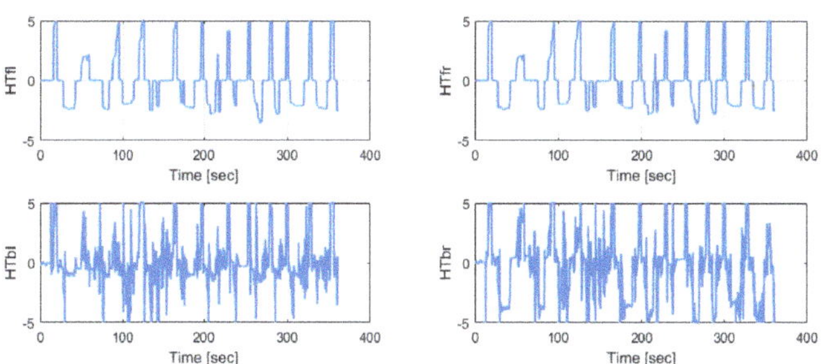

Figure 11. Control inputs for four of horizontal thrusters in the maximum forward speed test.

Figure 12. The test in the circulating water channel.

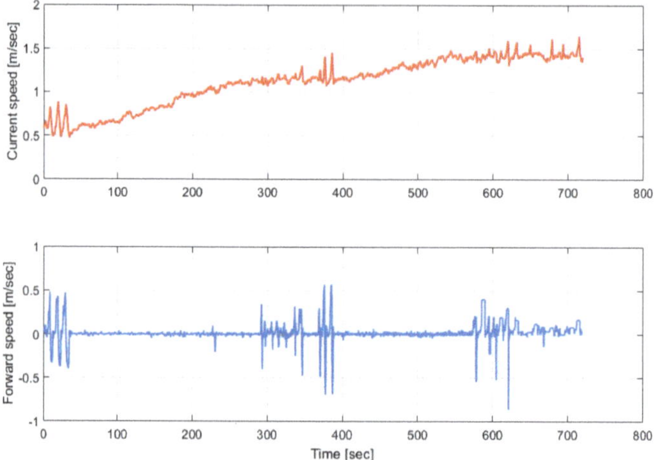

Figure 13. Vehicle's forward motion test results in the circulating water channel.

7. Conclusions

This paper has presented the development of a UUV platform and its control method to overcome strong current. The vehicle has been designed to have a flattened ellipsoidal exterior to minimize the hydrodynamic damping. In addition, vectored thrust force control algorithm has been developed to maximize its horizontal speed. All of these technologies have been evaluated through both simulation and experimental tests. Especially, through the experimental studies carried out in the water tank and circulating water channel, it is found that, for the current version of vehicle platform, more strong roll and pitch moments might be needed to guarantee the vehicle's horizontal stabilization.

During the next step of research works, the current version of the platform will be upgraded to have four vertical thrusters mounted symmetrically and each of the thrusters, similar to the horizontal thrusters, has the same forward and reverse dynamics. Moreover, each thruster will be more powerful compared to the current version and mounted more farther away from the center point. All of these are supposed to significantly increase the vehicle's capability of stabilizing horizontal motion in the strong current environment.

Author Contributions: Conceptualization by J.-H.L.; methodology by J.-H.L., foremost, and then G.R.C. and M.-G.K.; software by H.K.; validation by M.-J.L. H.K. and M.-G.K.; formal analysis by G.R.C.; writing–original draft preparation by J.-H.L.; writing–review and editing by J.-H.L.; research supervision by J.-H.L. Authorship must be limited to those who have contributed substantially to the work reported.

Funding: This research was supported by the project titled "Development of Underwater Robot Platform and its Control Technology to Overcome up to 3.5 knots of Sea Current," funded by the Ministry of Oceans and Fisheries (MOF) and Korea Institute of Marine Science and Technology Promotion (KIMST), Korea (20160148); It was also partially supported by the project No. 17-CM-RB-16 titled "Development of Multi-sensor Fusion based AUV's Terminal Guidance and Docking Technology," funded by the Agency for Defense Development (ADD) in the Korea.

Conflicts of Interest: The authors declare no conflict of interest.

References

1. Hur, M.Y. Revisiting the Cheonan sinking in the Yellow Sea. *Pac. Rev.* **2017**, *30*, 349–364. [CrossRef]
2. Kim, S.K. The Sewol Ferry Disaster in Korea and Maritime Safety Management. *Ocean Dev. Int. Law* **2015**, *46*, 345–358. [CrossRef]
3. Sands, T.; Bollino, K.; Kaminer, I.; Healey, A. Autonomous Minimum Safe Distance Maintenance from Submerged Obstacles in Ocean Currents. *J. Mar. Sci. Eng. Spec. Issue Intell. Mar. Robot. Model. Simul. Appl.* **2018**, *6*, 98.
4. Yoo, S.Y.; Shim, H.W.; Jun, B.H.; Park, J.Y.; Lee, P.M. Design of walking and swimming algorithms for a multi-legged underwater robot crabster CR200. *Mar. Technol. Soc. J.* **2016**, *50*, 74–87. [CrossRef]
5. Pyo, J.H.; Yu, S.C. Development of AUV (MI) for strong ocean current and zero-visibility condition. In Proceedings of the IEEE AUV 2016 Conference, Tokyo, Japan, 6–9 November 2016; pp. 54–57.
6. Eldred, R.A. Autonomous Underwater Vehicle Architecture Synthesis for Shipwreck Interior Exploration. Master's Thesis, Department of Systems Engineering, Naval Postgraduate Scholl, Monterey, CA, USA, 2015.
7. Bennett, S. A brief history of automatic control. *IEEE Control Syst. Mag.* **1996**, *16*, 17–25.
8. Astrom, K.J.; Murray, R.M. *Feedback Systems*; Princeton University Press: Princeton, NJ, USA, 2008.
9. Fossen, T.I. *Marine Control Systems: Guidance, Navigation and Control of Ships, Rigs and Underwater Vehicles*; Marine Cybernetics: Trondheim, Norway, 2002.
10. Newman, J.N. *Marine Hydrodynamics*; The MIT Press: Cambridge, MA, USA, 1977.
11. Prestero, T. Verification of a Six-Degree of Freedom Simulation Model for the REMUS Autonomous Underwater Vehicles. Master's Thesis, Department of Ocean Engineering and Mechanical Engineering, MIT, Cambridge, MA, USA, 2001.
12. Kim, J.; Chung, W.K. Accurate and practical thruster modeling for underwater vehicles. *Ocean Eng.* **2006**, *33*, 566–586. [CrossRef]
13. Tecnadyne Technical Report. Available online: http://tecnadyne.com/wp-content/uploads/2017/06/Model-1040-Brochure-1.pdf (accessed on 18 May 2018).
14. Lobo, K.; Lang, J.; Starks, A.; Sands, T. Analysis of Deterministic artificial Intelligence for Inertia Modifications and Orbital Disturbance. *Int. J. Control Sci. Eng.* **2018**, *8*, 52–62.
15. Sands, T. Improved Magnetic Levitation via Online Disturbance Decoupling. *Phys. J.* **2015**, *1*, 272–280.
16. Sands, T. *Physics-Based Control Methods*; Chapter in Advancements in Spacefraft Systems and Orbit Determination; InTech: London, UK, 2012; pp. 29–54.
17. Fossen, T. Comments on 'Hamiltonian Adaptive Control of Spacecraft. *IEEE Trans. Autom. Control* **1993**, *38*, 671–672. [CrossRef]
18. Nakatani, S.; Sands, T. Autonomous Damage Recovery in Space. *Autom. Control Intell. Syst.* **2016**, *2*, 23–36.
19. Nakatani, S.; Sands, T. Battle-damage tolerant automatic controls. *Electr. Electron. Eng.* **2018**, *8*, 10–23.
20. Sands, T.; Bollino, K. *Autonomous Underwater Vehicle Guidance, Navigation, and Control*; Chapter in Autonomous Vehicle; InTech: London, UK, 2019.
21. Slotine, J.J.E.; Benedetto, M.D. Hamiltonian Adaptive Cotnrol of Spacecraft. *IEEE Trans. Autom. Control* **1990**, *35*, 848–852. [CrossRef]

22. Sands, T.; Kim, J.; Agrawal, B. Improved Hamiltonian adaptive control of spacecraft. In Proceedings of the IEEE Aerospace Conference, Big Sky, MT, USA, 7–14 March 2009.

23. Astrom, K.J.; Wittenmark, B. *Adaptive Control*; Addison Wesley Longman: Boston, MA, USA, 1995.

24. Farrell, J.; Barth, M. *The Global Positioning System and Inertial Navigation*; McGraw-Hill, Two Penn Plaza: New York, NY, USA, 1996.

25. Hong, S.; Lee, M.H.; Chun, H.U.; Kwon, S.H.; Speyer, J.L. Experimental study on the estimation of lever arm in GPS/INS. *IEEE Trans. Veh. Technol.* **2006**, *55*, 431–448. [CrossRef]

26. Titterton, D.H.; Weston, J.L. *Strapdown Inertial Navigation Technology*; Peter Peregrinus Ltd.: London, UK, 1997.

27. Tecnadyne Technical Report. Available online: http://tecnadyne.com/wp-content/uploads/2017/06/Model-300-Brochure-3.pdf (accessed on 18 May 2018).

28. Li, J.H.; Kim, M.G.; Kang, H.; Lee, M.J.; Cho, G.R. Development of UUV platform and its control method to overcome strong currents: simulation and experimental studies. In Proceedings of the 11th IFAC Conference on Control Applications in Marine Systems, Robotics, and Vehicles, Opatija, Croatia, 10–12 September 2018.

29. Kim, J.H.; Sitorus, P.E.; Won, B.R.; Le, T.Q.; Ko, J.H.; Kim, D.Y.; Jang, I.S. A flow-visualization study of a multiple hydrofoils duct with particle image velocimetry equipment in KIOST. In Proceedings of the 12th International Symposium on Particle Image Velocimetry, Busan, Korea, 18–22 June 2017.

Journal of
*Marine Science
and Engineering*

Article

Autonomous Minimum Safe Distance Maintenance from Submersed Obstacles in Ocean Currents

Timothy Sands [1,*], **Kevin Bollino** [2], **Isaac Kaminer** [3] and **Anthony Healey** [3]

[1] Department of Mechanical Engineering, Stanford University, Stanford, CA 94305, USA
[2] Department of Electrical and Computer Engineering, George Mason University, Fairfax, VA 22030, USA; kbollino@gmu.edu
[3] Department of Mechanical and Aerospace Engineering, Naval Postgraduate School, Monterey, CA 93943, USA; kaminer@nps.edu (I.K.); healey@nps.edu (A.H.)
* Correspondence: dr.timsands@stanford.edu; Tel.: +1-3954-656-3954

Received: 27 June 2018; Accepted: 20 August 2018; Published: 22 August 2018

Featured Application: Submersed obstacle avoidance in unknown ocean currents via guidance, navigation, and control for autonomous underwater vehicles.

Abstract: A considerable volume of research has recently blossomed in the literature on autonomous underwater vehicles accepting recent developments in mathematical modeling and system identification; pitch control; information filtering and active sensing, including inductive sensors of ELF emissions and also optical sensor arrays for position, velocity, and orientation detection; grid navigation algorithms; and dynamic obstacle avoidance, amongst others. In light of these modern developments, this article develops and compares integrative guidance, navigation, and control methodologies for the Naval Postgraduate School's *Phoenix* submerged autonomous vehicle, where these methods are assumed available. The measure of merit reveals how well each of several proposed methodologies cope with known and unknown disturbances, such as currents that can be constant or harmonic, while maintaining a safe passage distance from underwater obstacles, in this case submerged mines. Classical pole-placement designs establish nominal baseline behaviors and are subsequently compared to performance of designs that are optimized to satisfy linear quadratic cost functions in regulators as well as linear-quadratic Gaussian designs. Feed-forward architectures and integral control designs are also evaluated. A noteworthy contribution is a very simple method to mimic optimal results with a "rule of thumb" criteria based on the design's time constant. Since the rule-of-thumb method uses the assumed system model for computation of the control, it is particularly generic. Cited references each contain methods for online system parameter identification (with a motivation of use in the finding the control signal), permitting the rule of thumb's generic applicability, since it is expressed in terms of the system parameters. This proposed method permits control design at sea where significant computation abilities are not available. Very simple waypoint guidance is also introduced to guide a vehicle along a preplanned path through a field of obstacles placed at random locations. The linear-quadratic Gaussian design proves best when augmented with integral control, and works well with reduced-order equations, while the "rule of thumb" design is seen to closely mimic the optimal performance. Feed-forward augmentation proves particularly efficient at rejecting constant disturbances, while augmentation with integral control is necessary to counter periodic disturbances, where the augmentations are also optimized in the linear-quadratic Gaussian procedures, yet can be closely mimicked by the proposed "rule of thumb" technique.

Keywords: actuator constraints; kinematics; dynamics; intelligent control; automation systems; submersible vehicles; obstacle avoidance; ocean research; guidance, navigation and control; approximated optimal control

---

1. Introduction

The Naval Postgraduate School's consortium for robotics and unmanned systems education and research (CRUSER) uses three autonomous underwater vehicles, the *Remus, Aries* [1], and *Phoenix* [2] vehicles to enhance education and research. The oldest vehicle, *Phoenix* [3] is used in this study to investigate integrated methodologies [4] for vehicle guidance, navigation, and control through a field of obstacles amidst unknown ocean currents that can be approximated by steady state, fixed disturbance ocean velocities, and can also be represented by harmonically oscillating velocities. This integrated approach is a natural extension of the recent innovations. The *Phoenix* vehicle's nominal mathematical modeling was articulated in the 1988 article [5] using surge motion to perform system identification. Recent innovations [6–10] have extended and improved the nominal system identification resulting in high-confidence mathematically modeling in computer simulations. Such simulations permitted Wu et al. [11] to redesign the L1 adaptive control architecture for pitch-control with anti-windup compensation based on solutions to the Riccati equation to guarantee robust and fast adaption of the underwater vehicle with input saturation and coupling disturbances and the approach was applied to the pitch channel alone. Stability was emphasized in the single-channel approach to emphasize dynamic nonlinearities and measurement errors. The Ricatti equation is also utilized in this research and proves effective when applied to all six degrees of freedom per [4], where the approach is applied to instances of disturbances that are constant with simultaneous harmonic disturbances simulating unknown ocean currents and waves. In addition to these recent achievements in control, improvements have also been made to guidance and navigation. In recent years, Bo He et al. [12] demonstrated, in simulations and open water experiments, the ability to overcome weak data links and sparse navigation data using a technique called extended information filter (EIF) applied to simultaneous localization and mapping (i.e., "SLAM") that proved computationally easier to implement than the traditional extended Kalman filter (EKF) SLAM. Low computational cost is emphasized here to keep the vehicle size low, but also to exaggerate the laudable goal of achieving optimal or near optimal results with methods that are simple. Such is an overt goal of the new research presented here.

Just last year, Yan et al. [13] integrated the navigation system using a modified fuzzy adaptive Kalman filter (MFAKF) to combine traditional strap-down inertial navigation with OCTANS and Doppler velocity log (DVL) to navigate the challenging polar regions where rapidly converging earth meridians and challenging ocean environments filled with submersed obstacles. This benchmark achievement requires the research here to utilize similar challenging ocean conditions, and provide the motivation for selection of simultaneous steady-state ocean currents together with sinusoidal varying unknown wave conditions amidst an ocean filled with obstacles (where here the non-polar ocean is used, so mines are added to fulfill the role of malignant submersed obstacles). Furthermore, simplified waypoint guidance is derived, based on the onboard-calculated distance from the vehicle to a submerged obstacle. The simplified waypoint guidance is proven effective, and should be considered in situations where onboard operation of a modified fuzzy adaptive Kalman filter proves to be computationally prohibitive. The distance to an underwater obstacle was measured by Wang et al. [14] with a novel method: measuring extremely low frequency (ELF) emissions with onboard inductive sensors. Such emissions are produced by ship hulls with relatively pronounced amplitudes compared to small subsurface obstacles, but the harmonic line spectra and fundamental signal frequency relate directly to the closing speed of approach to the obstacle. Experiments proved that even such small signals were detectable at long range with high sensitivity and low-noise sensors of the current state of the art, thus closing the distance to obstacles may now be presumed to be known passively, permitting the simplified waypoint guidance proposed in this manuscript. Particularly after ELF queuing, position, orientation, and velocity of obstacles may be monitored optically, as developed by Eren et al. [15], and these states may be used as feedback signals together with the waypoint guidance (desired trajectory) permitting augmentation with linear quadratic Gaussian techniques, as performed in this manuscript where full-order state observers are together optimized with attitude controller gains, followed by demonstration that reduced-order observers may also be optimized allowing vehicle operators to compensate for individually failed or degraded sensors, or instances where optimally-estimated signals are superior to sensor signals in individual or multiple channels.

Integrating these latest technological developments was demonstrated last year by Wei et al. [16], who integrated the Doppler velocity methods for obstacle monitoring into a dynamic obstacle avoidance scheme for collision avoidance. Following data fusion, a collision risk assessment model is used to avoid collisions, and claims to be effective in unknown dynamic environments, although the experiments did not go so far as to stipulate near-constant ocean currents in addition to harmonic wave actions. These challenging dynamic environments are addressed in this manuscript as a natural extension of the current state of the art.

Autonomous vehicle angular momentum control of rotational mechanics may be achieved using control moment gyroscopes, one potential momentum exchange actuator with a long, historic legacy of actuating space vehicles, where mathematical singularities have just recently been overcome [17–23], permitting the use of the actuator for underwater vehicles as recently achieved by Thorton et al. [24,25], including combined attitude and energy storage control. These developments suffice to reveal that attitude control is not controversial and, thus, the remainder of this manuscript focuses on guidance and navigation with a residual necessity to implement nominal, effective pitch and yaw control.

2. Materials and Methods

Assuming the availability of the recent technologies cited in the Introduction, this section describes the proposed methods to use these technologies to guide a submersed vehicle along a preplanned path through a field of randomly-placed obstacles. The constituent technologies are investigated through this section of the manuscript, and then combined in a fully-assembled system demonstration in Section 3 (Results), where the figure of merit used to assess the efficacy of the proposed methods is the maintenance of the miss distance from submersible objects in ocean currents.

Submersible vehicles require control systems to guide the vehicle around obstacles that can present dangers to vehicle health and safety in the presence of ocean currents. The challenge addressed here is to navigate one of the two Naval Postgraduate School's submersible vehicle (Figures 1 and 2) through a simulated minefield whose dimensions are 200 m × 5100 m in the presence of 0.5 m/s ocean currents. The field will contain at least 30 mines placed at locations using a random number generator. The resulting controller structure has an inner-outer loop structure, and several technologies will be described including pole-placement designs, linear-optimal (quadratic) Gaussian techniques, full- and partial-order observers for online disturbance identification for ocean currents (both constant lateral underwater ocean currents and also sinusoidal varying currents), tracking systems and feed-forward control designed to counter open ocean currents, in addition to integral control. The outer loop controller uses line-of-sight (LOS) guidance to provide a heading command to the inner loop. The inner loop controller uses output heading feedback to track heading commands. The vehicle is simulated to traverse a minefield and successfully travels no closer than 5 m from any mine and arrives within from the commanded destination autonomously; while this overall system design requirement drives subsystem requirements for trajectory tracking.

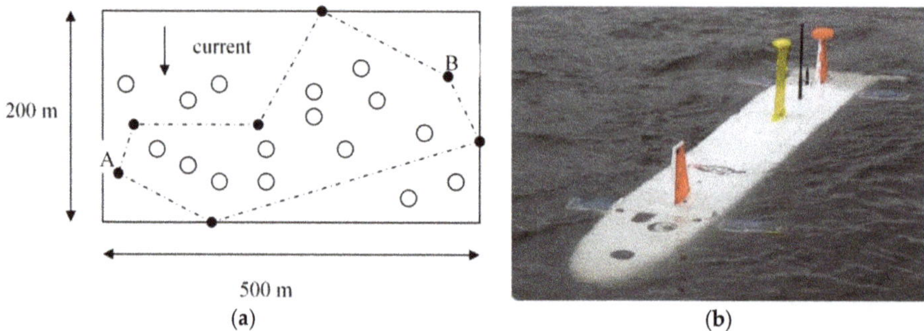

Figure 1. Submersible vehicle sample and notional minefield [1]. (**a**) Field of randomly-placed submersed mines to be avoided by the autonomous vehicle; and (**b**) *Aries* submersible in open ocean (illustrative sample is not simulated; the Figure 2 *Phoenix* vehicle is simulated).

2.1. System Dynamics

The equations of motion used to simulate the dynamic behavior of the autonomous submersible vehicle in a horizontal plane are listed in Equations (1)–(4). All variables in these equations are assumed to be in nondimensional form with respect to the vehicle length (7.3 feet) and constant forward speed (~3 ft/s). The vehicle weighs 435 lbs and is neutrally buoyant. Time is non-dimensionalized such that 1 s represents the time it takes to travel one vehicle length.

Figure 2. Vehicle geometry and reference axes. (**a**) *Phoenix* in open ocean [1]. The experimentally-determined dynamic model for this vehicle is listed in Equations (1)–(6) and forms the basis for the simulations in this manuscript; and (**b**) vehicle geometry and reference axis.

$$(m - Y_{\dot{v}})\dot{v} - (Y_{\dot{r}} - mx_G)\dot{r} = Y_{\dot{r}}v + (Y_r - m)r + Y_{\delta_s}\delta_s + Y_{\delta_b}\delta_b \tag{1}$$

$$(mx_G - N_{\dot{v}})\dot{v} - (N_{\dot{r}} - I_z)\dot{r} = N_v v + (N_r - mx_G)r + N_{\delta_s}\delta_s + N_{\delta_b}\delta_b \tag{2}$$

$$\dot{\psi} = r \tag{3}$$

$$\dot{y} = sin\psi + vcos\psi \tag{4}$$

In addition to the following dependent equation:

$$\dot{x} = cos\psi - vsin\psi \tag{5}$$

where the variables are		and the constants are	$Y_{\delta_s} = 0.01241$
v	Lateral (sway) velocity	$m = 0.0358$	$Y_{\delta_b} = 0.01241$
r	Turning rate (yaw)	$I_z = 0.0022$	$N_{\dot{r}} = -0.00047$
ψ	Heading angle (degrees)	$x_G = 0.0014$	$N_{\dot{v}} = -0.00178$
y	Lateral deviation (cross-track error)	$Y_{\dot{r}} = -0.00178$	$N_r = -0.00390$
δ_s	Stern rudder deflection	$Y_{\dot{v}} = -0.03430$	$N_v = -0.00769$
δ_b	Bow rudder deflection	$Y_r = 0.01187$	$N_{\delta_s} = -0.0047$
		$Y_v = -0.10700$	$N_{\delta_b} = 0.0035$

The constant definitions in the mass m, mass moment of inertia with respect to a vertical axis that passes through the vehicle's geometric center (amidships) I_z, position of the vehicle's center of gravity (measured positive forward of amidships) x_G, with the remaining terms referred to as the hydrodynamic coefficients. These constants are all presented in non-dimensional form.

Defining the state vector $\{x\} \equiv \{\ v\ \ r\ \ \psi\ \ y\ \}^T$ and the control $\{u\} \equiv \{\ \delta_s\ \ \delta_b\ \}^T$ and assuming small angles, the dynamics expressed in Equations (1)–(4) may be expressed in state space form as $\{\dot{x}\} = [A]\{x\} + [B][u]$ where:

$$[A] = \begin{bmatrix} -1.4776 & -0.3083 & 0 & 0 \\ -1.8673 & -1.2682 & 0 & 0 \\ 0 & 1 & 0 & 0 \\ 1 & 0 & 1 & 0 \end{bmatrix} \quad [B] = \begin{bmatrix} 0.2271 & 0.1454 \\ -1.9159 & 1.2112 \\ 0 & 0 \\ 0 & 0 \end{bmatrix} \tag{6}$$

The system may also be expressed in a transfer function ratio of outputs divided by inputs in Laplace form using Equation (7) where the observer matrix [C] is merely a proper identity matrix to this point of the manuscript. Equation (7) yields two transfer function relationships between each of the two possible rudder inputs as seen in Equations (8) and (9). Notice that both transfer functions have poles and zeros at the origin, while pole-zero cancellation is possible in the case of the stern rudder. On the other hand, even after pole-zero cancellation in the bow rudder Equation (9), there remains an open loop pole at the origin that must be dealt with during the control design, since it represents a potentially unstable element (at the very least, in the instance where the estimated constants are exactly correct, and these equations of motion exactly describe the system, an oscillatory element exists that will not decay). Nonetheless, the dynamics accord to nature. Consider trying to steer a row-boat using the rear rudder: it is much more stable than trying to steer the rowboat using a rudder in the front. This analogy applies to the submersible vehicle and is verified in these results.

$$G(S) = [C](s[I] - [A])^{-1}[B] \tag{7}$$

$$G(S)|_{\delta_s} \equiv \frac{Y(s)}{\delta_s(s)} = \frac{0.2271s^3 + 0.875s^2}{s^4 + 2.746s^3 + 1.298s^2} = \frac{s^2(0.2271s + 0.875)}{s^2(s^2 + 2.746s + 1.298)} \tag{8}$$

$$G(S)|_{\delta_b} \equiv \frac{Y(s)}{\delta_b(s)} = \frac{1.211s^2 + 1.518s}{s^4 + 2.746s^3 + 1.298s^2} = \frac{s(1.211s + 1.518)}{s^2(s^2 + 2.746s + 1.298)} \tag{9}$$

In Figure 3, the uncontrolled system is analyzed by merely performing a circular turn with each (and then both) rudders. The bow and stern rudders alone are each compared to the combined use of both bow and stern rudders. The bow rudder was deflected +15 degrees for about 21 s, while the stern rudder was deflected for -15 degrees for about 11 s. When both rudders were deflected the maneuver was completed in roughly 8 s. Two initial conditions for the sway velocity were investigated ($v(0) = 0$ and then $v(0) = \sqrt{8}$). In all cases, the bow rudder alone performed the poorest, with the stern rudder alone performing the turn in a smaller radius and shorter time. Furthermore, the combined use of both rudders resulting in tightest maneuver.

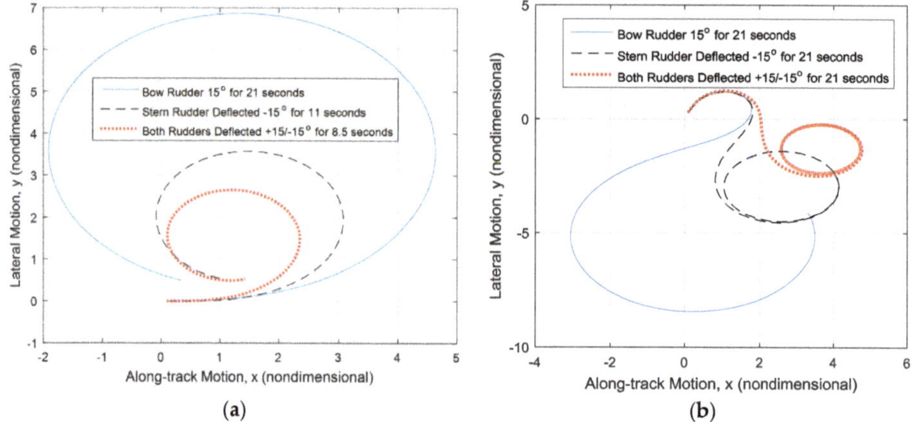

Figure 3. Analysis of uncontrolled system: comparison of rudder performance. (**a**) Counter-clockwise turn, $v(0) = 0$; and (**b**) initial sway velocity $v(0) = \sqrt{8}$.

Two simulation methodologies were used to investigate sensitivities to integration method. MATLAB was used with Euler integration, while SIMULINK was used with Runge-Kutte integration with identical timesteps, $\Delta t = 0.1$ s. Both software packages are manufactured and supplied by Mathworks®, Natick, Massachusetts, USA. The results were nearly negligible and are displayed in Table 1, from which insensitivity to the integration approach is established.

Table 1. Comparison of simulation integration methodologies.

Rudder Deflected	Euler: x-Distance [1]	Runge-Kutte: x-Distance [1]	Euler: y-Distance [1]	Runge-Kutte: y-Distance [1]
Bow	6.5471	6.5469	6.8647	6.8646
Stern	3.1665	3.1665	3.5768	3.5768
Both	2.4546	2.4546	2.6567	2.6567

[1] Distances calculated to traverse one circular path.

2.2. Control Law Design

In the system analysis, the optimal rudder implementation scheme was determined to be the application of both rudders, where the rudders were slaved to the same maneuver angle magnitude with the opposite sign, i.e., a "scissored-pair", per Equation (10). In the case where only variable y is to be measured, the new state space formulation of the system equation components are in Equation (11). Under the assumption of rudders constrained to behave as a scissored-pair the transfer function from rudder input to output y is given by Equation (12) whose poles and zeros are listed in Equation (13), with Equation (14) revealing the system's eigenvalues, noting the values are identical to the location of the poles in accordance with theory. The controllability and observability matrices ($[CO]$ and $[OB]$, respectively) are listed in Equation (15) (whose matrix product $[OC]$ is in Equation (16)) verifying these system equations are both controllable and observable, since these matrices are full-rank, while the determinant of the controllability matrix is 63.1778, a large value with a small value of the matrix condition number, 13.4513. The non-zero determinant of the controllability matrix proves controllability, but to see how close the system is to being uncontrollable, the matrix condition number proves more useful. These two figures of merit indicate the system equations are highly controllable and, accordingly, this manuscript will investigate and compare several options for navigation control: pole placement, linear quadratic optimal control, linear quadratic Gaussian, and time optimal control. The same holds true for observability and, thus, linear quadratic Gaussian. The matrix product $[OC]$ is the same for every definition of state variables for the given system.

A change in system parameters results in disparate system equation coefficients in Equations (1)–(5) which may be expressed in state space or transfer-function formulations and correlated modifications to Equations (6), (8), and (9). Nonetheless, these equations form the basis for finding linear-quadratic Gaussian optimal solutions or the simpler rule-of-thumb procedure driven by system time-constant. Since the process for finding the optimal solutions is inherently more challenging than simple control calculations using the time-constant, the rule of thumb technique is generically superior in regards to computational overhead, regardless of system parameters chosen.

$$\delta_b = -\delta_s \tag{10}$$

$$[A] = \begin{bmatrix} -1.4776 & -0.3083 & 0 & 0 \\ -1.8673 & -1.2682 & 0 & 0 \\ 0 & 1 & 0 & 0 \\ 1 & 0 & 1 & 0 \end{bmatrix} \quad [B] = \begin{bmatrix} 0.0816 \\ -3.1271 \\ 0 \\ 0 \end{bmatrix} \quad [C] = [0\ 0\ 0\ 1]\ [D] = [0] \tag{11}$$

$$G(S)|_\delta \equiv \frac{Y(s)}{\delta(s)} = \frac{0.08164s^2 - 2.06s - 4.773}{s^4 + 2.746s^3 + 1.298s^2} \tag{12}$$

$$poles\ at:\ s = 0,0,-0.6070,\ -2.1388;\quad zeros\ at:\ s = -6.1279 \times 10^{13},\ near-0,\ near-0 \tag{13}$$

$$eig(A) = \lambda = 0,0,-0.6070,\ -2.1388 \tag{14}$$

$$[CO] = \begin{bmatrix} 0.0816 & 0.8433 & -2.4216 & 5.5544 \\ -3.1271 & 3.8132 & -6.4105 & 12.6514 \\ 0 & -3.1271 & 3.8132 & -6.4105 \\ 0 & 0.0816 & -2.2838 & 1.3916 \end{bmatrix} \quad [OB] = \begin{bmatrix} 0 & 0 & 0 & 1 \\ 1 & 0 & 1 & 0 \\ -1.4776 & 0.6917 & 0 & 0 \\ 0.8917 & -0.4217 & 0 & 0 \end{bmatrix} \tag{15}$$

$$[OC] = \begin{bmatrix} 0 & 0.0816 & -2.838 & 1.3916 \\ 0.0816 & -2.2838 & 1.3916 & -0.8561 \\ -2.2838 & 1.3916 & -0.8561 & 0.5441 \\ 1.3916 & -0.8561 & 0.5441 & -0.3825 \end{bmatrix} \tag{16}$$

Diagonalizing the original system $[A]$ matrix, the spectral decomposition $[T][\Lambda] = [A][T] \rightarrow [\Lambda] = [T]^{-1}[A][T]$ in Equation (17) may be used to verify a diagonal matrix of eigenvalues $[\Lambda]$, and then write the system of equations in *normal-coordinate form* $\{\dot{x}'\} = [A']\{x'\} + [B'][u]; \{y'\} = [C'][x']$ using the following transformation: $[A'] = [\Lambda] = [T]^{-1}[A][T]$, $[B'] = [T]^{-1}[B]$, and $[C'] = [C][T]$ whose results are in Equation (18):

$$[\Lambda] = \underbrace{\begin{bmatrix} 0.4663 & -0.1074 & 0 & 0 \\ 1 & 0.3033 & 0 & 0 \\ -0.4676 & -0.4996 & 0 & 0 \\ 0.0006 & 1 & 1 & -1 \end{bmatrix}}_{T^{-1}}^{-1} \underbrace{\begin{bmatrix} -1.4776 & -0.3083 & 0 & 0 \\ -1.8673 & -1.2682 & 0 & 0 \\ 0 & 1 & 0 & 0 \\ 1 & 0 & 1 & 0 \end{bmatrix}}_{A} \underbrace{\begin{bmatrix} 0.4663 & -0.1074 & 0 & 0 \\ 1 & 0.3033 & 0 & 0 \\ -0.4676 & -0.4996 & 0 & 0 \\ 0.0006 & 1 & 1 & -1 \end{bmatrix}}_{T} \tag{17}$$

$$[A'] = \begin{bmatrix} -2.1388 & 0 & 0 & 0 \\ 0 & -0.6070 & 0 & 0 \\ 0 & 0 & 0 & 0 \\ 0 & 0 & 0 & 0 \end{bmatrix},\ [B'] = \begin{Bmatrix} -1.2502028 \\ -6.1888806 \\ -8.4625 \times 10^7 \\ -8.4625 \times 10^7 \end{Bmatrix},\ [C'] = \begin{Bmatrix} 0.0006 & 1 & 1 & -1 \end{Bmatrix} \tag{18}$$

$$\{u\}_{baseline} = \{\delta\} = -K_\upsilon\upsilon - K_r r - K_\psi\psi - K_y y \tag{19}$$

For the pole placement proportional-derivative (PD) controller articulated in Equation (19), the poles are set to have roughly the same time constant, while avoiding exactly coincident poles. Gains are iterated for various time constants as displayed in Figure 4, but the following rule of thumb is asserted as well to quickly achieve performance that closely mimics the performance of linear-quadratic optimal (LQR) gains where the control effort and tracking error are equally weighted in the cost function of the optimization. Since the rule-of-thumb method uses the assumed system model for computation of the control, it is particularly generic. References [7,9,10] each contain methods for online system parameter identification (with a motivation of use in the finding the control signal), permitting the rule of thumb's generic applicability, since it is expressed in terms of the system parameters.

RULE OF THUMB: Select unity time-constant t_c to roughly locate closed-loop poles per Equation (20). Then place other poles at slightly different locations (e.g., $s_p = s_1 \pm 0.01\ \forall\ p$)

$$Pole : s_1 = \frac{1}{t_c} \tag{20}$$

The gains achieved using the rule of thumb $K_{\text{R.O.T.}}$ = {0.5070 −0.3687 −0.7157 −0.1972} (see Table 2) have quite different values compared to the gains calculated through the matrix Ricatti equation in the linear-quadratic optimization K_{LQR} = {−0.0939 −1.2043 −2.2138 −1}, but, nonetheless, the resulting behaviors are indeed very similar.

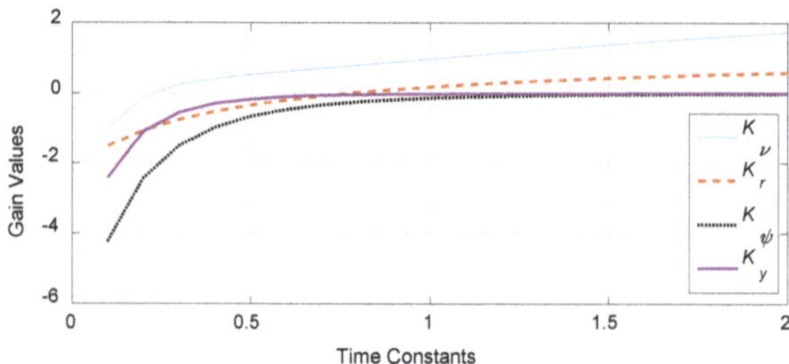

Figure 4. Gain values for each state iterated for various time constants.

Next, the initial feedback control design was evaluated in simulations where the ship is initially located off the desired track by one ship's length port side with zero heading, and rudder deflection was limited to 0.4 radians (~23 degrees). Next, another simulation was performed to test an initial heading angle of 30 degrees starboard where the initial $y(0) = 0$. The results are displayed in Figure 5a,b, respectively. All state variations were plotted in Figure 4, highlighting the fact that y converges to zero along with the other states. Furthermore, the results of rudder-limited simulations are displayed in Figure 6 for both scenarios, while the comparison of *rule of thumb* to LQR is shown in Figure 7.

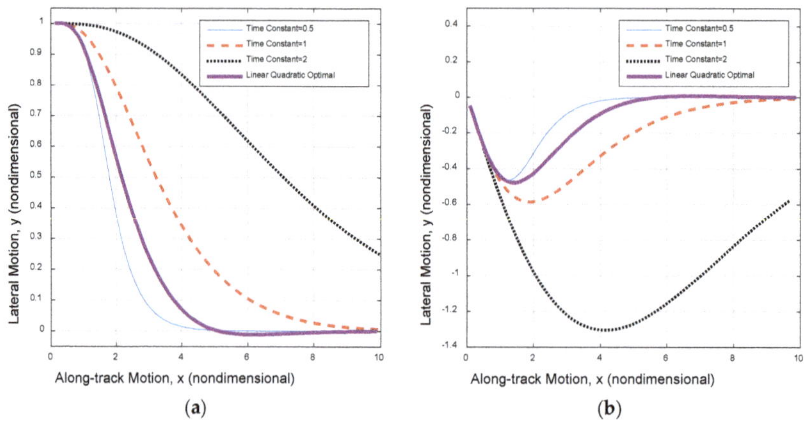

Figure 5. Simulations testing the initial baseline feedback controller in two scenarios. (**a**) Initially one ship's length port side; and (**b**) initial heading 30° starboard.

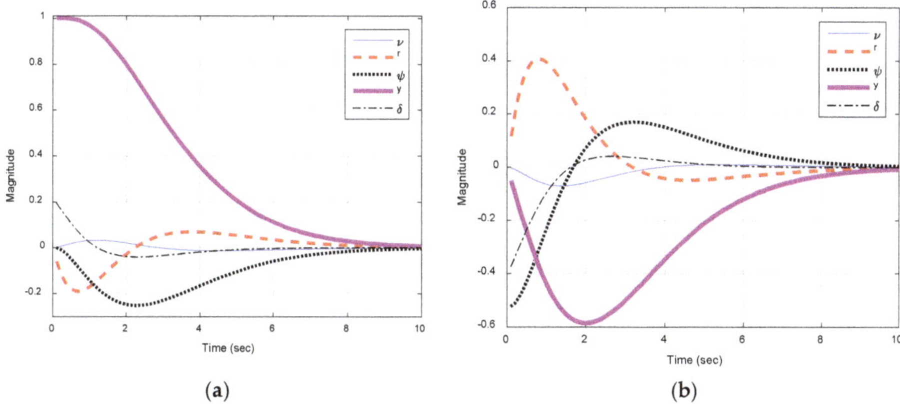

Figure 6. State variations for both scenarios simulated using pole-placement gains via *rule of thumb*. (a) Initially one ship's length port side; and (b) initial heading 30° starboard.

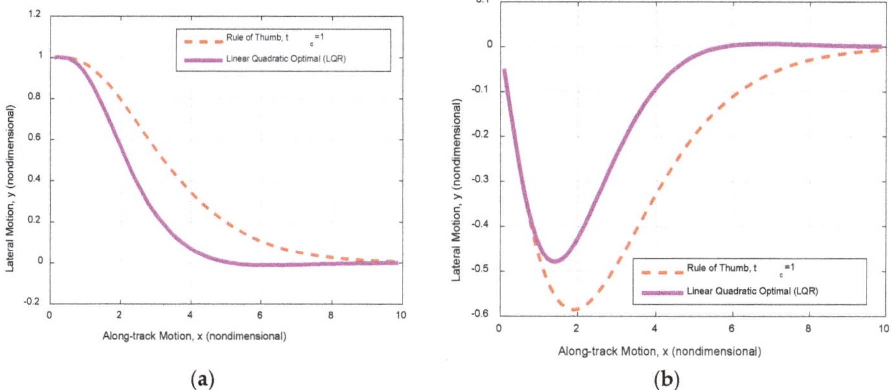

Figure 7. Rudder-limited trajectory track using pole-placement gains via *rule of thumb* and LQR. (a) Initially one ship's length port side; and (b) initial heading 30° starboard.

Table 2. Gains for various time constants and also solution to linear quadratic optimization.

Time Constant	K_v	K_r	K_ψ	K_y
0.5	-1.5135	-1.7005	-5.1508	-3.22524
1	0.5070	-0.3687	-0.7157	-0.1972
2	1.1248	0.2870	-0.0906	-0.0116
LQR	-0.0939	-1.2043	-2.2138	-1

[1] Reminder: state definition $\{x\} \equiv \{\; v \quad r \quad \psi \quad y \;\}^T$.

2.3. Observer Design

To design a state observer, the system must be observable [4], verifiable through examination of the observability matrix $[OB]$ per Equation (21), where for example $[C] = [v\ r\ \psi\ y] = [0\ 1\ 1\ 1]$, while several such examples will be iterated in this investigation. The condition of the observability matrix reveals the degree of observability, and it is defined by the ratio of maximum to minimal singular values:

$$[OB] = \begin{bmatrix} C \\ CA \\ CA^2 \\ \vdots \\ CA^{n-1} \end{bmatrix} \tag{21}$$

2.3.1. Full-Order Observer Design

$$\{\dot{\hat{x}}\} = [A]\{\hat{x}\} + [B][u] + [L](\{y\} - [C]\{\hat{x}\}) \tag{22}$$

$$\{\dot{x}\} - \{\dot{\hat{x}}\} = [A]\{x\} - [A]\{\hat{x}\} - [L]([C]\{x\} - [C]\{\hat{x}\}) \tag{23}$$

$$\{\dot{e}\} \equiv \{\dot{x}\} - \{\dot{\hat{x}}\} = ([A] - [L][C])(\{x\} - \{\hat{x}\}) \tag{24}$$

$$\{\dot{e}\} = ([A] - [L][C])\{e\} \tag{25}$$

Assuming that only v measurements are available, a mathematical model of the estimated system is shown in Equation (22) with a full-order observer design using the observer error Equation (23) leading to the error vector in Equation (24) allowing the re-expression of Equation (22) as Equation (25), where the dynamic behavior of the error vector is determined by the eigenvalues of matrix $[A] - [L][C]$, where $[L]$ gains of the observer may be chosen as desired for systems that prove observable, such that the error vector will converge to zero for any stable $[A] - [K_e][C]$. In the following paragraphs, $[L]$ is designed by solving the matrix Ricatti equation leading to linear quadratic optimal gains, and also by solving the *rule of thumb* relationship between gains and time constant as done for the controller gains resulting in Table 3.

Table 3. Full-order observer gains designed by *rule of thumb* for various time constants as multiple of controller time constant, t_c.

Multiple of the Controller Time Constant Used for the Observer	Observer Gain Matrix
$\frac{1}{2}t_c$ [1]	$\{\ \ -0.7464 \quad 1.8077 \quad 8.8270 \quad 5.1942 \ \ \}^T$
$10t_c$	$10^3 * \{\ \ -1.5909 \quad -3.4121 \quad 1.5953 \quad -0.0020 \ \ \}^T$

[1] Reminder: $\frac{1}{2}t_c$ is used in subsequent simulations.

Figure 8 displays the results of simulations revealing the accuracy of state estimation when $[L]$ is calculated by the *rule of thumb*, where the time constant is chosen to be half ($t_c = 1/2$) the time constant of the controller ($t_c = 1$), and the simulation is initialized with the heading angle 30 degrees off, while Figure 9 displays the simulation initialized at the one boat-length starboard position.

2.3.2. Reduced-Order Observer Design

Assuming that some measurements are available from sensors, this paragraph describes the possible iterations and reveals states that are relatively more important to measure with sensors. Four possible output matrices are used to investigate observability. Four options for output matrices $[C]_i$ for $i = 1, \ldots, 4$ result in four reduced-order observers $[OB]_i$ for $i = 1, \ldots, 4$ are detailed in Equations (26)–(29). The output matrix $[C]_1$ produces an observability matrix $[OB]_1$ with rank = 4 (observable) and determinant not nearly equal to zero. The output matrix $[C]_2$ produces an observability matrix $[OB]_2$ with rank = 4 (observable) and determinant not nearly equal to zero. The output matrix $[C]_3$ produces an observability matrix $[OB]_3$ with rank = 4 (observable) and determinant nearly equal to zero. The matrix condition number is very high indicating the system is barely observable. The output matrix $[C]_4$ produces an observability matrix $[OB]_4$ with rank = 3 (not observable) and determinant equal to zero with a matrix condition number equal to infinity. This means if all other states are measured by sensors, it is not possible to use an observer (even an optimal observer) to determine lateral deviation (cross-track error), y. It is a key state to measure with sensors. The sensor combinations that include y are observable. Using every other sensor, (except y) results in a system that is not observable. Furthermore, measuring y alone results in a barely observable system.

$$[C]_1 = \begin{Bmatrix} v \\ r \\ \psi \\ y \end{Bmatrix} = \begin{Bmatrix} 0 \\ 1 \\ 1 \\ 1 \end{Bmatrix} \rightarrow [OB]_1 = \begin{bmatrix} C \\ CA \\ CA^2 \\ \vdots \\ CA^{n-1} \end{bmatrix} = \begin{bmatrix} 0 & 1 & 1 & 1 \\ -0.8673 & -0.2682 & 1 & 0 \\ 1.7823 & 1.6074 & 0 & 0 \\ -5.6352 & -2.5879 & 0 & 0 \end{bmatrix} \tag{26}$$

$$[C]_2 = \begin{Bmatrix} v \\ r \\ \psi \\ y \end{Bmatrix} = \begin{Bmatrix} 0 \\ 1 \\ 0 \\ 1 \end{Bmatrix} \rightarrow [OB]_2 = \begin{bmatrix} C \\ CA \\ CA^2 \\ \vdots \\ CA^{n-1} \end{bmatrix} = \begin{bmatrix} 0 & 1 & 0 & 1 \\ -0.8673 & -1.2682 & 1 & 0 \\ 3.6496 & 2.8756 & 0 & 0 \\ -10.7624 & -4.7717 & 0 & 0 \end{bmatrix} \tag{27}$$

$$[C]_2 = \begin{Bmatrix} v \\ r \\ \psi \\ y \end{Bmatrix} = \begin{Bmatrix} 0 \\ 0 \\ 0 \\ 1 \end{Bmatrix} \rightarrow [OB]_3 = \begin{bmatrix} C \\ CA \\ CA^2 \\ \vdots \\ CA^{n-1} \end{bmatrix} = \begin{bmatrix} 0 & 0 & 0 & 1 \\ 1 & 0 & 1 & 0 \\ -1.4776 & 0.6917 & 0 & 0 \\ 0.8917 & -0.4217 & 0 & 0 \end{bmatrix} \tag{28}$$

$$[C]_2 = \begin{Bmatrix} v \\ r \\ \psi \\ y \end{Bmatrix} = \begin{Bmatrix} 1 \\ 1 \\ 1 \\ 0 \end{Bmatrix} \rightarrow [OB]_4 = \begin{bmatrix} C \\ CA \\ CA^2 \\ \vdots \\ CA^{n-1} \end{bmatrix} = \begin{bmatrix} 1 & 1 & 1 & 0 \\ -3.3450 & -0.5764 & 1 & 0 \\ 6.90190 & 1.7621 & 0 & 0 \\ -12.1843 & -4.0900 & 0 & 0 \end{bmatrix} \tag{29}$$

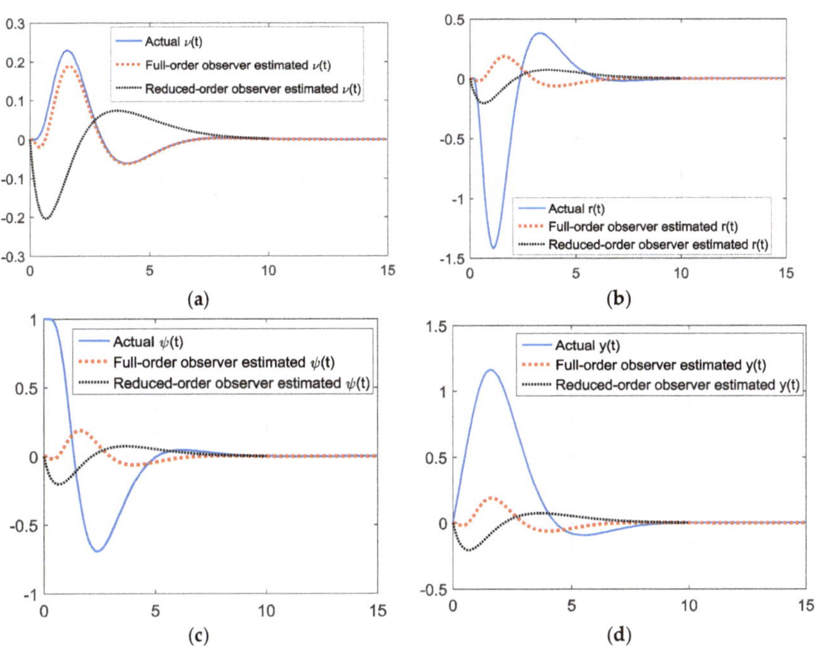

Figure 8. Simulations starting 30 degrees off heading with gains via *rule of thumb* state observer gains. (a) True and estimated sway velocity, $v(t)$; (b) true and estimated turning rate, $r(t)$; (c) true and estimated heading angle, $\psi(t)$; and (d) true and estimated cross track, $y(t)$.

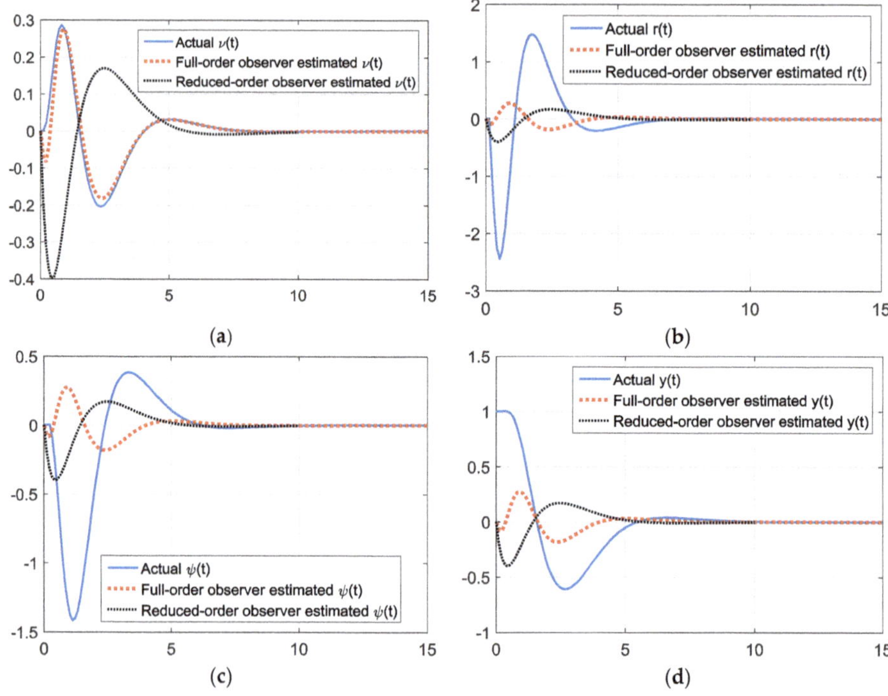

Figure 9. Simulations starting one boat-length starboard with gains via *rule of thumb*. (**a**) True and estimated sway velocity, $v(t)$; (**b**) true and estimated turning rate, $r(t)$; (**c**) true and estimated heading angle, $\psi(t)$; and (**d**) true and estimated cross track, $y(t)$.

Assuming y is to be measured by a sensor, Table 4 reveals that measuring v in addition to y produces the most observable system, and is recommended for designing reduced-order observers. The drawback is measuring v requires a Doppler sonar, which may not always be available. If all states are measureable except v the resulting reduced-order observer merely estimates v using gains on the measureable states displayed in Table 5. Figure 10 reveals a very good estimation of v when all other states are sensed, and this estimated value of v was fed to the motion controller in addition to the measured states (the poorly estimated states were neglected instead favoring the more-accurate measurements). State convergence to zero is achieved in the instance of state initialization 30 degrees off-heading. Figure 11 displays similar results for the instance of state initialization one boat-length starboard.

Table 4. Observability matrix condition number for options to supplement the y measurement.

Sensors Used to Measure States	Observability Matrix Condition Number [1]
y and v	8.8456
y and r	21.1306
y and ψ	31.2919

[1] Reminder: a high condition number means a less observable system.

Table 5. Reduced-order observer gains designed by *rule of thumb* for various time constants as a multiple of the controller time constant, t_c.

Multiple of the Controller Time Constant Used for Observer	Observer Gain Matrix
$\frac{1}{10}t_c$ [1]	$\{\ \ -0.2174 \quad 0 \quad 0.1164\ \ \}^T$
$2t_c$	$\{\ \ 0.4069 \quad 0 \quad -0.2179\ \ \}^T$
$10t_c$	$\{\ \ 0.5941 \quad 0 \quad -0.3182\ \ \}^T$

[1] Relatively faster $\frac{1}{10}t_c$ is used in subsequent simulations.

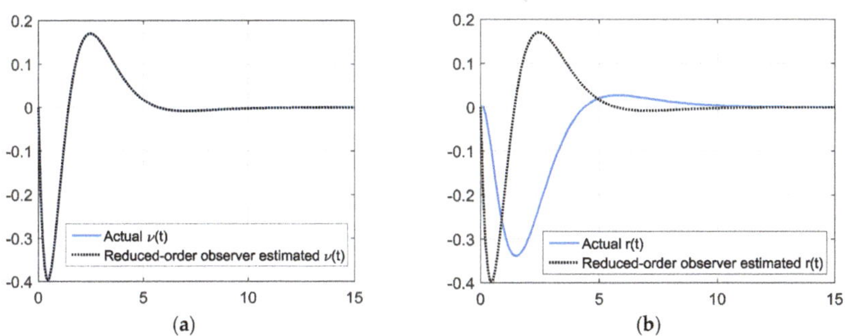

Figure 10. Simulations starting 30 degrees off heading gains via *rule of thumb* reduced-order state observer gains. (**a**) True and estimated sway velocity, $v(t)$ versus time (seconds); (**b**) true and estimated turning rate, $r(t)$ versus time (seconds); (**c**) true and estimated heading angle, $\psi(t)$ versus time (seconds); and (**d**) true and estimated cross track, $y(t)$ versus time (seconds).

Figure 11. *Cont.*

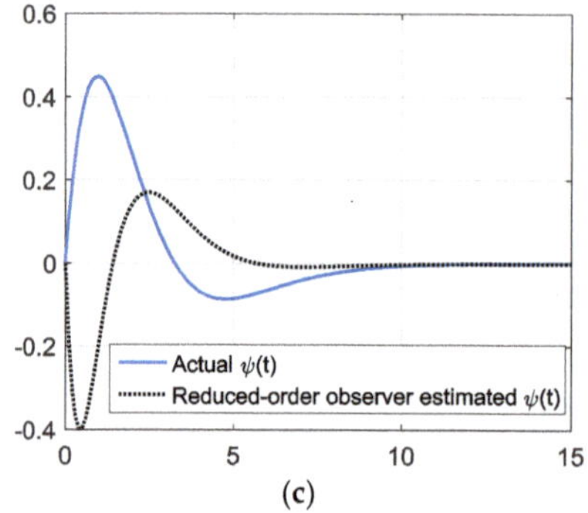
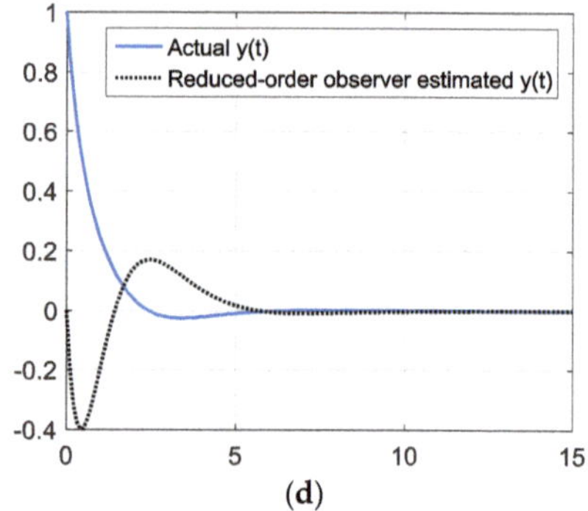

Figure 11. Simulations starting one boat-length starboard with gains via *rule of thumb* reduced-order state observer gains. (**a**) True and estimated sway velocity, $v(t)$ versus time (seconds); (**b**) true and estimated turning rate, $r(t)$ versus time (seconds); (**c**) true and estimated heading angle, $\psi(t)$ versus time (seconds); and (**d**) true and estimated cross track, $y(t)$ versus time (seconds).

2.3.3. Gain Margin and Phase Margin

Figure 12 compares the loop gains of the system with and without a compensator via the gain margin and phase margin with full-state feedback, while Figure 13 displays the loop gains when output-feedback via observers is used. Each has relative strengths. Full state (theoretical) feedback yields an infinite gain margin, yet a relatively lower phase margin (usually consider more important of the two), while output feedback (real-world) yields a good (but lesser) gain margin with an increased phase margin.

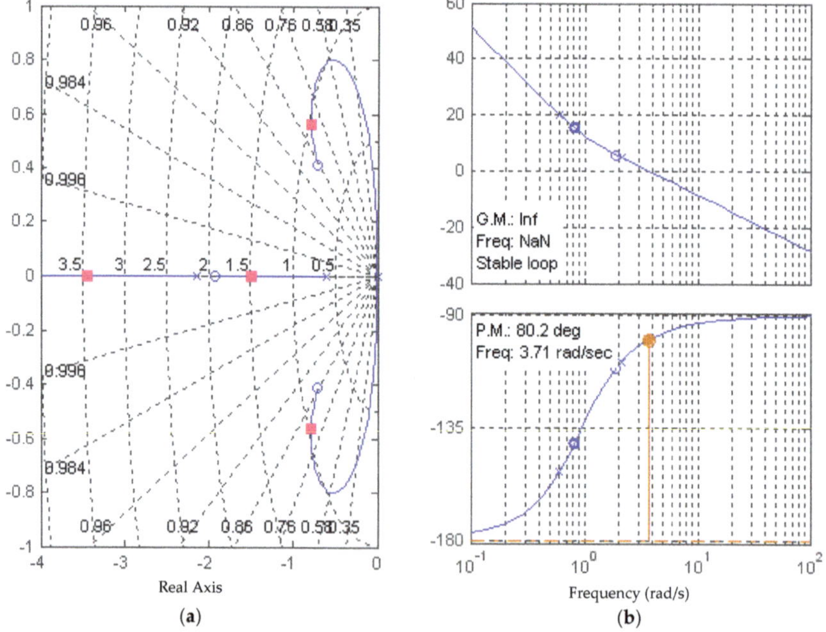

Figure 12. Infinite gain margin and 80.2° phase margin using full state feedback via full-ordered observer with rule of thumb controller gains. (**a**) Root locus; and (**b**) Bode plot.

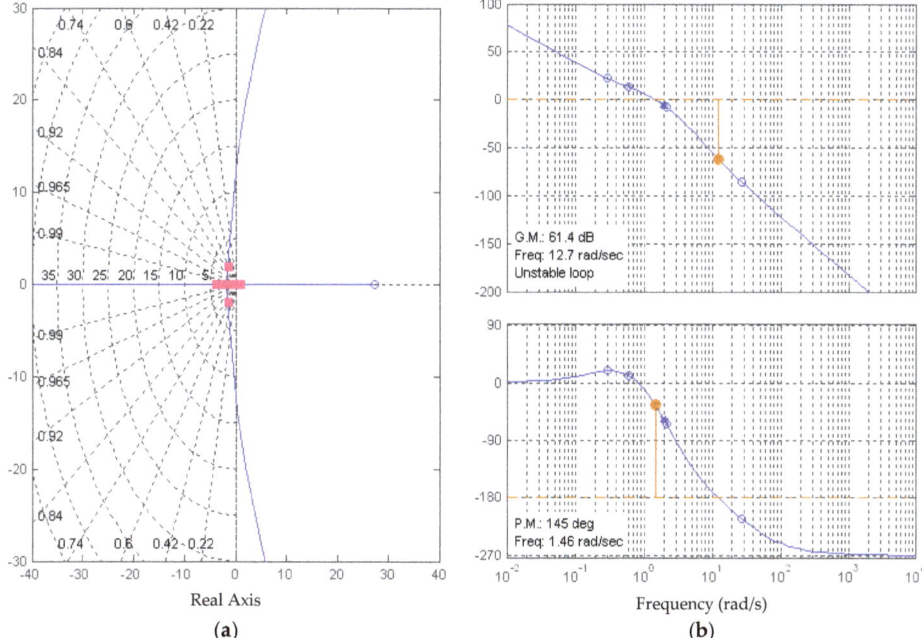

Figure 13. The 61.4 degree gain margin and 145 degree phase margin using a reduced-order observer (both rule of thumb gains for half-controller $t_c = 0.5$, and compensator with rule of thumb gains ($t_c = 1$). (**a**) Root locus; and (**b**) Bode plot.

2.4. Tracking Systems and Feed-Forward Control in the Presence of Constant Disturbance Currents

This section evolves the earlier developed system equations and performance analysis by adding non-quiescent conditions, in particular an introduction of a lateral underwater ocean current with an absolute velocity, v_0, requiring a modification of the system equations to add the lateral current to Equation (4) resulting in Equation (30):

$$\dot{y} = sin\psi + vcos\psi + v_0 \tag{30}$$

2.4.1. Analysis of Disturbed System in Ocean Currents via State Equations and Simulations

Using the controller (Equation (19)) and the modified system equations where Equation (4) is replaced by Equation (29), and applying the final value theorem: $f(t)_{t\to\infty}sF(s)_{s\to0}$, a steady state value $1/\omega+1$ has some variable quantity added to unity for various v_0. Thus, steady-state errors exist in all cases with such disturbances, which is verified by simulations depicted in Figures 14 and 15 using gain values from the *rule of thumb* (ROT) for the unity time constant. The steady-state errors are directly proportional to the disturbance magnitude. Figure 8 displays the maximum rudder deflection for the maximal lateral ocean current in the study (to verify the control design this continues to remain less than 0.4 radians).

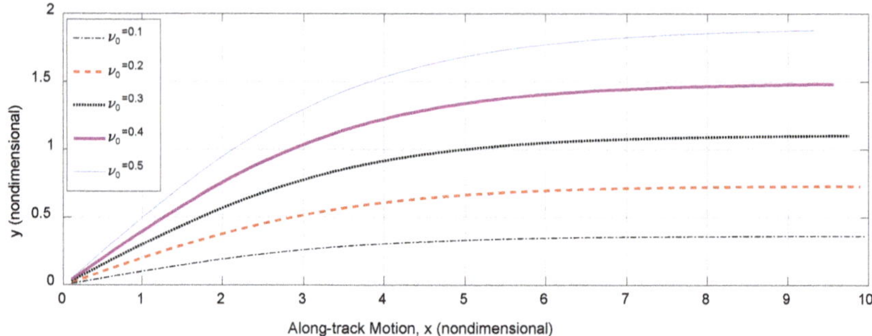

Figure 14. Steady-state position error for various lateral underwater ocean currents.

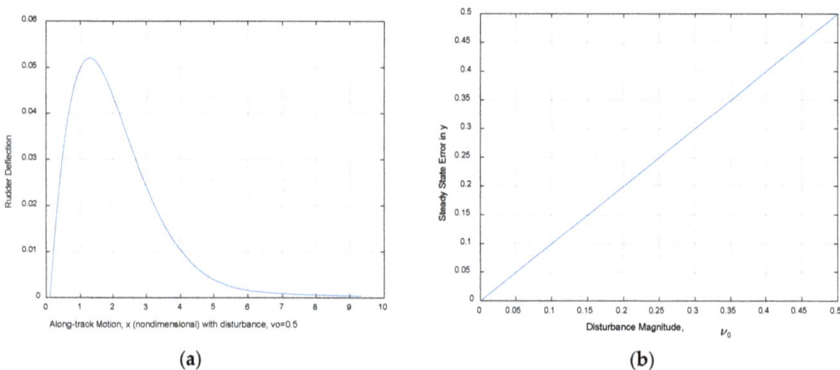

Figure 15. Feedback alone is unable to counter the constant lateral underwater ocean currents. (**a**) Rudder deflection, $v_0 = 0.5$; and (**b**) steady state error vs. v_0.

2.4.2. Elimination of Steady-State Error Using Feed-Forward Control

The control law is modified to $\{u\}_{feedforward} = \{\delta\} = -K_1 v - K_2 r - K_3 \psi - K_4 y - K_0$ in order to eliminate the steady-state error, where K_0 is chosen to ensure zero steady-state error, where the feedback gains are chosen by the *rule of thumb*, and the results are displayed in Figures 16 and 17.

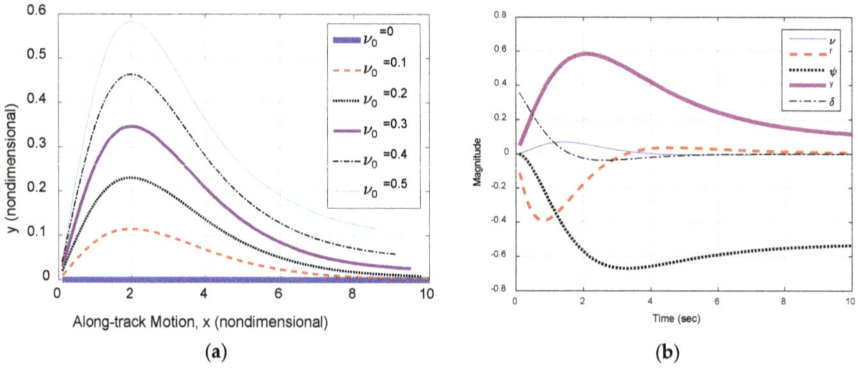

Figure 16. Feed-forward element included to counter constant lateral underwater ocean currents. (**a**) Rudder deflection, $v_0 = 0.5$; and (**b**) all states when $v_0 = 0.5$.

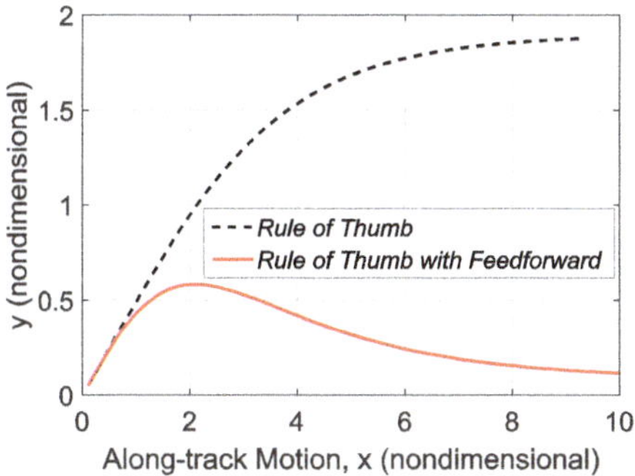

Figure 17. Comparison: feedback control with and without feed-forward ($v_0 = 0.5$).

2.5. Disturbance Estimation with Reduced-Order Observer and Integral Control

Section 2.4 demonstrated feed-forward control effectively countered the disturbance currents, but the current was presumed to be known. To be truly effective, the reduced-order observer is next augmented to include estimation of the unknown disturbance current velocity $\hat{v}_c$, where the observer now estimates the disturbance current velocity, the lateral sway velocity, v, the lateral deviation (cross-track error), y, and the heading angle, ψ. Figure 18a,b displays the estimates of the unknown current for two current velocity conditions: $\hat{v}_{c_1} = v_{est_1} = 0.1$ and $\hat{v}_{c_2} = v_{est_2} = 0.5$, respectively, while Figure 18c,d display the y and ψ states for each current velocity condition. Notice how large rudder deflections modify the heading angle to the command-tracking value which counters the disturbance current (sometimes referred to as "crabbing"), and after establishing the crab heading angle, the rudder deflection shifts towards zero, illustrating the effectiveness of command tracking.

Figure 19 displays all the states versus time in seconds and also the trajectory when a worst-case unknown disturbance current $v_c = 0.5$ is applied and estimated by the reduced-order observer where the observer gains are solutions to the linear quadratic Gaussian optimization. Meanwhile Figure 20 displays the results in cases utilizing command tracking with reduced order observer and with command: $\psi = -0.5$ and sinusoidal disturbance current $v_c(0) = A\sin(0.1t)$, but no disturbance estimation or feed-forward, while Figure 20 uses disturbance estimation, feed-forward, and rule of thumb gains. Figure 21 displays utilization of command tracking with reduced order observer, with command: $\psi = -0.5$, sinusoidal disturbance current $v_{c_0} = A\sin(0.1t)$, disturbance estimation, and feed-forward and rule of thumb gains. Lastly, Figure 22 displays the performance of reduced-order observers, which is especially useful in instances of limited at-sea computational capabilities.

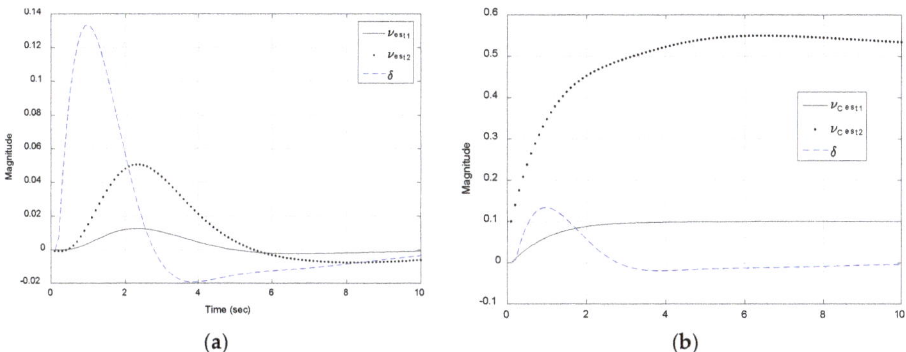

Figure 18. *Cont.*

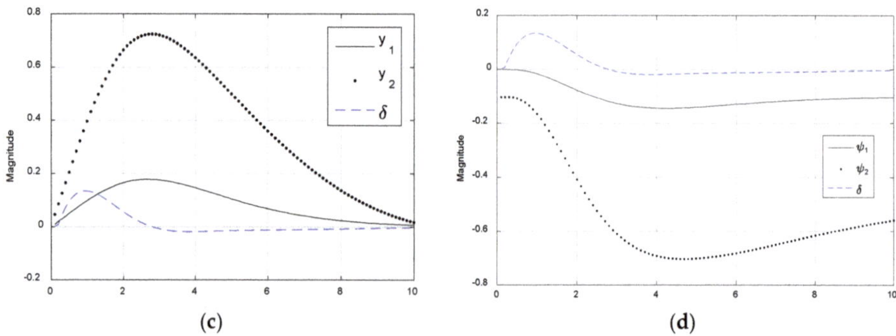

Figure 18. Reduced-order observer state estimates versus time (seconds) for two disturbance currents $v_{c_0} = \begin{bmatrix} 0.1 & 0.5 \end{bmatrix}$, where Δ is the rudder deflection using these estimates when the worst-case disturbance current is applied. (**a**) Sway velocity; (**b**) disturbance current; (**c**) lateral deviation (cross-track error); and (**d**) heading angle.

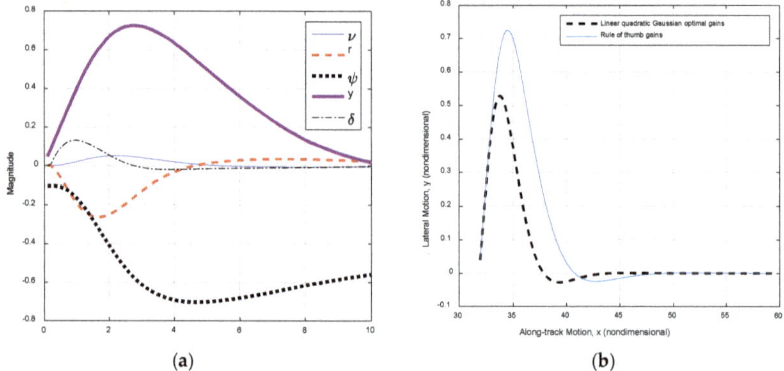

Figure 19. Performance with disturbance estimation and command tracking using LQR and *rule of thumb* gains in a reduced-order observer, and command tracking to $\psi = -0.5$ amidst a constant disturbance current $v_c = 0.5$. (**a**) States; and (**b**) trajectory.

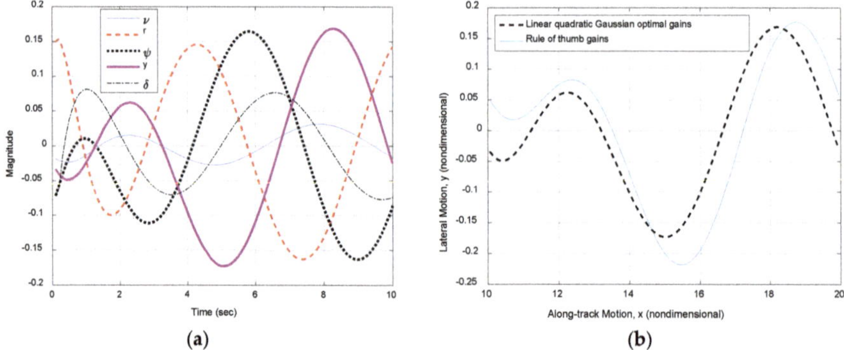

Figure 20. Utilization of command tracking with reduced order observer, with command: $\psi = -0.5$ and sinusoidal disturbance current $v_{c_0} = A sin(0.1t)$, but no disturbance estimation or feed-forward. (**a**) All states vs. time (seconds); and (**b**) trajectory.

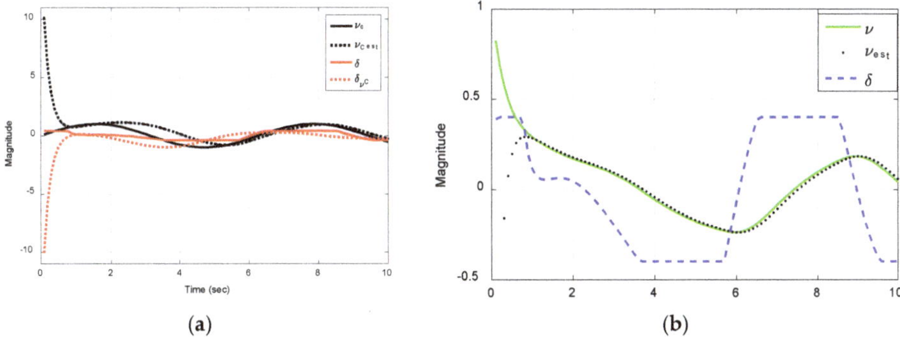

Figure 21. Utilization of command tracking with reduced order observer, with command: $\psi = -0.5$, sinusoidal disturbance current $v_{c_0} = Asin(0.1t)$, disturbance estimation, and feed-forward and rule of thumb gains.

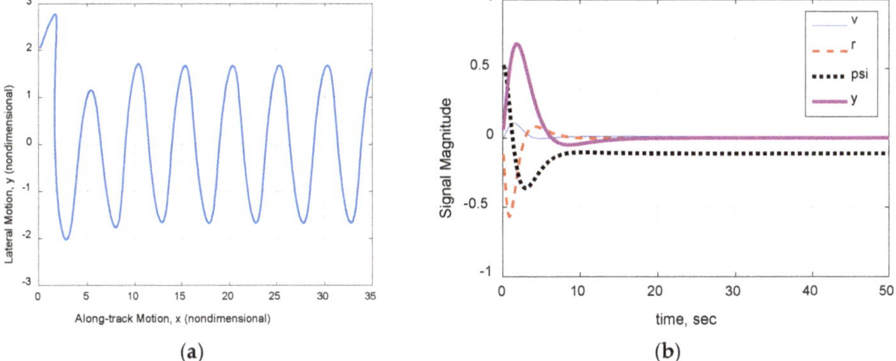

Figure 22. Utilization of command tracking with reduced order observer, with command: $\psi = -0.5$ and sinusoidal disturbance current $v_{c_0} = Asin(0.1t)$. (**a**) With disturbance estimation (and feed-forward), reduced order observer; and (**b**) with integral control, but no disturbance estimation or feed-forward.

2.6. Waypoint Guidance

A simple line-of-sight guidance routine was employed based on fixing waypoints through a minefield in order to navigate to a specified point and safely return home. The coordinates are fed to a logic determining when to turn per Equation (31), where d is the distance to the waypoint, and the heading command was autonomously calculated per Equation (32):

$$Turn \; if : \; \sqrt{(x_c - x)^2 + (y_c - y)^2} \leq d \tag{31}$$

$$\psi_{command} = K tan^{-1}\left(\frac{y_c - y}{x_c - x}\right) \tag{32}$$

Particular attention is brought to the inverse tangent calculation, since quadrants must be preserved in the calculation since the vehicle will navigate in 360 degrees.

3. Results

The following paragraphs mirror Section 2 above to provide a concise description of the simulation experiments to provide a concise and precise description of the experimental results, their interpretation, as well as the experimental conclusions that can be drawn in each sub-topic introduced and developed so far. Some new developments naturally follow in the paragraphs of the results in response to the lessons learned.

3.1. System Dynamics

Some basic lessons come from a brief analysis of the uncontrolled system dynamics. The open loop plant equations are potentially unstable (at least persistently oscillatory) with respect to only the bow rudder, while the relationship can be stable with respect to the stern rudder alone. *Can be stable* is exaggerated to emphasize the presence of pole-zero cancellation, which is an unwise practice (especially in this instance with both poles and zeros at the origin on the stability boundary) unless the estimates for the constants in the system equations are very well known. The analysis of the dynamics also revealed the bow rudder was least relatively-effective at maneuvering alone when compared to the stern rudder, however, the bow rudder does enhance vehicle maneuverability when used together with the stern rudder as a "scissored-pair" where the sign of the maneuver angle is opposite for each rudder. This "scissored-pair" constraint simplified the many-in-many-out (MIMO) control design, allowing the design engineer to treat the system as a single-in-single-out (SISO) design, since one rudder's deflection becomes a dependent variable constrained to the other rudder's deflection.

3.2. Control Law Design

Baseline proportional-derivative control designs effectively stabilized the dynamics, but were ineffective in the presence of a constant lateral open ocean current. Gains selected by *rule of thumb* performed similar to the linear-quadratic optimal control designs, so this underwater vehicle control could be designed at sea with rudimentary math in instances when higher level computational abilities are not available. Augmentation of the control including gains tuned to reject the constant current proved effective, but required the current to be measured to permit the control component to be properly tuned. Furthermore, when the lateral disturbance current had sinusoidal variation, the controller was rendered ineffective in rejecting the disturbance.

3.3. Observer Design

The submersible vehicle's system equations were verified observable by calculation of a full-ranked observability matrix in Section 2.3. A full sate observer was designed first to permit vehicle control with "full state feedback", yet without directly measuring velocity. Observer gains may be tuned using classical methods in the general spirit of duality between controller and observers. Their dual nature also permits the matrix Ricatti equation to produce optimal gains for a linear-quadratic cost function that exclusively emphasizes state estimation error, unlike the controller optimization where the cost function balanced control effort with state error. State observers permit the vehicle operator to have smooth, calculated estimates of all states at all times, which proves useful in the event of sensor interruptions or failures, and reduced-ordered observers may be used in instances where computations on-board the vehicle must be limited, for example to minimize computer size, weight, and/or power.

Especially in light of naturally occurring (roughly) sinusoidal variations in ocean current, the system equations were augmented to include the presumed-unknown disturbance as a state.

3.4. Tracking Systems and Feed-Forward Control in the Presence of Disturbance Currents

Simple feed-forward control elements proved effective against known or estimated constant lateral disturbance currents by allowing the vehicle to autonomously perform "set-and-drift" principles where a highly-trained helmsman would turn the bow of a ship into a current, but the simple feed-forward elements were ineffective at countering currents with sinusoidal variation. In the set and drift principle the heading is de facto non-zero, so the vehicle cannot simultaneously maintain center-pointing while countering the disturbance. If such a requirement were added, designers must decouple the scissored-pair rudder constraint and design the rudder commands separately to simultaneously counter the disturbance while maintaining centerline pointing.

3.5. Disturbance Estimation and Integral Control

Full-ordered observers effectively estimated constant and sinusoidal disturbance currents and proved useful in the control designs for feed-forward control, but, furthermore, reduced-ordered observers were applied in cases where disturbances were forces and moments and feed-forward control was not used. Integral control was used instead to drive the steady-state error to zero where sufficiently large time constants were used for the integrator, i.e., the fifth pole in the pole placement control must be less negative than the other poles.

3.6. Fully-Assembled System Demonstration

In light of all these results, a fully-assembled control system was used to navigate the proper mathematical models of the *Phoenix* autonomous submersible vehicle through a simulated 200 m × 500 m minefield in the presence of unknown ocean currents. The field was populated randomly with 30+ mines, and the vehicle successfully traversed the minefield in the presence of an unknown 0.5 m/s current with a miss distance from the nearest mine not less than 5 m, navigating from the starting point to pass within 0.5 m of a commanded en route point at sea, and then return to the start point. The outer loop controller used line-of-sight guidance to provide heading commands to the inner loop, and the inner loop controller was an output-feedback heading controller. Two control strategies both proved effective: linear-quadratic Gaussian, and approximate optimal pole-placement by *rule of thumb*. In the linear-quadratic Gaussian case, both the controller gains and observer gains were selected by optimization of the respective matrix Ricatti equation. Figure 23 displays the completed maneuver where each dot displays the location of a randomly-placed mine. Full state feedback was achieved with state observers via the certainly equivalence principle and the states were utilized in a proportional-derivative-integral feedback control architecture. Detailed outputs and figures of merit are plotted in Figures 24–28, including performance of a second transit of the minefield for validation purposes.

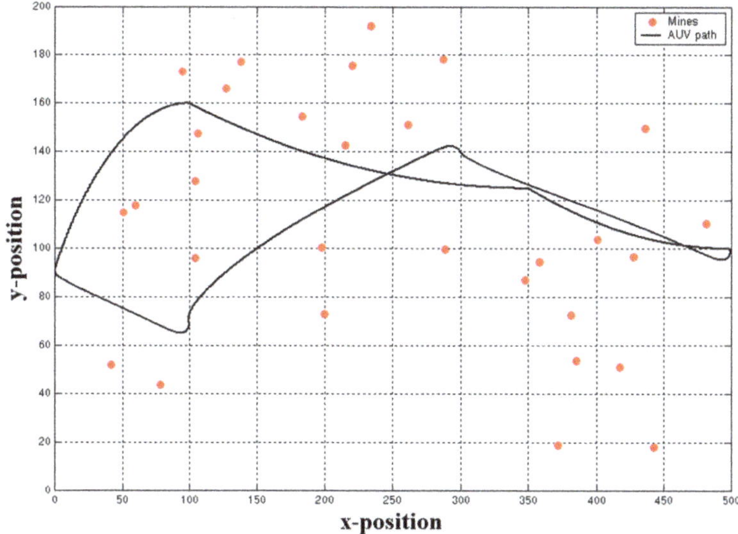

Figure 23. Navigation through simulated field of 30 randomly placed mines in −0.5 m/s current with linear quadratic Gaussian PID controller and full-state observer.

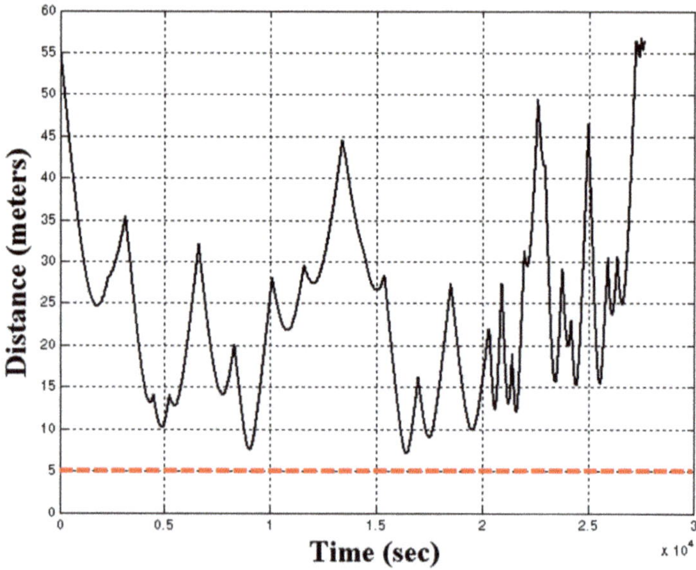

Figure 24. Continuous distance (meters) to closest mine with Linear quadratic Gaussian optimized *PID* controller and *full-state* observer versus time (seconds).

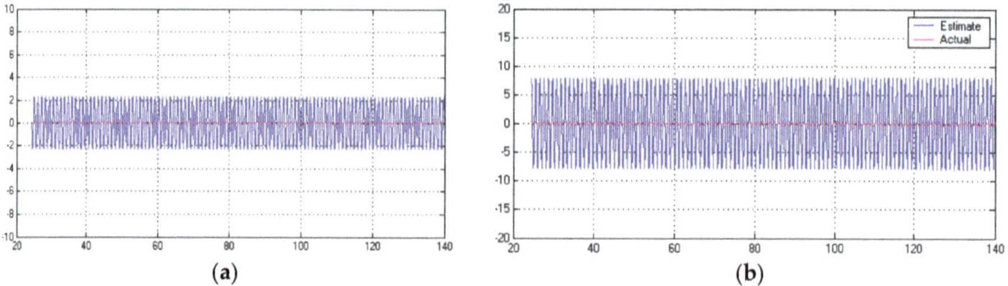

Figure 25. Linear quadratic (Gaussian) optimal observer convergence with actual value in light-pink near zero, while estimates are depicted oscillating in blue. (**a**) State v; and (**b**) state r.

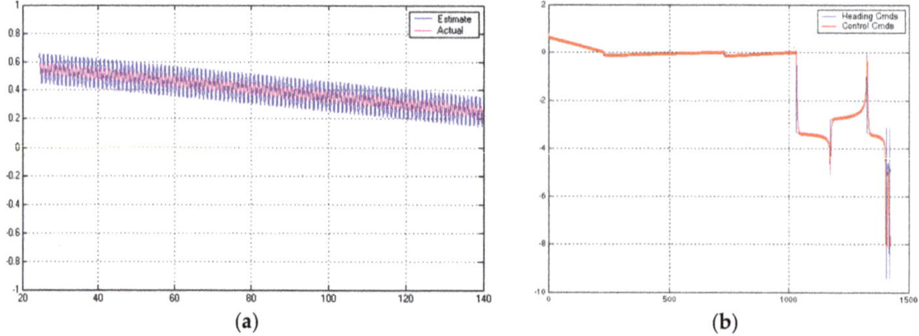

Figure 26. Linear quadratic (Gaussian) optimal observer convergence. (**a**) State ψ; and (**b**) command tracking (radians) versus time (seconds).

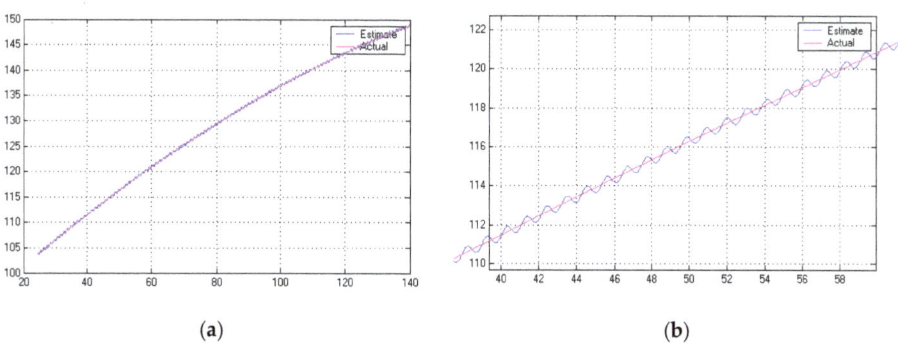

(a)   (b)

Figure 27. Linear quadratic (Gaussian) optimal observer convergence of y; (**a**) State y; (**b**) State.

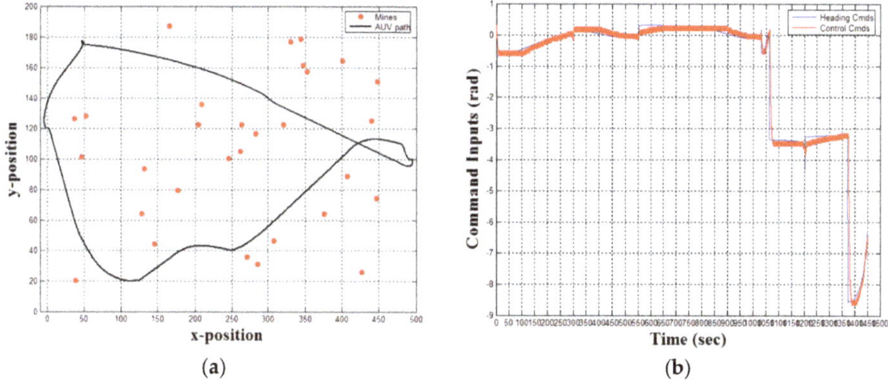

(a)   (b)

Figure 28. Validation trajectory through simulated field of 30 randomly placed mines in -0.5 m/s current with linear quadratic Gaussian optimized *PI* controller and *reduced-ordered* observer. (**a**) Second trajectory (results validation); and (**b**) heading command tracking.

4. Discussion

The results of this study establish both classical and modern control paradigms to guide autonomous submersible vehicles through obstacles in unknown ocean currents. Assuming the availability of the recent technologies cited in the Introduction, methods were investigate to use these technologies to guide a submersed vehicle along a preplanned path through a field of randomly place obstacles. The constituent technologies investigated in Section 2 were then combined in a fully-assembled system demonstration in Section 3 (Results), where the figure of merit used to assess the efficacy of the proposed methods is the maintenance of the miss distance from submersible objects in ocean currents. The conclusions from the experiments follow: both elegant and simplified autonomous controls proved effective (achieving consistent miss distance from mines greater than the goal of five meters), making this technology immediately accessible to low-end technology implementations. The results are consistent with the significant body of literature on motion mechanics in the presence of unknown disturbances with the added complication of restricted path planning due to randomly-placed obstacles, where mines were used in this study driving an additional requirement of minimum safe distance for obstacle passage. This consistency with the current literature leads to a natural direction for future research, since recent innovations in nonlinear idealized (and sometimes also adaptive) methods have recently proven to be natural extensions of technology in these fields.

These recent innovations stem from the imminent realization that the United States has been, for many years, preoccupied with low-intensity conflicts against technologically inferior opponents at a time when competitor nations have made great technological advancements in navigation and motion mechanics. The advancements in motion mechanics, as embodied in recent warhead maneuvering advancements for nuclear weapons, are exacerbated by a glaring reduction in the technical prowess of the American nuclear enterprise resulting in a very recent reinvigoration of the critical thinking abilities of the enterprise through education efforts designed to return the American enterprise to the forefront of technology in these areas. The idealized nonlinear methods are a part of the renewed education effort, and accompanying the renewed emphasis in this direction, the sequel to this manuscript should include an investigation of idealized nonlinear and adaptive methods with a direct comparison to the current state-of-the art including time-optimal control methods. The implications of the current American deficit expanded in this year's nuclear posture review indicate a steady funding source for such education in the near future.

Immediate future research will attempt to improve obstacle avoidance performance using nonlinear Feed-forward methods from the referenced literature when they show real promise. Another area of future research is the investigation of the limiting conditions of system parameter variation under which the rule-of-thumb technique remains effective, although it is assumed to be generically effective, since the system time constant method would be appropriately applied to any system equation. They key to effectiveness lies in online knowledge of the parameter variations, which is a natural extension of the nonlinear feed-forward methods to be investigated. The impetus for this research is the realization the United States has been preoccupied in the Middle East [26] and has lost the technical edge. In light of recent diplomatic failures [27] together with blatant signs of an insufficiently educated enterprise [28,29], and coupled with a resurgently aggressive defense posture [30,31], new efforts seek to elevate the critical thinking abilities of the enterprise with focused education including the topics elaborated in this manuscript amongst many others [32–36] in hopes that future enterprise members will be increasingly well prepared for highly technical defense missions [37–39]. A natural sequel to this manuscript would utilize most recent advancements in vehicle kinematics [40] in addition to the aforementioned methods ([32–36] in particular), which comprise nonlinear mathematical amplifications of the linear methods utilized here, and the impacts on critical thinking of enterprise members should be assessed after these methods are incorporated into a standard curriculum.

Author Contributions: Conceptualization by A.H., foremost, and then K.B. and T.S. afterwards; methodology by A.H.; software by K.B. and T.S.; validation by I.K., T.S., and K.B.; formal analysis by T.S. and K.B.; writing—original draft preparation by T.S.; writing—review and editing by T.S.; research supervision by I.K. Authorship has been limited to those who have contributed substantially to the work reported.

Funding: This research received no external funding. No sources of funding apply to the study. Grants were not received in support of our research work. No funds were received for covering the costs to publish in open access, instead this was an invited manuscript to the special issue.

Conflicts of Interest: The authors declare no conflict of interest.

References

1. Naval Postgraduate School Website for Consortium for Robotics and Unmanned Systems Education and Research (CRUSER). Available online: http://auvac.org/people-organizations/view/241 (accessed on 23 May 2018).
2. Brutzman, D.; Healey, A.J.; Marco, D.B.; McGhee, R.B. The Phoenix Autonomous Underwater Vehicle. In *AI Based Mobile Robots*; Kortenkamp, D., Bonasso, R.P., Murphy, R.R., Eds.; MIT/AAAI Press: Cambridge, MA, USA, 1998.
3. Naval Postgraduate School Website for Consortium for Robotics and Unmanned Systems Education and Research (CRUSER). Available online: https://my.nps.edu/web/cavr/auv (accessed on 23 May 2018).

4. Ogata, K. *Modern Control Engineering*, 4th ed.; Prentice-Hall: Upper Saddle River, NJ, USA, 2002; pp. 779–790, ISBN 0-13-060907-2.

5. Marco, D.B.; Healey, A.J. Surge Motion Parameter Identification for the NPS Phoenix AUV. In *Proceedings International Advanced Robotics Program IARP 98*; University of South Louisiana: Baton Rouge, LA, USA, 1998; pp. 197–210.

6. Kenny, T.; Sands, T. Experimental piezoelectric system identification. *J. Mech. Eng. Autom.* **2017**, *76*, 179–195. [CrossRef]

7. Sands, T. Nonlinear-adaptive mathematical system identification. *Computation* **2017**, *5*, 47–59. [CrossRef]

8. Kenny, T.; Sands, T. Experimental Sensor Characterization. *J. Space Explor.* **2017**, *7*, 140.

9. Sands, T. Space System Identification Algorithms. *J. Space Explor.* **2017**, *6*, 138.

10. Armani, C.; Sands, T. Analysis, correlation, and estimation for control of material properties. *J. Mech. Eng. Autom.* **2018**, *8*, 7–31. [CrossRef]

11. Wu, N.; Wu, C.; Ge, T.; Yang, D.; Yang, R. Pitch Channel Control of a REMUS AUV with Input Saturation and Coupling Disturbances. *Appl. Sci.* **2018**, *8*, 253. [CrossRef]

12. He, B.; Zhang, H.; Li, C.; Zhang, S.; Liang, Y.; Yan, T. Autonomous Navigation for Autonomous Underwater Vehicles Based on Information Filters and Active Sensing. *Sensors* **2011**, *11*, 10958–10980. [CrossRef] [PubMed]

13. Yan, Z.; Wang, L.; Zhang, W.; Zhou, J.; Wang, M. Polar Grid Navigation Algorithm for Unmanned Underwater Vehicles. *Sensors* **2017**, *17*, 1599. [CrossRef]

14. Wang, J.; Li, B.; Chen, L.; Li, L. A Novel Detection Method for Underwater Moving Targets by Measuring Their ELF Emissions with Inductive Sensors. *Sensors* **2017**, *17*, 1734. [CrossRef] [PubMed]

15. Eren, F.; Pe'eri, S.; Thein, M.; Rzhanov, Y.; Celikkol, B.; Swift, M.R. Position, Orientation and Velocity Detection of Unmanned Underwater Vehicles (UUVs) Using an Optical Detector Array. *Sensors* **2017**, *17*, 1741. [CrossRef] [PubMed]

16. Zhang, W.; Wei, S.; Teng, Y.; Zhang, J.; Wang, X.; Yan, Z. Dynamic Obstacle Avoidance for Unmanned Underwater Vehicles Based on an Improved Velocity Obstacle Method. *Sensors* **2017**, *17*, 2742. [CrossRef] [PubMed]

17. Kim, J.J.; Agrawal, B.; Sands, T. 2H Singularity free momentum generation with non-redundant control moment gyroscopes. In Proceedings of the 45th IEEE Conference on Decision and Control, San Diego, CA, USA, 13–15 December 2006; pp. 1551–1556. [CrossRef]

18. Kim, J.J.; Agrawal, B.; Sands, T. Control moment gyroscope singularity reduction via decoupled control. In Proceedings of the IEEE Southeastcon 2009, Atlanta, GA, USA, 5–8 March 2009; pp. 1551–1556. [CrossRef]

19. Sands, T.; Kim, J.J.; Agrawal, B. Nonredundant single-gimbaled control moment gyroscopes. *J. Guid. Control Dyn.* **2012**, *35*, 578–587. [CrossRef]

20. Kim, J.J.; Agrawal, B.; Sands, T. Experiments in Control of Rotational Mechanics. *Int. J. Autom. Control. Intell. Syst.* **2016**, *2*, 9–22.

21. Agrawal, B.; Kim, J.J.; Sands, T. Method and Apparatus for Singularity Avoidance for Control Moment Gyroscope (CMG) Systems without Using Null Motion. U.S. Patent 9,567,112, 2017.

22. Agrawal, B.; Kim, J.J.; Sands, T. Singularity Penetration with Unit Delay (SPUD). *Mathematics* **2018**, *6*, 23–38. [CrossRef]

23. Lu, D.; Chu, J.; Cheng, B.; Sands, T. Developments in angular momentum exchange. *Int. J. Aerosp. Sci.* **2018**, *6*, 1–7. [CrossRef]

24. Thornton, B.; Ura, T.; Nose, Y. Wind-up AUVs Combined energy storage and attitude control using control moment gyros. In Proceedings of the Oceans 2007, Vancouver, BC, Canada, 29 September–4 October 2007. [CrossRef]

25. Thornton, B.; Ura, T.; Nose, Y. Combined energy storage and three-axis attitude control of a gyroscopically actuated AUV. In Proceedings of the Oceans 2008, Quebec City, QC, Canada, 15–18 September 2008. [CrossRef]

26. Sands, T. Strategies for combating Islamic state. *Soc. Sci.* **2016**, *5*, 39. [CrossRef]

27. Mihalik, R.; Sands, T. Outcomes of the 2010 and 2015 nonproliferation treaty review conferences. *World J. Soc. Sci. Humanit.* **2016**, *2*, 46–51. [CrossRef]

28. Camacho, H.; Mihalik, R.; Sands, T. Education in Nuclear Deterrence and Assurance. *J. Def. Manag.* **2017**, *7*, 166. [CrossRef]

29. Mihalik, R.; Camacho, H.; Sands, T. Continuum of learning: Combining education, training and experiences. *Education* **2017**, *8*, 9–13. [CrossRef]

30. Camacho, H.; Mihalik, R.; Sands, T. Theoretical Context of the Nuclear Posture Review. *J. Soc. Sci.* **2018**, *14*, 124–128. [CrossRef]

31. Mihalik, R.; Camacho, H.; Sands, T.U.S. Nuclear Posture Review—Kahn vs. Schelling. *Soc. Sci.* **2018**, *14*, 145–154.

32. Sands, T. Physics-based control methods. In *Advances in Spacecraft Systems and Orbit Determination*; InTech: London, UK, 2012.

33. Nakatani, S.; Sands, T. Simulation of spacecraft damage tolerance and adaptive controls. In Proceedings of the 2014 IEEE Aerospace Conference, Big Sky, MT, USA, 1–8 March 2014; pp. 1–16. [CrossRef]

34. Nakatani, S.; Sands, T. Autonomous damage recovery in space. *Int. J. Autom. Control. Intell. Syst.* **2016**, *2*, 22–36.

35. Cooper, M.; Heidlauf, P.; Sands, T. Controlling Chaos—Forced van der Pol Equation. *Mathematics* **2017**, *5*, 70–80. [CrossRef]

36. Nakatani, S.; Sands, T. Battle-damage tolerant automatic controls. *Elec. Electr. Eng.* **2018**, *8*, 10–23. [CrossRef]

37. Sands, T. Satellite electronic attack of enemy air defenses. In Proceedings of the IEEE Southeastcon 2009, Atlanta, GA, USA, 5–8 March 2009; pp. 434–438. [CrossRef]

38. Sands, T. Space mission analysis and design for electromagnetic suppression of radar. *Int. J. Electromagn. Appl.* **2018**, *8*, 1–25. [CrossRef]

39. Nakatani, S.; Sands, T. Eliminating the existential threat from North Korea. *Sci. Technol.* **2018**, *8*, 11–16. [CrossRef]

40. Smeresky, B.; Rizzo, A.; Sands, T. Kinematics in the Information Age. *Mathematics* **2018**, accepted.

Journal of
*Marine Science
and Engineering*

MDPI

Article

Coupled and Decoupled Force/Motion Controllers for an Underwater Vehicle-Manipulator System

Corina Barbalata [1,*,†], **Matthew W. Dunnigan** [2] **and Yvan Petillot** [2]

[1]   Naval Architecture and Marine Engineering, University of Michigan, Ann Arbor, MI 48109, USA
[2]   School of Engineering and Physical Sciences, Heriot-Watt University, Edinburgh EH14 4AS, UK;
     m.w.dunnigan@hw.ac.uk (M.W.D.); y.r.petillot@hw.ac.uk (Y.P.)
*   Correspondence: corinaba@umich.edu; Tel.: +1-734-548-1770
†   A large part of this work was completed when Corina Barbalata was at Heriot-Watt University,
     School of Engineering and Physical Sciences, Edinburgh, UK.

Received: 10 July 2018; Accepted: 17 August 2018; Published: 21 August 2018

Abstract: Autonomous interaction with the underwater environment has increased the interest of scientists in the study of control structures for lightweight underwater vehicle-manipulator systems. This paper presents an essential comparison between two different strategies of designing control laws for a lightweight underwater vehicle-manipulator system. The first strategy aims to separately control the vehicle and the manipulator and hereafter is referred to as the decoupled approach. The second method, the coupled approach, proposes to control the system at the operational space level, treating the lightweight underwater vehicle-manipulator system as a single system. Both strategies use a parallel position/force control structure with sliding mode controllers and incorporate the mathematical model of the system. It is demonstrated that both methods are able to handle this highly non-linear system and compensate for the coupling effects between the vehicle and the manipulator. The results demonstrate the validity of the two different control strategies when the goal is located at various positions, as well as the reliable behaviour of the system when different environment stiffnesses are considered.

Keywords: underwater vehicle-manipulator system; autonomy; low-level control; position control; force control; parallel control; dynamic modelling

---

1. Introduction

In a world where only 5% of the oceans has been explored, the challenges of underwater exploration is driving the development of new technologies. The barrier of deep-water exploration, not reachable by humans, has been removed by the emergence of underwater robotics. Remotely operated vehicles (ROVs) were one of the first robotics systems used to survey underwater environments. Using artificial intelligence, autonomous underwater vehicles (AUVs) are one of the main systems being developed and constantly improved on to explore and survey deep-waters. The real challenge is to interact with the environment. To solve this a robotic manipulator is added to the underwater vehicle, the complete system being referred to as an underwater vehicle-manipulator system (UVMS).

Underwater vehicle-manipulator systems are highly complex systems, characterized by a large number of degrees-of-freedom, coupled and non-linear dynamics [1]. The dynamics of the system are highly affected by the dry mass ratio between the two subsystems that form the UVMS. In the case when the vehicle has a considerable mass with respect to the manipulator, e.g., SAUVIM UVMS [2], the effects of the manipulator motion are not significant. This is not true if the manipulator is attached to a light vehicle. This type of structure is known as a lightweight underwater vehicle-manipulator system [3]. Using this type of system presents challenges regarding the stability of the robot and simple control laws are not sufficient to perform the required tasks. While extensive research for vehicle-manipulator systems in aerial domain has been made in recent years [4–6], studies of underwater vehicle-manipulator systems are still limited due to the high operational costs and need of large infrastructure facilities.

Interaction tasks in the underwater environment using a lightweight vehicle-manipulator system are highly challenging and research in this area is slowly developing. Most of the available literature is based on classic force control approaches: impedance, hybrid or parallel control. Impedance control is based on the dynamic relationship between the position and the force variable, controlling one of them through the other [7]. Hybrid control is based on the assumption that ideal conditions are available in the robot space and the task that has to be performed by the robot can be defined in two separate orthogonal and complementary directions covering the 3D space [8]. The parallel control approach combines a motion controller and a force controller. It is claimed to increase the robustness of the force/position control as it incorporates the advantages of both the impedance and hybrid control. It is as simple and robust as the impedance control and enables the control of the position and force separately [9]. The difficulties encountered in the interaction between the UVMS and the environment include uncertainties in the UVMS model knowledge, the hydrodynamic effects, redundancy of the system, the coupling effects between the manipulator and the vehicle and the effects on the vehicle stability when interacting with the environment. Most of these challenges have been analyzed in the available literature. In [10] the authors present the impedance controller for an UVMS considering it as a single dynamic system. An adaptive impedance controller is used together with a hybrid controller in [11]. The system switches between the two controllers by using a fuzzy logic approach. The authors argue that using both types of controllers is beneficial for systems where uncertainties are present in the system.

Underwater vehicle-manipulator systems can be controlled at a low-level either in the operational space or in the vehicle/joints space. Two main categories of control strategies for complex systems can be identified. One approach investigates a separate type of controller for the vehicle and another type of controller for the manipulator. McLain et al. [12] presents a coordinated-control approach for a UVMS having a single link manipulator. A separate feedback controller for the vehicle and a different controller for the manipulator are designed and the hydrodynamic model of the manipulator is used to coordinate and reduce the effects of the manipulator on the vehicle. Vehicle control can be achieved with advance strategies such as a sliding-mode control system using a direction-based genetic algorithm and fuzzy inference mechanism [13,14]. Among the control structures specific to underwater environments Simetti et al. proposes the use of task priority control [15]. Wit et al. [16] approaches the issue of different bandwidth properties for the vehicle and manipulator. A Proportional-Derivative controller is implemented for the manipulator whose gains are limited according to the bandwidth of the manipulator. A similar problem is solved in the paper of Kim et al. [17] where the UVMS is presented as a decentralized system. A proportional vehicle controller is designed in operational space and a feedback linearised controller is used for the manipulator. A different group of control strategies for a UVMS includes a single type of controller designed for the overall system. In most cases, the control law is designed in the operational space. Antonelli et al. [18] proposes the use of a sliding-mode controller (SMC) to track a desired trajectory with the UVMS. The method is advantageous for the system as there is no need to have an exact dynamic model as the method handles uncertainties and disturbances. In [19] the authors present a comparison between an operational-space sliding mode controller and a classical Proportional-Derivative Operational Space Controller. The advantages of the SMC can be observed through the simulation results.

The contribution of this paper is the first extensive comparative analysis between the two main types (coupled and decoupled) of implementing force/position control laws for a lightweight vehicle-manipulator system. The evaluation aims to study the differences between the case when the proposed controller is applied in a centralized (coupled) strategy with the case when a descentalized (decoupled) strategy is used. The coupled strategy allows coordinated movement of the vehicle and manipulator while the decoupled strategy represents a method where movement of the manipulator is restricted during the motion of the vehicle. Moving the vehicle and manipulator simultaneously in the deacoupled strategy would be valid only if at any moment in time the object to be reached is in the workspace of the manipulator. In most real case scenarious of mobile manipulation, the object to be reached is outside of the workspace of the manipulator. This results in the impossibility of controlling just the manipulator, in a deacoupled strategy, before the vehicle brings the manipulator in an area where the object is its manipulation space. Furthermore, if simultaneous movement is desired, this would lead to a coupled control structure as presented in Section 3.2.

Details of both strategies are discussed in this paper. The performances of underwater manipulation and the area of manipulability are improved by joining together an underwater vehicle with a manipulator. The additional degrees-of-freedom of the vehicle represent an extension to the system that can be used to compensate for the oscillations and disturbances caused by the underwater environment while maintaining good end-effector pose keeping. This paper aims to use these benefits of the UVMS system for the coupled strategy, while for the decoupled strategy the systems are considered separate and the coupling effects are considered as disturbances. The low-level controller used in both strategies combines the theory of sliding mode control for force regulation and the integrative sliding mode control for position regulation in a parallel implementation. The method is robust to disturbances by incorporating the dynamic model of the underwater vehicle-manipulator system in the control architecture. Reliable and efficient UVMSs are not currently available for performing underwater tasks such as probe sampling and maintenance of underwater oil and gas platforms. Research in this field is scarce mostly due to high costs of developing and deploying these type of systems. Before experimental testing takes place with an UVMS, the concepts have to be tested and analysed based on a simulation environment. It is important to demonstrate the validity of the proposed methods in simulation as this step is essential in reducing the probability of failures and damaging the system. Furthermore, the authors argue that it is important to understand the benefits and limitations of the coupled and decoupled control strategies to be able to choose the appropriate approach based on the application the system has to perform.

The mathematical model of the UVMS is described in Section 2, followed by the presentations of the coupled and the decoupled control strategies in Section 3. A comparative evaluation of the two methods is presented in Section 4 and a discussion based on the comparative simulation results is made in Section 5. All the mathematical symbols used in this paper are listed in Appendix A.

2. System Model

In this section the underwater vehicle-manipulator system is presented, including the characteristics of the system, the kinematic relationship between different coordinate frames and the dynamic model used to describe the UVMS.

Multiple coordinate systems are available to represent the UVMS: vehicle (body) coordinates, joint coordinates, end-effector (operational/task space) coordinates and earth-fixed inertial coordinates. The kineamtic chain of the proposed UVMS is presented in Figure 1.

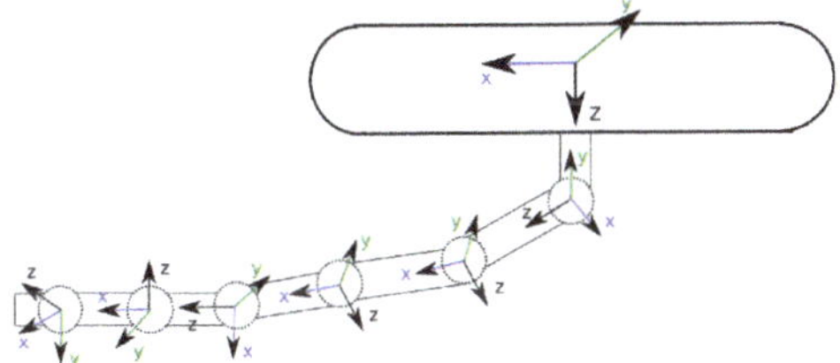

Figure 1. Kineamtic tree of the vehicle-manipulator system.

The system position is defined based on the manipulator joint position vector $q = [q_1 \cdots q_n]^T$, vehicle position $\eta_1 = [x, y, z,]^T$ and vehicle orientation $\eta_2 = [\phi, \theta, \psi,]^T$. Using the generalized coordinates of the system, $\rho = [\eta_1, \eta_2, q]^T$, the end-effector position and orientation $x_E^I \in \mathbb{R}^6$ with respect to the inertial frame is given by:

$$x_E^I = f(\rho) \tag{1}$$

where $f(\rho)$ is the general transformation dependent on the pose of the vehicle and joint positions. A single chain-representation is used to describe the dynamic model of the underwater-vehicle manipulator based on the work presented in [20]. This representation is characterised by considering the vehicle to be a part of the manipulator, an extra link with 6 DOFs: 3 prismatic joints and 3 revolute joints. A recursive implementation of the dynamic system as presented in [21] is used to compute the dynamic model of the system. The classical principle of Newton-Euler that transmits the velocities and forces between subsystem is used to calculate the bias forces. The Composite Rigid Body Algorithm is implemented for determining the inertia matrix of the system. A rigorous analysis of the hydrodynamic effects is performed and the approximated mathematical forces are included into the computation of the dynamic model. A study of the dynamic and hydrodynamic parameters has been made to accurately represent the underwater vehicle-manipulator system. This detailed description of the UVMS model and the corresponding parameters were previously presented in [21].

The dynamics of the UVMS can be further described in a matrix form by the following equation:

$$M(\rho)\dot{\zeta} + C(\rho, \zeta)\zeta + D(\rho, \zeta)\zeta + g(\rho) = \tau - J^T F \tag{2}$$

where $M(\rho) \in \mathbb{R}^{n \times n}$ is the inertia matrix, $C(\rho, \zeta)\zeta \in \mathbb{R}^n$ is the Coriolis and Centripetal vector, both consisting of rigid body terms and added mass terms, $D(\rho, \zeta)\zeta \in \mathbb{R}^n$ is the damping and lift forces vector, $g(\rho) \in \mathbb{R}^n$ represents the restoring forces, $\tau = [\tau_v, \tau_m] \in \mathbb{R}^n$ is the vector of total forces and moments applied to the vehicle $\tau_v \in \mathbb{R}^l$ and the torques applied to the manipulator $\tau_m \in \mathbb{R}^m$, $J \in \mathbb{R}^{6 \times n}$ is the Jacobian of the system, n is the total number of degrees-of-freedom of the system, l is the number of DOFs of the vehicle and m is the number of DOFs of the manipulator. $\tilde{F} \in \mathbb{R}^6$ is the external disturbance vector produced by the interaction with the environment, modelled by Equation (3).

$$F = K_e(x - x_e) \tag{3}$$

where $x_e \in \mathbb{R}^6$ is the point of a plane at rest, $x \in \mathbb{R}^6$ is the end-effector position and $K_e \in \mathbb{R}^{6 \times 6}$ is the stiffness matrix of the environment [22].

In mobile manipulation the tasks to be solved are naturally expressed in task space coordinates. The dynamic description of the system in operational space can be described by Equation (4) [8], where $x \in \mathbb{R}^6$ represents the independent parameters vector described in the operational space.

$$M(x)\ddot{x} + C(x)\dot{x} + D(x)\dot{x} + G(x) = T - F \tag{4}$$

where $M(x) \in \mathbb{R}^{6 \times 6}$ is a positive operational space inertia matrix, $C(x)\dot{x} \in \mathbb{R}^6$ is the vector of Coriolis and Centripetal forces, $D(x)\dot{x} \in \mathbb{R}^6$ is the damping vector, $G(x) \in \mathbb{R}^6$ is the vector of restoring forces, all defined in operational space coordinates and $T \in \mathbb{R}^6$ is the vector of generalized forces at the end-effector.

3. UVMS Position/Force Control Strategies

An underwater vehicle-manipulator system can be considered either as consisting of two separate parts where the coupling effect between the subsystems is seen as an external disturbance or as a single and unique system. For each of these representations a different design strategy for the low-level control structure can be employed referred to as the *decoupled* strategy and the *coupled* strategy. In this section the two strategies are detailed for position/force applications where the system first has to navigate to a goal and then the manipulator has to interact with the environment.

3.1. The Decoupled Strategy

In the decoupled approach the goal is to control the vehicle and the manipulator separately so that each of the components of the system performs the desired task. In most real-world applications, the underwater-vehicle manipulator system has to interact with the environment at a location outside of the workspace of the manipulator. To design an independent control structure for the manipulator, the goal has to be at all times in its manipulation space. For the proposed decoupled strategy this is translated to sending separate commands for the manipulator and vehicle, leading to a sequential movement of the two components.

The main focus of this work is on the control of lightweight vehicle-manipulator systems where the effects of the manipulator movement on the vehicle behaviour is significant as it was previously demonstrated in [21]. These interaction effects are considered as disturbances in the decoupled strategy and a reliable and good vehicle controller has to be employed to solve this challenge. In this case a separate controller is designed for the vehicle and another controller is needed for the manipulator. A separate task has to be defined for each part of the system at every time step. This leads to integrating a high-level component that decomposes the main task in separate tasks for the vehicle and the manipulator. A schematic representation of the overall strategy is presented in Figure 2.

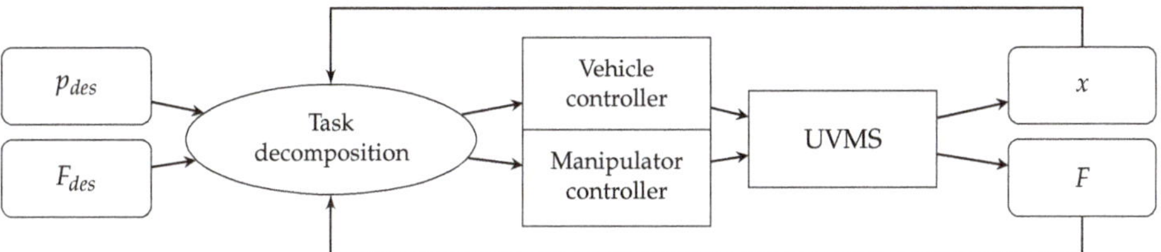

Figure 2. The *decoupled* strategy.

The task decomposition component of the decoupled strategy is responsible for defining the mission for each of the components of the UVMS. The current approach is based on the Euclidian distance between the center of mass of the vehicle and the final location where the system has to reach and interact with the environment. The distance between the two components is computed at every time step and if it is larger than the total length of the manipulator, the vehicle is required to move and the manipulator is required to be in station keeping mode. At the moment when the vehicle is close enough to the object the vehicle is in station keeping mode and the manipulator is commanded to move and interact with the object with the desired force. The approach is summarized in Algorithm 1, where p_{des} is the desired (object) position, x_v is the vehicle position, x_{des_v} is the vehicle desired position, x_{des_m} is the manipulator desired position, x_m is the current manipulation position, p_{des_i}, is the i-th component of the desired goal location ($i = 1, 2, 3$), x_{v_i} is the i-th component of the position of the vehicle ($i = 1, 2, 3$) and L is the threshold that decides if either the vehicle or the manipulator should move. All the notations and the overall strategy are defined in world coordinates. After defining the task decomposition the two low-level controllers are presented.

Algorithm 1 Task decomposition.

$$d(p_{des}, x_v) = \sqrt{\sum_{i=1}^{n}(p_{des_i} - x_{v_i})^2}$$

2: **if** $d(p_{des}, x_v) \geq L$ **then**

$\quad\quad x_{des_v} = p_{des}$

4: $\quad\quad x_{des_m} = x_m$

$\quad\quad$ **else**

6: $\quad\quad x_{des_v} = x_v$

$\quad\quad x_{des_m} = p_{des}$

8: **end if**

3.1.1. Vehicle Controller

A robust but simple and efficient strategy is used in the *decoupled* approach to control the degrees-of-freedom of the vehicle. The controller is introduced in detail in [23]. Considering that for underwater environments the operational speed of the AUV is relatively small leads to the assumption that each degree-of-freedom of the vehicle can be independently controlled. Nevertheless, in the case when a manipulator is added to a lightweight AUV the effects of the manipulator are noticeable on the DOFs of the vehicle. The controller has to be powerful enough to handle these coupling effects.

The vehicle controller, PILIM (Proportional-Integrative LIMited) is designed using two control loops one for position and one for velocity. The position controller is based on the error in position of the vehicle $e_{p_v} \in \mathbb{R}^6$ and the positive definite matrix of proportional gain $K_{p_p} \in \mathbb{R}^{6\times6}$. The velocity loop takes into account the error in vehicle velocity $e_{v_v} \in \mathbb{R}^6$ and has two components one proportional and another integrative with positive definite gain matrices $K_{p_v}, K_{i_v} \in \mathbb{R}^{6\times6}$. The system is characterized by the following equations:

$$
\begin{aligned}
e_{p_v} &= J_v^{\dagger}\left(x_{des_v} - x_v\right) \\
u_{p_v} &= K_{p_p}e_{p_v} \\
e_{v_v} &= \dot{x}_v - u_{p_v} \\
\tau_v &= K_{p_v}e_{v_v} + K_{i_v}\int_0^t e_{v_v}d\tau \\
\tau_v &= \mathbf{sat}(\tau_v, -l, l)
\end{aligned}
\tag{5}
$$

where $\dot{x}_v \in \mathbb{R}^6$ is the velocity of the vehicle, $x_{des_v} \in \mathbb{R}^6$ is the desired position, $u_{p_v} \in \mathbb{R}^6$ is the output forces and moments of the position loop and $\tau_v \in \mathbb{R}^6$ is the control output for the vehicle, taking into account the saturation limits $\pm l$. $J_v^{\dagger} \in \mathbb{R}^{6\times6}$ is the pseudo-inverse of the Jacobian matrix for the manipulator.

3.1.2. Manipulator Controller

The task that the system has to solve is a motion/force task where the system has to interact with the environment. A parallel force/position control law based on the sliding mode control (SMC) theory [24] defined in the operational space coordinates is proposed as presented in [25]. A parallel controller develops separate and independent control laws for position and force compensation. The two components are merged together as shown in Equation (6).

$$\tau_m = J_m^T\left(u_p + u_f\right) \tag{6}$$

where $u_p \in \mathbb{R}^6$ is the control signal from the position controller and $u_f \in \mathbb{R}^6$ is the control signal from the force controller. $J_m \in \mathbb{R}^{6\times m}$ is the Jacobian of the manipulator. The two components are presented in the following lines. According to Utkin [26] an Integral Sliding Mode Controller is able to maintain the order of the compensated system dynamics in the sliding mode, being advantageous where uncertainties, coupling effects and parameter variations are present in the system. This control structure consists of two sliding mode variables: one sliding variable accounts for the bounded disturbances and another sliding variable is responsible with driving the sliding dynamics to zero. Based on this theory, this paper presents an integral sliding mode controller for position regulation and a classical sliding mode control for force regulation.

The primary sliding mode variable for the position controller is defined by Equation (7) and its corresponding control component is defined in Equation (8).

$$\sigma = \dot{e}_p + c_1 e_p + c_2 \int_0^t e_p d\tau, \quad c_1, c_2 > 0 \tag{7}$$

$$u_1 = kM^{-1}(x)\sigma, \quad k > 0 \tag{8}$$

where e_p is the error in the end-effector position, $e_p = p_{des} - x$, defined based on the difference between the desired position $p_{des} \in \mathbb{R}^6$ and the current end-effector position $x \in \mathbb{R}^6$. $M^{-1}(x)$ is the inverse of the manipulator's inertia matrix in end-effector coordinates and $k \in \mathbb{R}^{6\times6}$ is a positive matrix. The auxiliary sliding variable for the position controller is designed by Equation (9) and the corresponding control component is defined by Equation (10).

$$\begin{cases} s = \sigma - z \\ \dot{z} = M(x)u_1 \end{cases} \tag{9}$$

$$u_2 = \rho_1 \mathbf{sign}(s) \tag{10}$$

The primary sliding mode controller is responsible to compensate for disturbances and uncertainties in the system while the auxiliary control law is responsible for driving to zero in finite time the position error. The total control law for position is:

$$u_p = u_1 + u_2 = kM^{-1}(x)\sigma + \rho_1 \mathbf{sign}(s) \tag{11}$$

For the force controller a sliding mode control law is chosen. The sliding variable is expressed by Equation (12).

$$\delta = c_3 e_f + \int_0^t e_f d\tau, \quad c_3 > 0 \tag{12}$$

where $c_3 \in \mathbb{R}^{6\times6}$ is a positive matrix and e_f is the force error defined as $e_f = F_{des} - F$. Choosing the control law as presented by Equation (13) drives $\delta \to 0$ in a finite time.

$$u_f = \rho_2 \mathbf{sign}(\delta) \tag{13}$$

where ρ_2 is the control gain.

The manipulator position/force controller is described by:

$$\tau_m = J_m^T(\rho_1 \mathbf{sign}(s) + kM^{-1}(x)\sigma + \rho_2 \mathbf{sign}(\delta)) \tag{14}$$

3.2. The Coupled Strategy

The coupled controller considers the UVMS as a unique system and the same type of controller is used for all degrees-of-freedom. Similar to the decoupled controller, a parallel force/position law with sliding mode dynamics is used for the coupled strategy:

$$\tau_x = u_p + u_f \tag{15}$$

where $\tau_x \in \mathbb{R}^6$ is the total control force in operational space. The position control law, u_p, is based on the integral sliding mode controller and is presented in Equation (16). The details of this formulation are described in Section 3.1.2.

$$u_p = kM^{-1}(x)\sigma + \rho_1 \mathrm{sign}(s) \tag{16}$$

where $M(x) \in \mathbb{R}^{6\times6}$ represents the operational space inertia matrix considering the full vehicle-manipulator system and k, σ, ρ_1, $\mathrm{sign}(s)$ are the sliding mode parameters defined in Section 3.1.2. The force control law, u_f, described through the sliding mode control law is defined based on Equation (13) from Section 3.1.2. The sliding mode controllers for the position and force regulation used for full UVMS control are responsible in handling the uncertainties in the system. Nevertheless, the coupling effects are significant and they do have to be considered in the controller implementation. To remove these coupling effects a feedback-linearisation technique is incorporated together with the parallel position/force controller. For the system defined by Equation (4), the feedback linearization control structure [27,28] is defined by:

$$T_x = M(x)\left[\ddot{p}_{des} + \tau_x\right] + C(x,\dot{x})\dot{x} + D(x,\dot{x})\dot{x} + G(x) \tag{17}$$

In this case the coupling effects between the vehicle and manipulator are incorporated in the dynamic model and have an active role in the control strategy. To compute the control forces and torques at vehicle and manipulator level the transformation from operational space to joint space is obtained using:

$$\tau = \bar{J}^T T_x \tag{18}$$

where $\bar{J}$ is the weighted Jacobian of the UVMS.

The final control law in the joint space is defined by Equation (19).

$$\tau = \bar{J}^T \left\{ \tilde{M}(x)\left[\ddot{p}_{des} + u_p + u_f\right] + \alpha \right\} \tag{19}$$

where:

$$
\begin{aligned}
u_p &= \rho_1\mathrm{sign}(s) + kM^{-1}(x)\sigma, \quad \rho_1 > 0,\ k > 0\\
u_f &= \rho_2\mathrm{sign}(\delta), \quad \rho_2 > 0\\
\sigma &= \dot{e}_p + c_1 e_p + c_2 \int_0^t e_p dt, \quad c_1, c_2 > 0\\
s &= \sigma - z\\
\dot{z} &= M(x)u_2\\
\delta &= c_3 e_f + \int_0^t e_f dt, \quad c_3 > 0\\
\alpha &= \tilde{C}(x,\dot{x})\dot{x} + \tilde{D}(x,\dot{x})\dot{x} + \tilde{G}(x)
\end{aligned}
\tag{20}
$$

$\tilde{M}(x)$ is an estimate of the inertia term, $\tilde{C}(x,\dot{x})$, $\tilde{D}(x,\dot{x})$, $\tilde{G}(x)$ are estimates of the real values of the system defined in operational space coordinates according to the boundary errors. As mentioned in Section 2 in an UVMS mathematical model it is really difficult to have accurate estimates, uncertainties in the model and unmodelled disturbances are always present. This is the main reason that the estimates of the parameters of the system are used in the control law presented in Equation (19). By not matching exactly the model used for the simulation of the UVMS and the model used in the control strategy a realistic environment is created, compensating for unmodelled dynamics and underwater currents.

In both the coupled and decoupled system the chattering effect in the sliding mode controllers tend to affect the behaviour of the system. To remove the chattering effect, a continuous/smooth control function should be designed. This leads to approximate the discontinuous function by a smooth function by replacing the `sign` function with the sigmoid function.

4. Simulation Results

The evaluation of the two strategies is presented through the simulation results highlighting the benefits and drawbacks of both methods. The core analysis is on the two different strategies (coupled vs. decoupled).

The simulation environment, Figure 3, is based on an accurate model of the two robotics systems available in the Ocean Systems Laboratory: Nessie VII an autonomous underwater vehicle developed as a research platform and a commercially available underwater manipulator, HDT-MK3-M. Nessie VII AUV is a torpedo shaped 5 degrees-of-freedom vehicle with a mass of 60 kg, a length of 1.1 m and a diameter of 0.15 m. The vehicle nominal velocity in the translational degrees of freedom is 1 m/s and the angular velocity of the vehicle is 0.5 rad/s. The vehicle *roll* degree-of-freedom is not controlled. The manipulator has 6 revolute joints and a total weight of 9 kg. The length of the extended arm is 0.8 m and the radius of each link is 0.07 m. The manipulator maximum joint velocity for each degree-of-freedom is 0.75 rad/s. The system represents a lightweight UVMS characterized by significant effects on the vehicle caused by the manipulator movement.

Figure 3. UVMS model.

The system is implemented in Python 2.7.0 using the mathematical model developed in Section 2 and the control strategies presented in Section 3. The working frequency for the two systems is 10 Hz, both the vehicle and the manipulator having approximately the same bandwidth. This corresponds to the real hardware systems: both the HDT-MK-3 and Nessie VII cannot be commanded at a higher frequency. As the proposed system operates at low speeds and a reliable model of the system is available, the proposed working frequency is sufficient to generate appropriate control commands. The control parameters used in the implementation of the controller are given in the Appendix B of the paper.

The problem to be solved is described by the following scenario: the UVMS has to reach a desired object in the underwater environment and interact with it at a predefined force. A compliant, frictionless point-contact at the x-axis of the end-effector describes the interaction force between the end-effector and the object. When the desired final goal is reached the interaction should take place. In these simulations different objects with a variety of stiffness coefficients are used to validate the behaviour of the system. The translational degrees-of-freedom: x, y and z axes of the end-effector are under control. The world coordinates are represented at the point of contact between the base of the manipulator and the vehicle. The end-effector initial position is at $p_{init} = (0.0,\ 0.0,\ 0.97)$ m.

The end-effector trajectory between the initial position and object location is described by a cycloid function as defined in Equation (21). The function represents an interpolation for the position $p_{des}(t)$ starting with an initial value $p_{init}(0)$ and a desired final value $p_{des}(t_f)$.

$$p_{des}(t) = p_{init}(0) + \Delta/2\pi[\omega t - \sin(\omega t)] \tag{21}$$

where

$$\omega = 2\pi/t_f, \quad \Delta = p_{des}(t_f) - p_{init}(0), \quad 0 \leq t \leq t_f$$

4.1. Flexible Environments

In the first testing scenario it is desired to interact with an object that has a $K_e = 10^3$ N/m stiffness coefficient. This stiffness coefficient corresponds to flexible environments, an example of this can be a rubber ball as the one presented in Figure 4a.

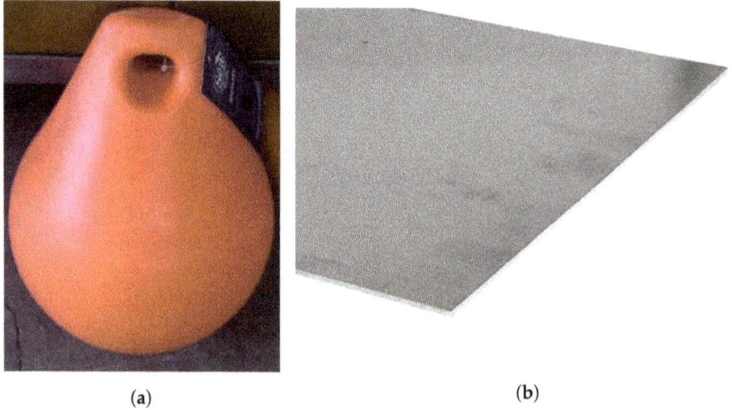

(a)   (b)

Figure 4. Objects representative for experimental set-up. (**a**) Rubber ball [29], (**b**) Aluminum plate [30].

In this case the final goal is (4.0, 2.0, −3.0) m and the desired interaction force is $F_{des} = 100$ N. The end-effector trajectory tracking is presented in Figure 5. The desired trajectory starting point is (0.0, 0.0, 0.0) m in world coordinates that represents the point where the manipulator is attached to the vehicle. Starting at this point, the end-effector is commanded to move towards the goal. The decoupled and coupled strategy behaviour can be observed based on the 3D trajectory of the end-effector.

By analysing Figures 6 and 7a clearer analysis of the behaviour of the end-effector axes can be made. The results are given in the world coordinate system and present the behaviour of the two strategies. Due to the mobile platform the end-effector is able to reach a goal that is placed outside of its fixed workspace. The decoupled approach computes separate goals for the vehicle and manipulator at each time step based on the current location of the vehicle with respect to the goal. This creates a sudden change in the requested end-effector trajectory, Figure 6c, and is responsible in creating a slower trajectory generation. This represents the main difference in the system behaviour compared with the coupled approach. In the decoupled case, the nature of the system is described by the following behaviour: the vehicle is under control while the manipulator is in station keeping mode until the system reaches the vicinity of the object which the system has to interact with. At the moment when the object is reached the vehicle is in station keeping mode while the manipulator is commanded to move and interact with the object. Using this strategy the end-effector passes beyond the object before it is commanded to move and this causes a different behaviour on the z-axis in the decoupled strategy compared with the coupled approach. The sudden jump in the z-trajectory is caused by the selection of the threshold in the decoupled approach. During the vehicle movement, the manipulator is kept in the same configuration as the one seen in Figure 3. If the vehicle overshoots in z-axis at the moment when the manipulator starts to move, the trajectory generation module will ask the manipulator to compensate for this overshoot, hence the sudden z-axis jump in the case of the decoupled approach. Nevertheless, using any of the two strategies, the end-effector trajectory is accurately followed and reliable behaviour is obtained. In the coupled approach, using a single model based controller that handles uncertainties presents good results and the output is comparable with the case when specific controllers are used for each of the systems.

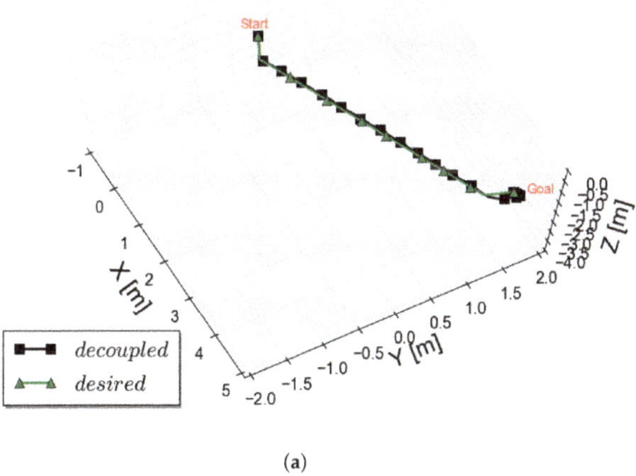

(**a**)

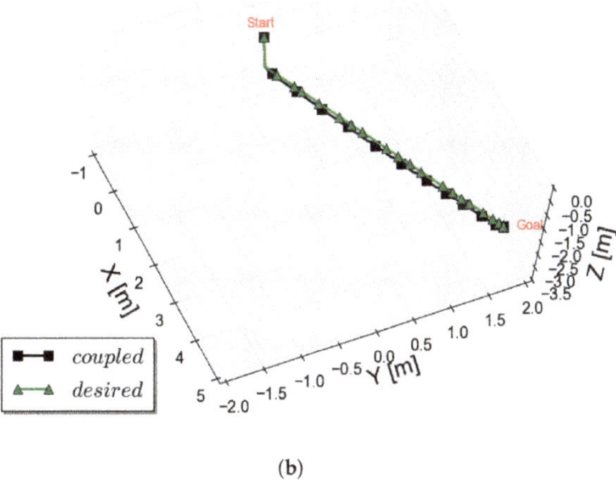

(**b**)

Figure 5. End-effector 3D position for goal at $(x, y, z) = (4.0, 2.0, -3.0)$ m. (**a**) Decoupled strategy, (**b**) Coupled strategy.

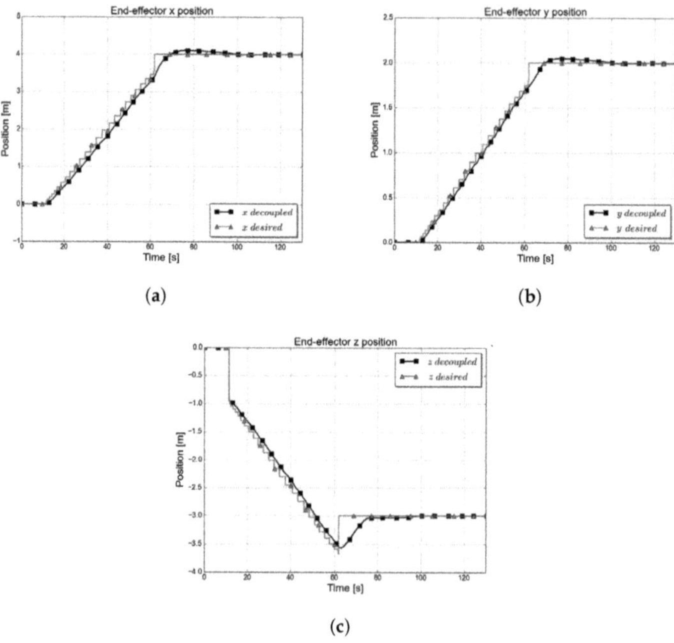

Figure 6. Decoupled strategy UVMS end-effector position tracking, goal at $(x, y, z) = (4.0, 2.0, -3.0)$ m and $K_e = 10^3$ N/m. (**a**) x-position decoupled, (**b**) y-position decoupled, (**c**) z-position decoupled.

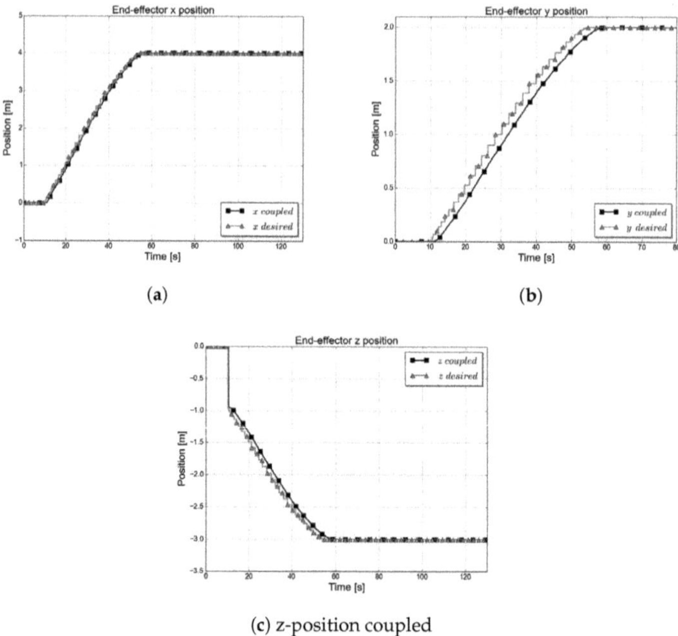

Figure 7. Coupled strategy UVMS end-effector position tracking, goal at $(x, y, z) = (4.0, 2.0, -3.0)$ m and $K_e = 10^3$ N/m. (**a**) x-position coupled, (**b**) y-position coupled, (**c**) z-position coupled.

The interaction with the environment, Figure 8, takes place as soon as the end-effector is within one centimetre of the centre of the object. In the decoupled strategy it can be noticed that the contact with the environment starts at around 90 s, although the system has reached the vicinity of the goal in a similar time as in the coupled case (60 s). This 30 s gap is explained by the overshoot and large time to obtain zero steady-state error. The settling time is large in this approach due to the fact that for a very short time both the vehicle and the manipulator are moving towards the goal. This is caused by a small overshoot in the response when the vehicle controller is used, leading to an overshoot in the overall behaviour of the system.

The desired force is achieved and maintained using both strategies. Small oscillations can be visible in the decoupled approach. For the coupled case an initial larger overshoot is observed but smaller oscillations are seen in Figure 8b. At the moment when contact with the environment takes place, the manipulator compensates for the force and is trying not to lose position while maintaining contact. This will drive the manipulator to apply a larger force that results in an overshoot.

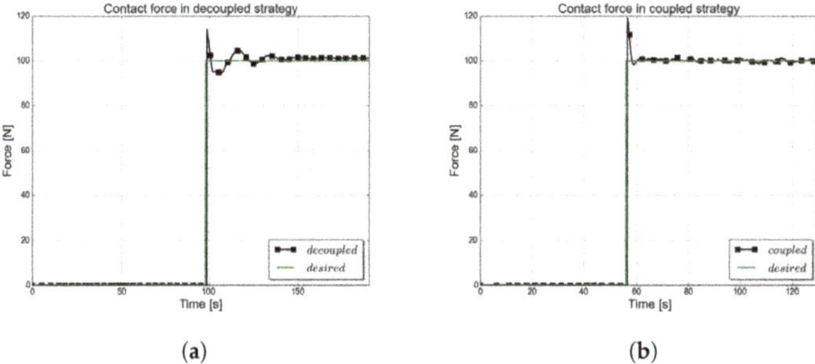

(a) (b)

Figure 8. Interaction with the environment when goal is at $(x, y, z) = (4.0, 2.0, -3.0)$ m and $K_e = 10^3$ N/m. (**a**) Decoupled strategy, (**b**) Coupled strategy.

4.2. Stiff Environments

The second set of experiments presents the interaction between the end-effector and an object with a higher stiffness coefficient, $K_e = 10^5$ N/m. This stiffness coefficient corresponds to an aluminum plate sheet as the one presented in Figure 4b. Based on the results in Figure 9 it can be observed that the overshoot in the force response increases with the stiffness of the environment. As an integral sliding mode controller is used for the position component, this has priority over the force component and the end-effector position is maintained leading to the overshoot in the force behaviour. Increasing the stiffness of the environment is an additional factor that contributes to the large overshoot.

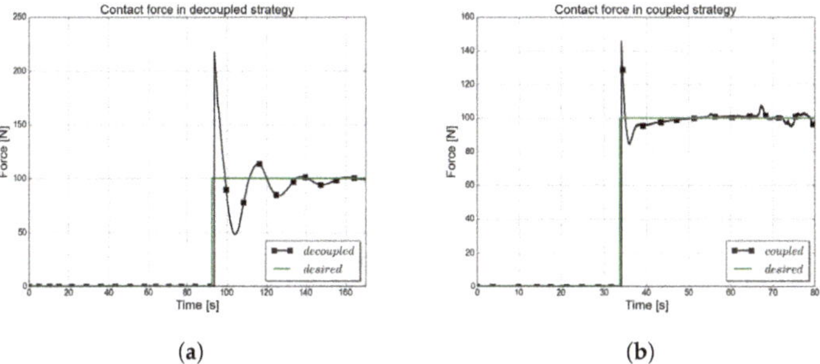

(a) (b)

Figure 9. Interaction with the environment when goal is at $(x, y, z) = (2.0, 0.0, -2.0)$ m and $K_e = 10^5$ N/m. (**a**) Decoupled strategy, (**b**) Coupled strategy.

The desired location where contact with an object should take place is set at $(2.0, 0.0, -2.0)$ m. Looking at the end-effector trajectory tracking, Figure 10, before the system reaches a steady-state, an overshoot is present in the x and y axes. These are caused by using two different controllers for the system. At the moment when the end-effector interacts with a stiff environment large forces affect the system. In the case when the vehicle and manipulator are controlled separately, the vehicle receives these large disturbances and reacts to them by generating a larger control command that will keep the vehicle's position. Furthermore, the manipulator controller tries to enforce zero steady-state error in position due to the integral term in the control loop and generates large torques. These two elements combined lead to the overshoot in the decoupled strategy behaviour. The coupled approach, Figure 11, similarly tries to enforce zero-steady state but the large control forces are distributed across the whole system taking into account the full UVMS. Overshoot is present only on a single axis.

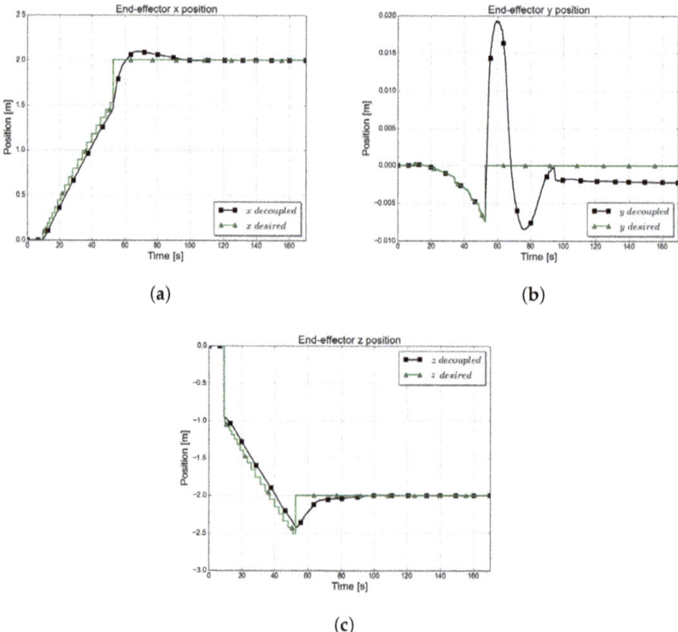

Figure 10. Decoupled strategy UVMS end-effector position tracking for goal at $(x, y, z) = (2.0, 0.0, -2.0)$ m. (**a**) x-position decoupled, (**b**) y-position decoupled, (**c**) z-position decoupled.

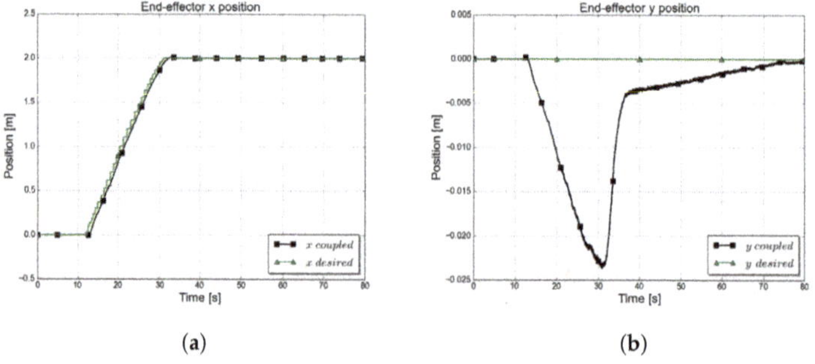

Figure 11. *Cont.*

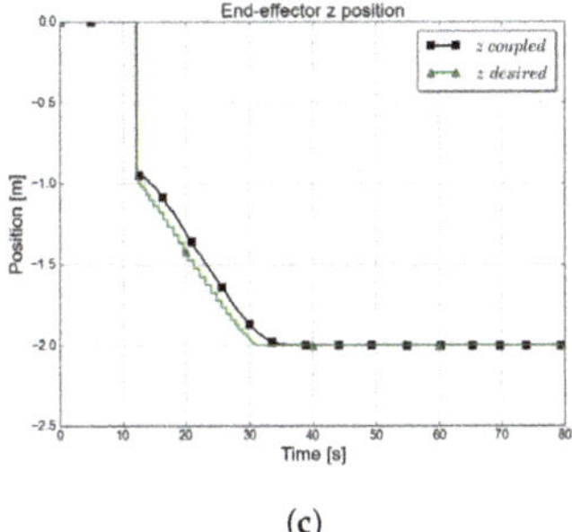

(**c**)

Figure 11. Coupled strategy UVMS end-effector position tracking for goal at $(x, y, z) = (2.0, 0.0, -2.0)$ m. (**a**) x-position coupled, (**b**) y-position coupled, (**c**) z-position coupled.

In Figure 12 the 3D behaviour is presented for the case when the final goal is $(2.0, 0.0, -2.0)$ m. The goal is reached using any of the two strategies with slightly different behaviour being observed. The coupled strategy is a straightforward approach as the desired trajectory is generated dependent on the end-effector location. In the decoupled strategy the trajectory is generated taking into account the position of the vehicle and the configuration of the manipulator.

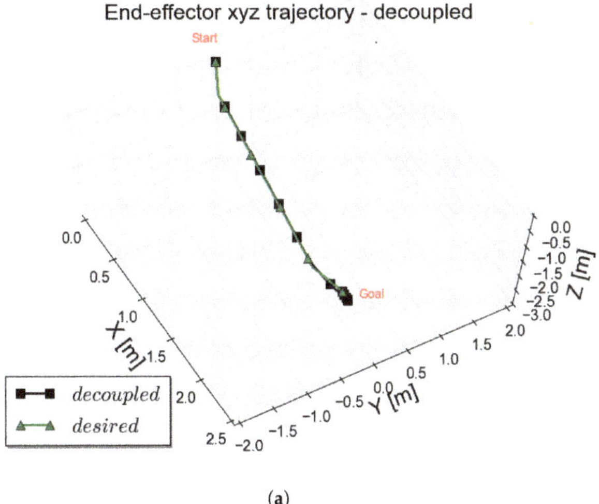

(**a**)

Figure 12. *Cont.*

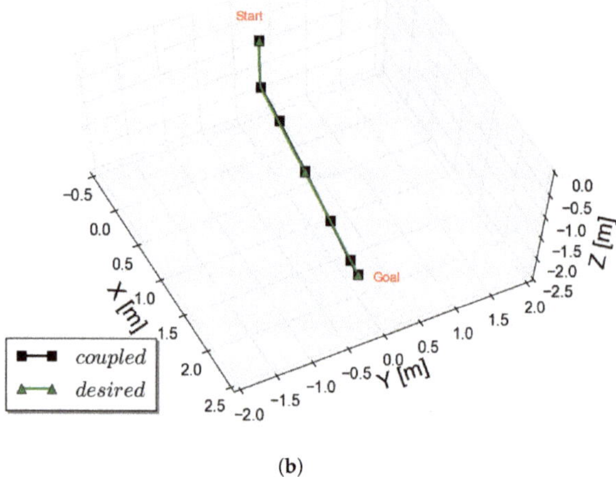

(**b**)

Figure 12. End-effector 3D position, goal at $(x, y, z) = (2.0, 0.0, -2.0)$ m. (**a**) Decoupled strategy, (**b**) Coupled strategy.

5. Discussion

In this paper an operational space parallel position/force controller is used based on the Sliding Mode Control theory and an estimate of the system's model. The main benefit in using this type of control structure are the reduction of the coupling effects between the light vehicle and the manipulator and a robust response regardless of the uncertainties in the mathematical model or the underwater environment. Using a dynamical sliding-mode control and a continuous function in its implementation leads to a smooth behaviour without chattering effects. The oscillations characteristics to integral sliding mode controllers for the position component of the strategy have been removed with appropriate tuning. Nevertheless, using an integral sliding mode controller gives higher priority to the position component over the force component, resulting in oscillations at the moment of contact when both position and force components are under regulation. These oscillations have a small effect in the first instance of the contact with the environment, but this is rapidly compensated by the control structure producing steady contact and station keeping. The control structure is used in a lightweight vehicle-manipulator system using two different strategies: a centralized method (the coupled approach) and a decentralised structure (the decoupled strategy). Comments are further made regarding the two proposed strategies.

In [12] the decoupled approach is presented as a classical control strategy for underwater vehicle-manipulator systems where different control laws are used for the vehicle and manipulator. Having a straightforward implementation and being easy to design are the key advantages of this method. By using a robust vehicle control law the disturbances caused by the coupling effects between the vehicle and manipulator are handled as well as the effects of the interaction with the environment. Good trajectory tracking for the end-effector is obtained as well as the desired interaction force with the environment is maintained. One of the key aspects of this strategy is the task decomposition component responsible with distributing what component is active performing movement and which one is kept in station keeping.

When the vehicle is in station keeping it is common that the movement of the manipulator is large, being dependent on the threshold used in the task decomposition component. Setting a very small threshold leads for the vehicle to be in very close proximity to the object while a large threshold may lead for the end-effector to not interact with the environment as the object might be out of reach.

Improvements to the decoupled approach can be made using a better tuning for the vehicle controller or by increasing the frequency used for the control loops. Nevertheless, in the results presented, due to the characteristics of the real systems and to obtain a reliable comparison between the two strategies, the same frequency is used for both controllers.

The behaviour of the system using an ill-setted threshold is presented in Figure 13. As can be seen, the manipulator joints reach their physical limits and a constant error in the steady-state response of the system, Figure 14, is seen.

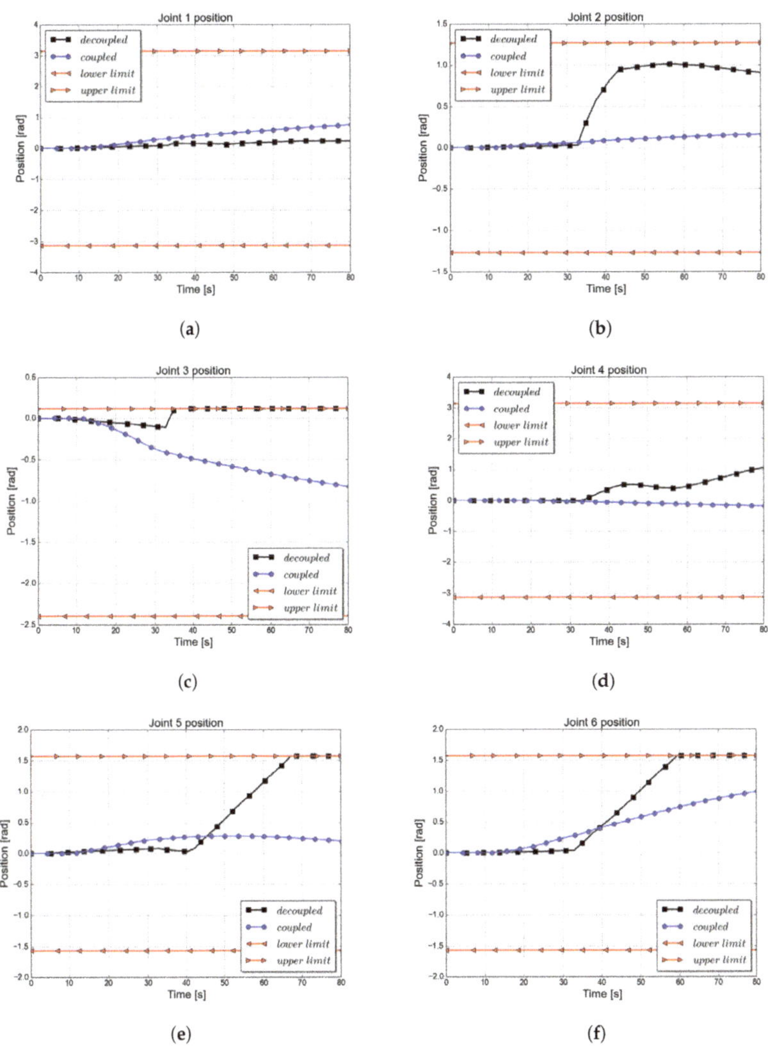

Figure 13. Joint position for goal at $(x, y, z) = (2.0, 0.0, -2.0)$ m. (**a**) Joint 1 position, (**b**) Joint 2 position, (**c**) Joint 3 position, (**d**) Joint 4 position, (**e**) Joint 5 position, (**f**) Joint 6 position.

Figure 13 presents the behaviour of the joint positions. The physical minimum and maximum limits of the joints are also presented. It was shown in the previous section that this goal, $p_{des} = (2.0, 0.0, -2.0)$ m, is reachable. In this case it is intended to demonstrate that setting the threshold to half of the manipulator's length does not lead to the desired behaviour. Furthermore, it is aimed to show that the strategy is sensitive and highly dependent on the threshold used for the task decomposition. Three out of six joints reach their limits, Figure 13c,e,f and this prevents the end-effector reaching the final goal. The direct connection between the threshold and the success of the task represents a disadvantage of the decoupled approach. Nonetheless, this represents only a naive strategy to decide which subsystem is in station keeping. If a planning strategy such as an optimal trajectory generation approach would be used instead of the cycloid function for waypoints generation improvements in the the system response are expected. Using the dynamic model of the system in the optimal trajectory generation, suitable trajectories can be generated for both vehicle and manipulator, taking into account the interactions between the two subsystems. A detailed description of this type of optimal trajectory generator for UVMS is presented in [31]. Nevertheless, this is out of the scope of this paper and represents a topic of itself that the authors aim to explore in a future paper.

Figure 13 shows also the behaviour of the joints when the coupled strategy is used. It can be observed that this shows a more restrictive movement of the arm. The joint movements presented in Figure 13 lead to the end-effector error presented in Figure 14. It can be seen that the end-effector is locked and a constant x-axis error is present for the simulation time. In this case, the exact location of the goal is not reached and the manipulator is not in contact with the object with no force being requested to the end-effector.

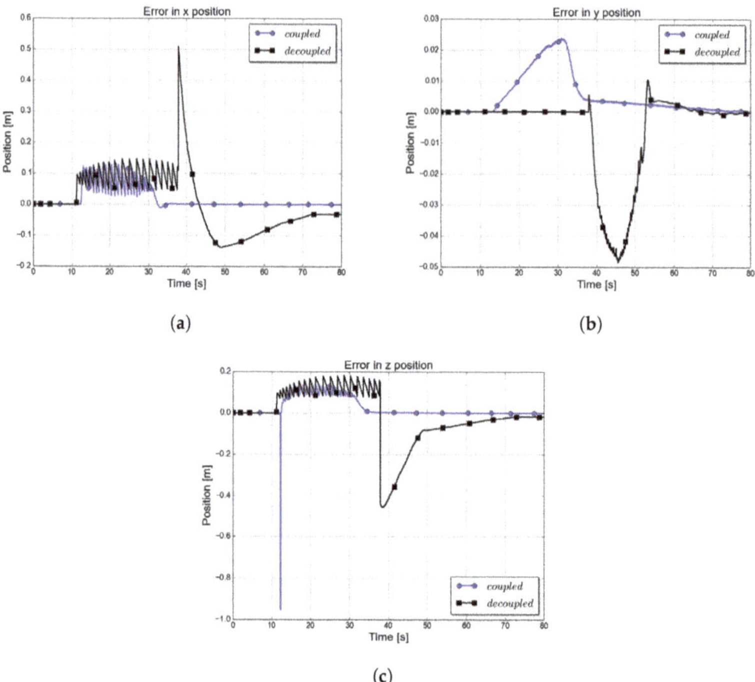

Figure 14. End-effector position errors when goal is at $(x, y, z) = (2.0, 0.0, -2.0)$ m. (**a**) x-axis error, (**b**) y-axis error, (**c**) z-axis error.

In the coupled strategy the control law is designed in operational space and the vehicle control forces and the joint torques are generated based on the inverse transformation from task space to joint space. The control system is designed in operational space. The effects of the manipulator movement on the vehicle and the interactions with stiff environments are compensated by incorporating a linearisation technique based on an estimate of the dynamic model. The main advantage of this strategy is the simplicity of the strategy, by using a single controller for the UVMS. In this case there is no need to design an interaction strategy and decide on an appropriate threshold for it. This makes the coupled strategy less sensitive to failure, guaranteeing the success of the task. Another important characteristic is that when using this method the execution time for the task is reduced as the system does not have to evaluate at each time step the distance to the goal and plan accordingly as happens in the decoupled strategy.

Different stiffnesses, from $K_e = 10^3$ N/m and $K_e = 5 \times 10^5$ N/m and ten different locations of the goal are used to test the two control strategies. To evaluate the results, the generalized root mean square error in end-effector coordinates, Equation (22), is used. The evaluation metric is computed separately for the position and force.

$$GRMS = \sqrt{\frac{1}{N} \sum_{k=1}^{N} e_k^2} \tag{22}$$

N is the number of total measurements and e is the generalized error. From Table 1 it can be observed that the performance of the trajectory tracking/position control is independent of the environment stiffness. The overall error in position is improved for the coupled approach, compared with the decoupled strategy. Nevertheless the difference is not significant and the decoupled approach provides accurate trajectory tracking results. For the force error, the coupled approach provides better results than the decoupled strategy. The oscillatory behaviour and the large overshot at high stiffness environments degrades the performance of the decoupled strategy. One specific case is represented by the most compliant environment where the decoupled strategy offers better results than the coupled approach. The object in this case does not oppose high contact forces and the desired force value is reached without any overshoot.

Table 1. Performance errors for decoupled and coupled strategies.

Characteristic	Strategy							
	Decoupled				Coupled			
K_e (N/m)	1×10^3	5×10^3	1×10^4	5×10^4	1×10^3	5×10^3	1×10^4	5×10^4
Position error (m)	0.13	0.13	0.13	0.13	0.10	0.10	0.10	0.10
Relative position error (%)	4.6	4.6	4.6	4.6	3.54	3.54	3.54	3.54
Force error (N)	1.98	10.25	13.48	47.74	2.80	4.07	8.89	39.97
Relative force error (%)	1.32	6.68	8.66	31.8	1.38	2.71	5.92	26.64

Based on the generalized root mean square error and the relative position error it can be said that the coupled approach performs better in terms of position and force tracking. The error reduction is a result of the use of the overall UVMS dynamic model in the control law and not only for the manipulator. The coupling effects between the manipulator and the vehicle when the system is in contact with the object are handled in this case. The vehicle successfully reacts to these forces facilitating a better position keeping for the end-effector.

One common disadvantage for both strategies rests in the sensitive tuning of the controllers. Not setting the parameters accurately can lead to an oscillatory system or in having a large steady-state error. Tuning the controllers can be challenging as both position and force behaviour have to be accurately obtained. As presented in [2] tuning operational space controllers is difficult as mapping the control effort from operational space to joint space can produce overshoot in the position-force response or producing a constant-steady state error. Nevertheless, during our simulation it was observed that the most sensitive parameter of both coupled and decoupled methods is the gain, c_2 corresponding to the integral sliding mode controller. Very small variations of this parameter could lead to large oscillatory behaviour and stability loss. Furthermore for having an accurate tracking behaviour the position sliding mode gains ρ_1 and the force sliding mode gains ρ_2 were chosen with similar values.

It can be concluded that both strategies can be used on the underwater vehicle-manipulator system. The decoupled strategy represents a controlled method where movement of the manipulator is restricted during the movement of the vehicle. This reduces the risk of collisions with the environment. Using appropriate vehicle and manipulator control structures and reliable interaction strategies, the decoupled method produces similar results to the coupled approach. The main advantage of the coupled approach is the use of a single controller for the overall system as this reduces the coupling effects between the vehicle and manipulator.

It has to be highlighted that the thruster model is not incorporated into this study. According to [32] this represents a simplification of the system and is a valid approach when the thrusters are used below the critical velocity of the vehicle. This is ensured in our approach as the vehicle that we based our work operates only at low speeds. As mentioned in Section 4, the simulation system used in this paper is highly representative of a real underwater vehicle and manipulator available in the Ocean System Laboratory, Heriot-Watt University. The control design and operation capabilities have been considered using the physical capabilities of the real vehicle and manipulator system. Implementation of this controller on the real system will only require minor parameter changes, everything else being directly transferable. The underwater currents and the thruster model are the only simplifications made to the real system in the simulation environment. Nevertheless, using an estimate model of the system in the control structure and as sliding mode controllers handle uncertainties and disturbances in the environment ensures that the proposed control structure will be viable on the real system in real underwater environments. Both the vehicle and the manipulator controllers have a sampling a frequency of 10 Hz, the response of the vehicle's position and orientation transient response being consistent with the transient response from the manipulator.

6. Conclusions

This paper presented a detailed investigation into two possible strategies to control a lightweight underwater vehicle-manipulator system that interacts with an object in the underwater environment. A parallel position/force control based on sliding mode control is used for this study. The simulation results present how the control method can be used on an UVMS either using a coupled or a decoupled strategy. The decoupled method incorporated the proposed control law for the manipulator while a different control law is used for the vehicle. A task decomposition strategy is used to decide which component is in station keeping and which one is requested to move. In the coupled strategy, the underwater vehicle-manipulator system is controlled by the parallel position/force law designed in the operational space. The joint and vehicle commands are computed from the control law output based on the full Jacobian of the system. The evaluation of the system is focused on the two control strategies and analysing the differences between them. Based on the simulation results, it can be concluded that both control strategies provide accurate position and force tracking. The desired interaction force is achieved both in compliant and stiff environments and the steady state is maintained. A detailed and extended comparison between the coupled and decoupled strategies when contact with the environment takes place is presented for the first time for autonomous underwater systems, to the best knowledge of the author. Future work aims to extend the current work by incorporating an optimal trajectory generation that it is expected to improve both the coupled and decoupled control stratedies. Furthermore, upon hardware availability the two architectures aim to be implemented on a real experimental set-up.

Author Contributions: Conceptualization, C.B. and M.W.D.; Methodology, C.B.; Software, C.B.; Validation, C.B., M.W.D. and Y.P.; Formal Analysis, C.B.; Investigation, C.B.; Resources, C.B., M.W.D. and Y.P.; Data Curation, C.B.; Writing—Original Draft Preparation, C.B.; Writing—Review & Editing, C.B. and M.W.D.; Visualization, C.B.; Supervision, M.W.D. and Y.P.; Project Administration, M.W.D. and Y.P.; Funding Acquisition, M.W.D.

Funding: This research received no external funding

Acknowledgments: The authors would like to acknowledge the support provided by the ETP program Scotland and SeeByte Ltd.

Appendix A. List of Mathematical Symbols

In this appendix the mathematical symbols used in this paper are defined:

- $q = [q_1 \cdots q_n]^T$—joint positions
- $\eta_1 = [x, y, z]^T$—vehicle positions
- $\eta_2 = [\phi, \theta, \psi]^T$—vehicle orientation
- $\rho = [\eta_1, \eta_2, q]^T$—vehicle-manipulator generalized position coordinates
- ξ—vehicle-manipulator generalized velocity coordinates
- x_E^I—position and orientation of the end-effector
- $f(\rho)$—generalized transformation from vehicle-joint coordinates to end-effector coordinates
- $M(\rho)$—inertia matrix
- $C(\rho, \xi)$—Coriolis and centripetal vector
- $D(\rho, \xi)$—damping and lift forces vector
- $g(\rho)$—restoring forces
- τ—total forces, moments and torques applied to the vehicle-manipulator
- F—external forces applied to the UVMS
- J—Jacobian of the system
- n—number of degree-of-freedom of the UVMS
- K_e—stiffness matrix of the environment
- x—end-effector position
- x_e—environment position
- $M(x)$—inertia matrix in operational space
- $C(x)$—Coriolis and centripetal vector in operational space
- $D(x)$—damping and lift forces vector in operational space
- $G(x)$—restoring forces in operational space
- p_{des}—desired position of the end-effector
- x_v—vehicle position
- x_m—manipulator position
- J_v—vehicle Jacobian
- $K_{P_p}, K_{p_v}, K_{i_v}$—vehicle controller gains
- τ_v—vehicle control output
- τ_m—manipulator control output
- u_p—position control output
- u_f—force control output
- e_p—error in position
- σ—primary sliding mode variable
- s—secondary sliding mode variable
- $c_1, c_2, c_3, k, \rho_1, \rho_2$—controller parameters

Appendix B. Control Parameters

The controllers were designed in operational space, assuming a decoupled system as a consequence of using a feedback-linearisation technique. The control parameters for x, y, z, roll (*phi*), pitch (θ) and yaw (ψ) are presented in the following lines:

Position gains:

- $\rho_1 : [x : 6.6, y : 1.0, z : 1.6, \phi : 0.1, \theta : 0.1, \psi : 0.1]$
- $k : [x : 50, y : 50, z : 130, \phi : 0.003, \theta : 0.003, \psi : 0.003]$
- $c_1 : [x : 11.3, y : 17.2, z : 8.2, \phi : 0.1, \theta : 0.2, \psi : 0.12]$
- $c_2 : [x : 0.5, y : 0.1, z : 0.6, \phi : 0.0, \theta : 0.0, \psi : 0.0]$

Force gains:

- $\rho_2 : [x : 1.4, y : 1.0, z : 1.0, \phi : 0.0, \theta : 0.0, \psi : 0.0]$
- $c_3 : [x : 0.9, y : 0.9, z : 0.9, \phi : 0.0, \theta : 0.0, \psi : 0.0]$

References

1. Schempf, H.; Yoerger, D. Coordinated Vehicle/Manipulator Design and Control Issues For Underwater Telemanipulation. In Proceedings of the International Federation of Automatic Control (CAMS'92), Genova, Italy, 8–10 April 1992.
2. Yuh, J.; Choi, S.; Ikehara, C.; Kim, G.; McMurty, G.; Ghasemi-Nejhad, M.; Sarkar, N.; Sugihara, K. Design of a Semi-Autonomous Underwater Vehicle for Intervention missions (SAUVIM). In Proceedings of the 1998 International Symposium on Underwater Technology, Tokyo, Japan, 17 April 1998; pp. 63–68.
3. Sanz, P.J.; Ridao, P.; Oliver, G.; Melchiorri, C.; Casalino, G.; Silvestre, C.; Petillot, Y.; Turetta, A. TRIDENT: A Framework for Autonomous Underwater Intervention Missions with Dexterous Manipulation Capabilities. In Proceedings of the 7th IFAC Symposium on Intelligent Autonomous Vehicles IAV-2010, Lecce, Italy, 6–8 September 2010.
4. Ruggiero, F.; Trujillo, M.A.; Cano, R.; Ascorbe, H.; Viguria, A.; Peréz, C.; Lippiello, V.; Ollero, A.; Siciliano, B. A Multilayer Control for Multirotor UAVs Equipped with a Servo Robot Arm. In Proceedings of the 2015 IEEE International Conference on Robotics and Automation (ICRA), Seattle, WA, USA, 26–30 May 2015; pp. 4014–4020.
5. Baizid, K.; Giglio, G.; Pierri, F.; Trujillo, M.A.; Antonelli, G.; Caccavale, F.; Viguria, A.; Chiaverini, S.; Ollero, A. Behavioral control of unmanned aerial vehicle manipulator systems. *Auton. Robots* **2017**, *41*, 1203–1220. [CrossRef]
6. Tummers, R.; Fumagalli, M.; Carloni, R. Aerial Grasping: Modeling and Control of a Flying Hand. *arXiv* **2017**, arXiv:1706.00036.
7. Hogan, N. Impedance control: An approach to manipulation: Part II Implementation. *J. Dyn. Syst. Meas. Control* **1985**, *107*, 8–16. [CrossRef]
8. Khatib, O. A unified approach for motion and force control of robot manipulators: The operational space formulation. *IEEE J Robot. Autom.* **1987**, *3*, 43–53. [CrossRef]
9. Siciliano, B. Parallel Force/Position Control of Robot Manipulators. In *Robotics Research*; Springer: London, UK, 1996; pp. 78–89.
10. Cui, Y.; Podder, T.K.; Sarkar, N. Impedance Control of Underwater Vehicle-Manipulator Systems (UVMS). In Proceedings of the 1999 IEEE/RSJ International Conference on Intelligent Robots and Systems, Kyongju, Korea, 17–21 October 1999; Volume 1, pp. 148–153.
11. Cui, Y.; Yuh, J. A Unified Adaptive Force Control of Underwater Vehicle-Manipulator Systems (UVMS). In Proceedings of the 2003 IEEE/RSJ International Conference on Intelligent Robots and Systems, Las Vegas, NV, USA, 27–31 October 2003; Volume 1, pp. 553–558.
12. McLain, T.W.; Rock, S.M.; Lee, M.J. Experiments in the Coordinated Control of an Underwater Arm/Vehicle System. In *Underwater Robots*; Springer: Boston, MA, USA, 1996; pp. 139–158.
13. Chin, C.S.; Lin, W.P. Robust Genetic Algorithm and Fuzzy Inference Mechanism Embedded in a Sliding-Mode Controller for an Uncertain Underwater Robot. *IEEE/ASME Trans. Mechatron.* **2018**, *23*, 655–666. [CrossRef]
14. Lin, W.P.; Chin, C.S.; Looi, L.C.W.; Lim, J.J.; Teh, E.M.E. Robust design of docking hoop for recovery of autonomous underwater vehicle with experimental results. *Robotics* **2015**, *4*, 492–515. [CrossRef]
15. Simetti, E.; Casalino, G.; Torelli, S.; Sperinde, A.; Turetta, A. Floating underwater manipulation: Developed control methodology and experimental validation within the TRIDENT project. *J. Field Robot.* **2014**, *31*, 364–385. [CrossRef]
16. Canudas de Wit, C.; Olguin Diaz, E.; Perrier, M. Robust Nonlinear Control of an Underwater Vehicle/ Manipulator System with Composite Dynamics. In Proceedings of the 1998 IEEE International Conference on Robotics and Automation, Leuven, Belgium, 20 May 1998; Volume 1, pp. 452–457.
17. Kim, J.; Chung, W.K.; Yuh, J. Dynamic Analysis and Two-Time Scale Control for Underwater Vehicle-Manipulator Systems. In Proceedings of the 2003 IEEE/RSJ International Conference on Intelligent Robots and Systems (IROS 2003), Las Vegas, NV, USA, 27–31 October 2003; Volume 1, pp. 577–582.
18. Antonelli, G.; Chiaverini, S. Singularity-Free Regulation of Underwater Vehicle-Manipulator systems. In Proceedings of the 1998 American Control Conference, Philadelphia, PA, USA, 26 June 1998; Volume 1, pp. 399–403.

19. Santhakumar, M.; Jinwhan, K.; Yogesh, S. A Robust Task Space Position Tracking Control of an Underwater Vehicle Manipulator System. In Proceedings of the 2015 ACM Conference on Advances in Robotics, Goa, India, 2–4 July 2015; p. 2.

20. Featherstone, R. *Rigid Body Dynamics Algorithms*, 1st ed.; Springer: New York, NY, USA, 2008.

21. Barbalata, C.; Dunnigan, M.W.; Pétillot, Y. Dynamic Coupling and Control Issues for a Lightweight Underwater Vehicle Manipulator System. In Proceedings of the Oceans-St. John's, St. John's, NL, Canada, 14–19 September 2014; pp. 1–6.

22. Siciliano, B.; Villani, L. *Robot Force Control*; The Springer International Series in Engineering and Computer Science: Boston, MA, USA, 1999; Volume 540.

23. Barbalata, C.; De Carolis, V.; Dunnigan, M.W.; Ptillot, Y.; Lane, D. An Adaptive Controller for Autonomous Underwater Vehicles. In Proceedings of the 2015 IEEE/RSJ International Conference on Intelligent Robots and Systems (IROS), Hamburg, Germany, 28 September–2 October 2015; pp. 1658–1663.

24. Edwards, C.; Spurgeon, S. *Sliding Mode Control: Theory and Applications*; Series in Systems and Control; Taylor & Francis: London, UK, 1998.

25. Barbalata, C.; Dunnigan, M.W.; Petillot, Y. Position/force operational space control for underwater manipulation. *Robot. Auton. Syst.* **2018**, *100*, 150–159. [CrossRef]

26. Utkin, V.; Shi, J. Integral Sliding Mode in Systems Operating under Certain Conditions. In Proceedings of the 35th IEEE Conference on Decision and Control, Kobe, Japan, 13 December 1996; Volume 4, pp. 4591–4596.

27. Piltan, F.; Keshavarz, M.; Badri, A.; Zargari, A. Design Novel Nonlinear Controller Applied to RobotManipulator: Design New Feedback Linearization Fuzzy Controller with Minimum Rule Base Tuning Method. *Int. J. Robot. Autom.* **2012**, *3*, 1–12.

28. Henson, M.A.; Seborg, D.E. Feedback Linearizing Control. In *Nonlinear Process Control*; Prentice Hall PTR: Upper Saddle River, NJ, USA, 1997; Volume 149.

29. Action Outdoors. Available online: https://www.actionoutdoors.kiwi/Fishing-Net-Accessorie (accessed on 21 August 2018).

30. Inventables: Raw Aluminum Sheet. Available online: https://www.inventables.com/categories/materials/metal/raw-aluminum-sheets (accessed on 21 August 2018).

31. Podder, T.K.; Sarkar, N. A unified dynamics-based motion planning algorithm for autonomous underwater vehicle-manipulator systems (UVMS). *Robotica* **2004**, *22*, 117–128. [CrossRef]

32. Steenson, L.V.; Turnock, S.R.; Phillips, A.B.; Harris, C.; Furlong, M.E.; Rogers, E.; Wang, L.; Bodles, K.; Evans, D.W. Model predictive control of a hybrid autonomous underwater vehicle with experimental verification. *Proc. Inst. Mech. Eng. Part M J. Eng. Marit. Environ.* **2014**, *228*, 166–179. [CrossRef]

Journal of
Marine Science and Engineering

Article

A Novel Gesture-Based Language for Underwater Human–Robot Interaction

Davide Chiarella [1,*,†], Marco Bibuli [1], Gabriele Bruzzone [1], Massimo Caccia [1], Andrea Ranieri [1], Enrica Zereik [1], Lucia Marconi [2] and Paola Cutugno [2]

[1]   Institute of Intelligent Systems for Automation—National Research Council of Italy, Via E. De Marini 6, 16149 Genova, Italy; marco.bibuli@ge.issia.cnr.it (M.B.); gabriele.bruzzone@ge.issia.cnr.it (G.B.); massimo.caccia@ge.issia.cnr.it (M.C.); ranieri@ge.issia.cnr.it (A.R.); enrica.zereik@ge.issia.cnr.it (E.Z.)

[2]   Institute of Computational Linguistics—National Research Council, Via E. De Marini 6, 16149 Genova, Italy; lucia.marconi@ilc.cnr.it (L.M.); paola.cutugno@ilc.cnr.it (P.C.)

*   Correspondence: davide.chiarella@cnr.it; Tel.: +39-010-6475220

†   Main author.

Received: 30 May 2018; Accepted: 27 July 2018; Published: 1 August 2018

Abstract: The underwater environment is characterized by hazardous conditions that make it difficult to manage and monitor even the simplest human operation. The introduction of a robot companion with the task of supporting and monitoring the divers during their activities and operations underwater can help to solve some of the problems that usually arise in this scenario. In this context, a proper communication between the diver and the robot is imperative for the success of the dive. However, the underwater environment poses a set of technical challenges which are not readily surmountable thus limiting the spectrum from which possibilities can be chosen. This paper presents the design and development of a gesture-based communication language which has been employed for the entire duration of the European project CADDY (Cognitive Autonomous Diving Buddy). This language, the *Caddian*, was built upon consolidated and standardized underwater gestures that are commonly used in recreational and professional diving. Its use and integration during field tests with a remotely operated underwater vehicle (ROV) is also shown.

Keywords: marine robotics; underwater human–robot interaction; gesture-based language; field trials

---

1. Introduction

Recreational and professional divers generally work in environments that are difficult to monitor and are characterized by severe conditions. In such a context, it is difficult to monitor the status of divers: any sudden episode causing them to deal with an emergency, such as technical problems or human error, may jeopardize the underwater work or even lead to dramatic consequences, which may involve the divers' safety.

In order to avoid and decrease the probability of such events, standard procedures recommend adherence to well-defined rules, for example, to pair up divers. Nevertheless, during extreme diving campaigns, divers' current best practices may not be sufficient to avoid dangerous episodes.

The EU-funded Project CADDY (Cognitive Autonomous Diving Buddy) has been developed with the aim of transferring the robotic technology into the diving world to improve the safety levels during the dives. The main objective of the project is to develop a pair of companion/buddy robots—an Autonomous Underwater Vehicle (AUV) and its counterpart, an Autonomous Surface Vehicle (ASV) (see Figure 1)—to monitor and support human operations and activities during the dive.

In this scenario, one of the major challenges consists in the development of a communication protocol, which allows the diver and the underwater robot to actively interact and cooperate for the fulfillment of the objectives of the mission.

Given the extreme attenuation of high-frequency electro-magnetic waves underwater, medium-to long-range WiFi/radio communication becomes unreliable already at low depths (i.e., 0.5 m), while optical communications are limited by the reverberation of the water and by the scattering caused by suspended debris [1]. The most used and reliable solution for underwater communication is the exploitation of acoustics, with two main disadvantages: the high prices of the devices and the very low data transmission rates [2,3].

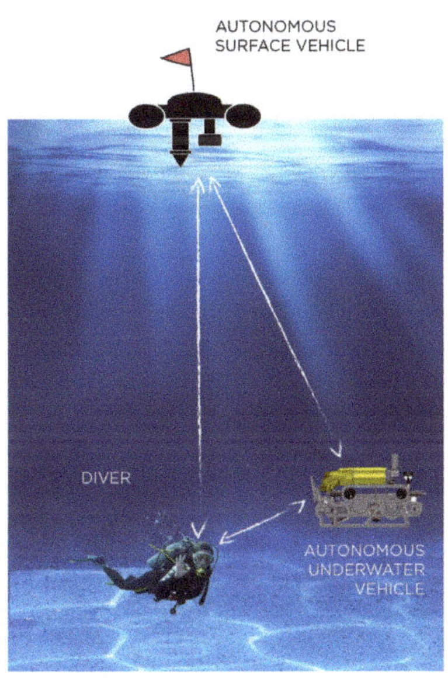

Figure 1. The CADDY (Cognitive Autonomous Diving Buddy) concept.

For all the aforementioned reasons, the solution adopted during the CADDY project has been to develop a novel communication framework with the specific purpose of letting the diver communicate through the most "natural" method available underwater: the gestures. This language, called *Caddian*, has been created as an extension to the established and universally accepted gestures employed by divers worldwide [4–7]. The choice of making the *Caddian*, for all intents and purposes, backward compatible with the current method of communication used by divers has been made in the hope of fostering a widespread adoption among communities of divers.

The field of gesture-based communication between a robotic vehicle and a human being, especially underwater, represents an open challenge in robotics. Very little work has been done.

This paper, with additional results, also extends and completes preliminary work described in [8].

2. State of the Art

The literature on human–robot interaction on dry land shows many languages based on natural language processing and gestures. For example, the authors in [9] use a finite state machine to develop a speech interaction system. Conversely, the choice of the authors in [10] has fallen on a gesture-based human–robot interface. In this second case, however, the limited set of gestures (only five) limits the usability of the language. The work in [11] uses gestures with two alternative methods: a template-based approach and a neural network approach. As can be seen, the literature on human–robot interaction (HRI) on dry land is abundant and presents several ways to make humans and robots communicate. On the other hand, the literature on HRI languages for underwater environments is not as rich and few works present a formal language described by a formal grammar [12,13]. Among those regarding HRI in underwater environments, we cite authors in [14,15], who developed the Robochat language providing Backus–Naur form (BNF) productions. However, the language developed was based on fiduciary markers such as ARTags [16] and Fourier tags [17], lacking the simplicity and instinctivity of gestures [18,19]. Furthermore, authors in [20] developed a programming language for AUV with essential instructions for mission control with the given grammar similar to the assembly language: in this case, the interaction between divers and robots is missing and the use of assembly language seems to be overly complex and hard to remember.

Changing perspectives and, instead of focusing on the language, focusing on the robot's ability to understand, a large amount of research (see, for example, [21,22]) has been done in order to provide robots with robust perceiving capabilities, with the aim of making them reactive to the external world and its occurring events. One of the main goals of the robotic field is to obtain a more natural interaction between men and machines without compromising the efficiency and the robustness of the "system" as a whole. To this aim, gesture recognition seems to be the most promising technique, since it is judged to be almost effortless by humans [4–7].

From the gesture recognition point of view, the literature shows many works focused on the problem of exploiting hand gesture recognition algorithms within different contexts [23], such as robotics or computer science. However, these works have almost always been developed for dry land applications, where the environment is simpler than the harsh underwater scenario (e.g., with bubbles, turbid water, etc.) or the task and the working environment are highly simplified. For example, some of them assume that the hand is the only moving object or make sure that the gestures are performed in front of a very neutral and uniform background. Furthermore, "in-air" applications can exploit cameras with extra sensors such as IR systems that are not employable in water or directly exploit RGBD (Red Green Blue Depth) cameras, like in [24].

In this branch of research (i.e., the "in-air" one) many interesting techniques are proposed; approaches based on adaptive skin detection [25–28], motion analysis [29–31], pose classification [32,33], and others are investigated.

Likewise, there are many distinctive problems essentially in relation to robustness and repeatability of the recognition procedure; for many in-air applications, a simple and structured background (e.g., uniform gray or white without other objects in view) is often considered, since many algorithms fail in more complex scenarios (e.g., in the presence of objects similar to the ones to be detected). Moreover, changes in illumination and light highly affect vision techniques, decreasing their robustness and usability in real-world situations. Conditions are even worse in underwater applications: for instance, the in-air consolidated techniques based on skin detection, or more in general relying on color detection (recall that usually divers wear gloves and masks that totally or partially cover their bodies or their faces), are not suitable in water because of hue attenuation. Due to such a phenomenon, color appearance is very different and the usually employed algorithms lose their effectiveness and robustness. Thus, a more complex approach has to be adopted; as an example, a model for light attenuation is considered in [34] and a color registration technique is tested to demonstrate the effectiveness of the overall approach.

In the specific case of divers, since their suite, gloves and all garments are usually black, difficulties can arise while segmenting their body parts: for a posture with the diver hand right in front of the chest, algorithms can be easily deceived and fail in correctly detecting the hand. To this aim, approaches based on stereocamera systems can improve the detection, exploiting the depth information: in an image like the one above described, the hand and arm of the diver will be slightly closer to the cameras. Indeed, this difference in depth is very small, so the algorithm should be very precise and able to overcome problems badly affecting measurements, such as distortion. Furthermore, another problem strictly related to underwater perception in the presence of humans consists in occlusions due to bubbles generated by divers' breath: the object to be detected can be concealed for many frames, so some sort of prediction and tracking algorithms should be considered. These are some of the additional problems related to the underwater environment that have to be faced and solved by research on underwater perception and that can undermine robustness and repeatability of the robot behavior.

A wide number of different techniques can be exploited and has to be tailored to the specific application: geometric classifiers, Principal Component Analysis (PCA) [35–37], silhouette recognition [38], feature extraction [39], Haar classifiers [40], learning algorithms [41] and so on.

The large variety of works about gesture recognition in the relevant literature testifies to the importance of the development of a robust and effective natural HRI system and indicates that the solution to this problem is still far from being found. Moreover, most of the works presented in the literature are about in-air applications, where the operational conditions present few difficulties: the underwater environment poses many further problems, such as visibility, cloudy water, bubbles occluding the captured scene, illumination, and constraints in the range to be kept between the diver and the robot.

The hostile and harsh underwater environment and the few works addressing the problems of communication in it underline the innovation and utility of the system proposed. Moreover, in a scenario where robots will help divers in their tasks, there is a need for defining and developing a rich language to enable divers to communicate complex commands to their robotic buddies.

This article explains the first implementation and evaluation of the *Caddian* language, namely the phase following the creation of the language from alphabet, syntax and semantics, and the communication protocol that must be followed by the divers to communicate with the AUV. The work is structured as follows. Section 3 presents the definition of *Caddian* language and its communication protocol. In Section 4, the subset of the language gestures used for trials is described, while in Section 5 the robotic platform employed is outlined. Section 6 contains the description of the missions' trials and the BNF syntax of the trial language. Section 7 combines the results from the individual trials. Section 8 describes a study about the language learning curve. Section 9 presents our conclusions.

3. A Gesture-Based Language for Underwater Environments: The Caddian Language

3.1. Human–Robot Interaction Based on Gestures

The development of the HRI language *Caddian* is based on divers' sign language. Given the fact that a language has to be easy to learn and to be taught, *Caddian* signs have been mapped with easily writable symbols such as the letters of the Latin alphabet: this bijective mapping function translates signs to our alphabet and vice versa, as depicted in Figure 2.

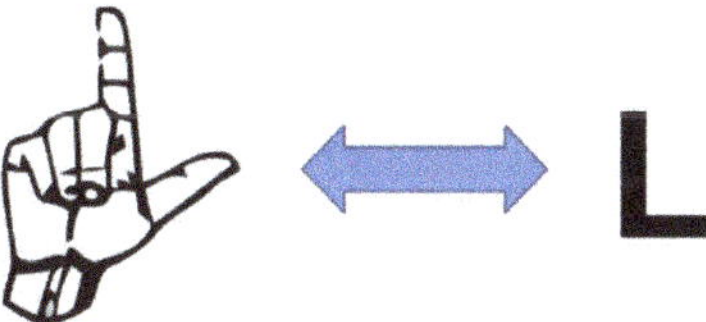

Figure 2. Bijective function mapping from gestures to letters.

Sequences of *Caddian* gestures and the corresponding sequences of characters of the written alphabet (i.e., Σ) are mapped to a semantic function that translates them into commands/messages.

A classifier encodes/decodes gestures, which should be feasible in the underwater environment and as intuitive as possible to cope both with the learning aspect of the language and divers' acceptance. The more dimensions can be discerned by the classifier, the more gestures can be used:

- if it is able to extract and match features from both hands, the amount of recognizable gestures increases (for an example of two-handed gesture, see the "boat" signal in [5]);
- if it is able to extract features and match them in the time domain, thus being able to classify hand gestures with motion, the gestures alphabet becomes richer (for an example of motion gesture, see the "something is wrong" gesture in [5]).

In the creation of the language, the issued sentences have been sequentially defined to allow the synchronization of the recipient with the issuer, and have been defined with boundaries to ensure efficient interpretation: the "start communication" and "end of communication" gestures enclose, as the name says, the communication, while the "start communication" gesture is also used as delimiter between a message and the following one during a complex communication.

3.2. A Specialized Language

Caddian is a language for communicating between divers and underwater robots, in particular autonomous underwater vehicles (AUVs), so the list of commands/messages defined by the language in the scope of the project is strictly context-dependent. Currently, there are 40 implemented commands. The commands/messages (see Table 1 and [8]) are separated into six groups: Problems (9), Movement (5), Setting Variables (10), Interrupt (4), Feedback (3), and Works/Tasks (9).

Table 1. List of commands as seen in [8].

Group	Commands/Messages	
Problems	I have an ear problem	I'm out of breath
	I'm out of air [air almost over]	Something is wrong [diver]
	I depleted air	Something is wrong [environment]
	I'm cold	I have a cramp
	I have vertigo	
Movement	Take me to the boat	You lead (I follow you)
	Take me to the point of interest	I lead (you follow me)
	Go X Y	Return to/come X
	$X \in Direction$	$X \in Places$
	$Y \in \mathbb{N}$	
Interrupt	Stop [interruption of action]	Abort mission
	Let's go [continue previous action]	General evacuation
Setting Variables	Keep this level (actions are carried out at this level)	Free level ("Keep this level" command does not apply any more)
	Level Off (AUV cannot fall below this level)	Slow down/Accelerate
	Set point of interest	Give me air (switch on the on board oxygen cylinder)
	Give me light (switch on the on board lights)	No more air (switch off the on board oxygen cylinder)
	No more light (switch off the on board lights)	
Feedback	No (answer to repetition of the list of gestures)	I don't understand (repeat please)
	Ok (answer to repetition of the list of gestures)	
Works	Wait n minutes $n \in \mathbb{N}$	Tessellation X * Y area $X, Y \in \mathbb{N}$
	Tell me what you're doing	Photograph of X * Y area $X, Y \in \mathbb{N}$
	Carry a tool for me	Stop carrying the tool for me [release]
	Do this task or list of task n times $n \in \mathbb{N}$	Photograph of point of interest/boat/here
	Tessellation of point of interest/boat/here	
$Direction = \{ahead, back, left, right, Up, Down\}$		
$Places = \{point\,of\,interest, boat, here\}$		

3.3. Communication Protocol and Error Handling

The *Caddian* language is used inside a communication protocol that guarantees error handling. The protocol is designed having in mind a strict cooperation between the diver and the AUV: for example, we can mention the possibility for the diver to query the robot at any time about the progress of a task with the purpose of understanding whether it has been executed.

For these reasons, AUV is equipped with three light emitters (red, green, and orange):

- green = IDLE STATE — everything is ok, all tasks have been accomplished and I'm waiting for orders;
- orange = BUSY STATE — everything is ok and I'm working the last mission received;
- red = FAILURE STATE — a system failure has been detected or an emergency has been issued.

The *Caddian* protocol handles these three possible types of error:

1. the AUV does not recognize a gesture inside a sequence: the robot shows an error message and the diver repeats the gesture, but the sequence is not aborted. This allows the diver to make mistakes and, in such cases, to save time avoiding to repeat the whole sequence.
2. the AUV recognizes the sequence, but the resulting command is not semantically correct. When the whole sequence is issued, a semantic error message is shown: the sequence of gestures is aborted and must be repeated.
3. the AUV recognizes the sequence and the resulting command is semantically correct, but it is not what the diver intended. This type of error is more subtle, because it involves a semantical evaluation that only the diver is able operate with the necessary swiftness.

To deal with this last category of errors, we have to introduce the definition of "mission." As a sequence is issued to the buddy AUV, before it can turn into a real mission, the AUV repeats the sequence, writing it in plain text on its screen and waiting for the diver's confirmation. At this point, the diver simply accepts the sequence showing thumbs up, thus letting the mission start, or he refuses the sequence with thumbs down: that sequence won't turn into a mission and thus it will never be translated into a series of actions. Moreover, it would be very hard (or even impossible) for the AUV to guess where the error was in the sequence; therefore, if the diver does not confirm the sequence, the only feasible approach is to make the diver repeat it from the beginning.

Regarding the robot's mission status, two accessibility aspects were also considered crucial while creating the *Caddian* language:

1. Divers should always understand if the assigned mission has been terminated.
2. Divers should always be able to know the progress of a mission.

In the first case, the buddy AUV just turns on the idle status (i.e., green light) and remains stationary. In the second case, the proposed behavior is the following:

- the buddy is in operation executing a task;
- the diver approaches the buddy, facing it, to be clearly visible on both the camera and the sonar;
- for safety reasons, the diver always remains outside a predefined safety range (e.g., 2 m). If closer, the buddy is programmed to automatically back off from the diver.
- if all the above conditions are met, the buddy AUV suspends the current action, remaining however in the BUSY state.

In this situation, a diver can

- query the AUV on the mission's progress with the "Check" command;
- erase the current mission with the "Abort mission" command;
- report an emergency using a command belonging to the "Problems" subset;
- leave the range of safety, letting the AUV return to the assigned mission.

3.4. Language Definition

The diver–robot language has been defined as a formal language. A formal grammar can describe a formal language [12,13] and can be represented as a quadruple $< \Sigma, N, P, S >$ as follows:

-   a finite set Σ of terminal symbols (disjoint from N), the alphabet, which are assembled to form the sentences of the language;

- a finite set N of non-terminal symbols or variables or syntactic categories, which represents some collection of subphrases of the sentences;
- a finite set P of rules or productions which describe how each non-terminal is defined in terms of terminal symbols and non-terminals. Each production has the form $B \rightarrow \beta$, where B is a non-terminal and β is a string of symbols from the infinite set of strings $(\Sigma \, UN)$;
- a differentiated non-terminal S, the start symbol, which specifies the principal category being defined, such as a sentence, a program, or a mission.

This said, the language L_G (i.e., generated by grammar G) can be formally defined as the set of strings composed of terminal symbols that can be derived through productions from the start symbol S.

$$L_G = \{s/s \in \Sigma^* \text{ and } S \rightarrow^* s\}. \tag{1}$$

For the *Caddian* language, the signs of the alphabet Σ are the set of letters belonging to the Latin alphabet mixed with some complete words, Greek letters, math symbols, and natural numbers (also used as subscripts) defined as follows:

$$\Sigma = \{A, \ldots, Z, ?, const, limit, check, \ldots, 1, 2 \ldots\}. \tag{2}$$

By definition, the grammar of *Caddian* is a context-free grammar because on the left side of the productions only non-terminal symbols and no terminal symbols can be found [42,43]. In addition, the resulting language is an infinite language given that the first production (i.e., $\langle S \rangle$) uses recursion and the dependency graph of the non-terminal symbols contains a cycle.

3.5. Syntax

Syntax has been given through the following BNF productions:

$\langle S \rangle$ ::= A $\langle \alpha \rangle$ $\langle S \rangle$ | $\forall$

$\langle \alpha \rangle$ ::= $\langle agent \rangle$ $\langle m\text{-}action \rangle$ $\langle object \rangle$ $\langle place \rangle$ | ƀ$\langle feedback \rangle$ $\langle p\text{-}action \rangle$ $\langle problem \rangle$ | $\langle set\text{-}variable \rangle$ | $\langle feedback \rangle$ | $\langle interrupt \rangle$ | $\langle work \rangle$ | $\varnothing$ | $\triangle$

$\langle agent \rangle$ ::= I | Y | W

$\langle m\text{-}action \rangle$ ::= T | C | D | F | G $\langle direction \rangle$ $\langle num \rangle$

$\langle direction \rangle$ ::= forward | back | left | right | up | down

$\langle object \rangle$ ::= $\langle agent \rangle$ | Λ

$\langle place \rangle$ ::= B | P | H | Λ

$\langle problem \rangle$ ::= E | C_1 | B_3 | P_g | A_1 | K | V | Λ

$\langle p\text{-}action \rangle$ ::= H_1 | B_2 | D_1 | Λ

$\langle feedback \rangle$ ::= ok | no | U | Λ

$\langle set\text{-}variable \rangle$ ::= S $\langle quantity \rangle$ | L $\langle level \rangle$ | P | L_1 $\langle quantity \rangle$ | A_1 $\langle quantity \rangle$

$\langle quantity \rangle$ ::= + | -

$\langle level \rangle$ ::= const | limit | free

$\langle interrupt \rangle$ ::= Y $\langle feedback \rangle$ D

$\langle work \rangle$ ::= Te $\langle area \rangle$ | Te $\langle place \rangle$ | Fo $\langle area \rangle$ | Fo $\langle place \rangle$ | wait $\langle num \rangle$ check | $\langle feedback \rangle$ carry | for $\langle num \rangle$ $\langle works \rangle$ end | Λ

$\langle works \rangle$ ::= $\langle work \rangle$ $\langle works \rangle$ | Λ

$\langle area \rangle$ ::= $\langle num \rangle$ $\langle num \rangle$ | $\langle num \rangle$

$\langle num \rangle$ ::= $\langle digit \rangle$ $\langle num \rangle$ | Ψ

$\langle digit \rangle ::= 1 \mid 2 \mid 3 \mid 4 \mid 5 \mid 6 \mid 7 \mid 8 \mid 9 \mid 0$

With the given syntax, we can translate the previously identified messages and commands and obtain a translation table (Table 2) [8].

Table 2. Translation table as seen in [8].

	Message/Command	Caddian
Problems	Ear problem	$A\,bH_1E\forall$
	Out of breath	$A\,bB_2B_3\forall$
	Out of air [air almost over]	$A\,bB_2A_1\forall$
	Something is wrong [diver]	$A\,bH_1P_g\forall$
	Air depleted	$A\,bD_1A_1\forall$
	Something is wrong [environment]	$A\,bP_g\forall$
	I'm cold	$A\,bH_1C_1\forall$
	I have a cramp	$A\,bH_1K\forall$
	I have vertigo	$A\,bH_1V\forall$
Movement	Take me to the boat	$AYTMB\forall$
		$AIFYAYCB\forall$
	You lead (I follow you)	$AIFY\forall$
	Take me to the point of interest	$AYTMP\forall$
		$AIFYAYCP\forall$
	I lead (you follow me)	$AYFM\forall$
	Go X Y	$AYG\,Directions\,n\forall$
	$X \in Directions$ and $Y \in \mathbb{N}$	$n \in \mathbb{N}$
	Return to/come X	$AYCP\forall$
	$X \in Places$	$AYCB\forall$
		$AYCH\forall$
Interrupt	Stop [interruption of action]	$AY\,no\,D\forall$
	Let's go [continue previous action]	$AY\,ok\,D\forall$ or $AYD\forall$
	Abort mission	$A\varnothing\forall$
	General evacuation	$A\,\triangle\,\forall$
Setting Variables	Slow down	$AS-\forall$
	Accelerate	$AS+\forall$
	Set point of interest	$AP\forall$
	Level Off	$AL\,limit\,\forall$
	Keep this level	$AL\,const\,\forall$
	Free level	$AL\,free\,\forall$
	Give me air	$AA_1+\forall$
	No more air	$AA_1-\forall$
	Give me light	$AL_1+\forall$
	No more light	$AL_1-\forall$
Feedback	No	$A\,no\,\forall$
	Ok	$A\,ok\,\forall$
	I don't understand (repeat please)	$AU\forall$
Works	Wait n minutes $n \in \mathbb{N}$	$A\,wait\,n\,\forall$
	Tessellation X * Y area	$ATe\,n\,m\forall$
	$X, Y \in \mathbb{N}$	$ATe\,n\forall$ [square]
	Tessellation of point of interest/boat/here	$ATe\,P\forall$
	Tell me what you're doing	$A\,check\forall$
	Photograph of X * Y area $X, Y \in \mathbb{N}$	$AFo\,n\,m\forall$
		$AFo\,n\forall$ [square]
	Take a picture of point of interest/boat/here	$AFoP\forall$
	Carry a tool for me	$A\,carry\forall$
	Stop carrying the tool for me [release]	$A\,no\,carry\forall$
	Do this task or list of task n times $n \in \mathbb{N}$	$A\,for\,n\,\ldots\,end\forall$
	$Directions = \{ahead, back, left, right, Up, Down\}$	
	$Places = \{point\,of\,interest, boat, here\}$	

3.6. Semantics

The communication scenarios and relating commands can be divided into six groups. In the following paragraphs, these commands sets are briefly explained.

Problems—This category of messages refers to issues happening to the diver or to the environment around the operating area. All productions contain the ♭ symbol, which intrinsically denotes that there is an emergency and that any action is being executed needs to be suspended to take care of the issue.

Movement—This category of commands makes the robot move or tells the robot how to move.

Interrupt—This category of commands makes the robot stop the current task/mission. The "general evacuation" command has a special meaning: in fact, this command makes the buddy AUV abort the mission and makes it emit any possible warning signal both to the surface and to any diver in the operation area (e.g., it turns on flashing red lights and sends emergency messages through the acoustic link towards the surface vehicle).

Setting Variables—These commands set internal variables inside the robot. At this moment, there are eight of them, but only seven can be set directly by the diver (see below).

- *Speed*: the robot speed has discrete values. With the "+" or "-" signs, the diver increases or decreases this variable by a quantum.
- *Level*:

    - constant: any following command is carried out at the current level of depth;
    - off: the buddy AUV cannot move below the actual depth: if a subsequent command tries to force the buddy AUV to break this rule, the robot interrupts the mission. This behavior has been thought as a safety measure mainly for the buddy AUV (and for the diver as a direct consequence) but it may be also useful in specific scenarios, for example, during the exploration of archaeological sites;
    - free: clears previous statuses set by other commands which refer to the level of depth: the AUV is now free to move up and down underwater.

- *Point of Interest*: set a single point of interest, to be recalled later within other commands.
- *Light*: this is a binary variable which switches on or off the vehicle lights.
- *Air*: this is a binary variable which toggles on or off the vehicle's onboard oxygen cylinder.
- *Here*: store the actual 3D coordinates of the position in memory (i.e., where the buddy is located while the command is being issued, and its yaw angle—where the buddy is facing).
- *Boat*: boat or base position, which cannot be set by the diver.

Communication Feedback—These commands refer to the communication feedback between the diver and the AUV. The diver can accept or reject a previously issued command (see Section 3.3) and can also ask the AUV to repeat the command if he did not comprehend it (by accident or by distraction).

Works—These commands refer to tasks the robot is able to do. The "tell me what you're doing" command (i.e., check the mission progress) can be used when the diver approaches the robot (and the robot consequently pauses anything it is doing). The "wait X minutes" command instructs the robot to float and wait X minutes then proceed with the next command (useful to pause the mission or to let the seafloor dust settle down). The "carry a tool for me" command instructs the robot to carry equipment upon diver request: after the equipment has been placed into the robot compartment, the AUV waits for a physical confirmation (i.e., a button to press to give physical feedback).

4. Outline of Gestures Used during Trials

A diver's underwater gestures are not formalized in any international standard. In fact, all organizations and diving agencies worldwide teach divers their own subset of diving hand signals, causing some gestures to vary from region to region: in this paper, the most famous and common ones [4–7] were chosen, carefully picking both from the ones used by largest diving organizations and from gestures akin to natural or instinctive meaning. *Caddian* gestures were also chosen following the two most important requirements of the language: all gestures should be feasible and as easy as possible to perform underwater and they should be as intuitive as possible to make them easy to remember and to render the language truly effective.

In some cases, whenever possible, basic mnemonic techniques have been exploited, associating a gesture to objects related to the action it expresses: for example, in the "take a picture" gesture, the diver shows just three fingers, which can be mnemonically associated to the tripod used to stabilize and elevate a camera.

The list in Figures 3 and 4 contains a subset of gestures used during CADDY trials. As can be seen, a significant amount of task gestures and all the natural numbers are recognized, but we eventually decided not to introduce dynamic gestures, because the effort to recognize them through computer vision was too high with respect to the expected benefits.

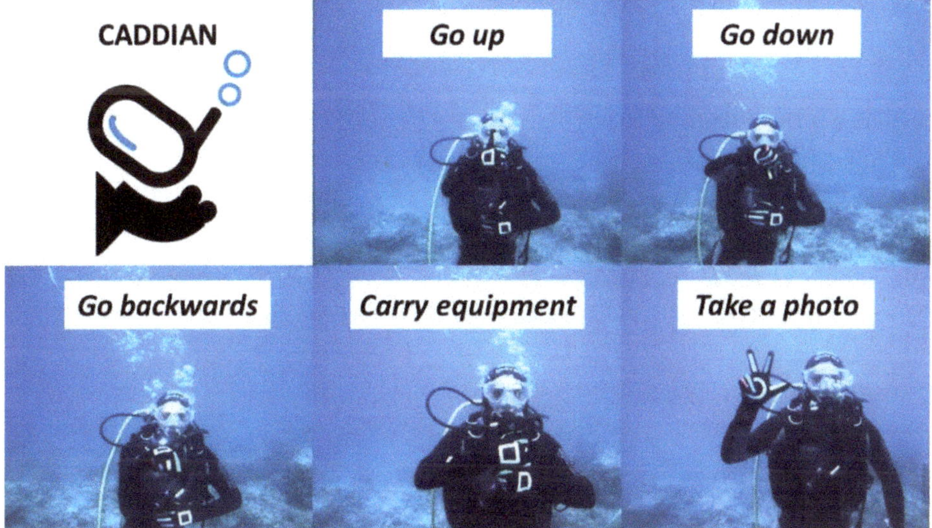

Figure 3. A subset of the *Caddian* gestures tested during 2015 validation trials.

Figure 4. Trials gestures: numbers from 1 to 5.

Mapping Gestures to Syntax; Syntax to Semantics

As already said, a bijective mapping function translates from the domain of signs to our alphabet and vice versa (Figure 5). Accordingly, one or more gestures and the corresponding characters are also mapped to a semantic function that translates them into commands/messages.

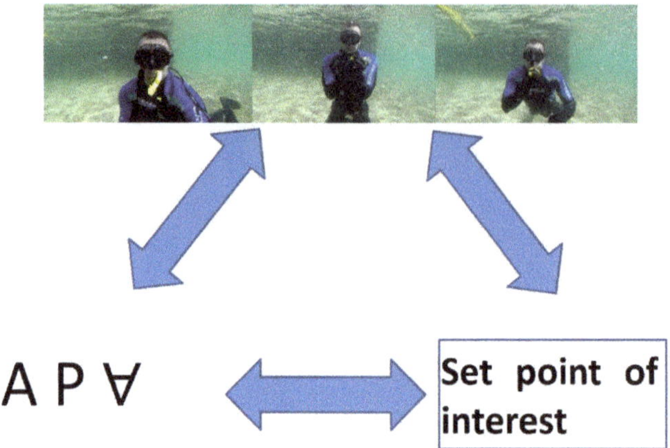

Figure 5. Gestures, written alphabet, and semantics.

5. Outline of Trial Vehicle: R2 ROV

At-field trials have been carried out employing the robotic platform R2 ROV/AUV design and developed by CNR-ISSIA (Figure 6); the R2 underwater vehicle is the product of a retrofit process of the former Romeo ROV, built and developed during the 1990s. R2 is characterized by an open-frame structure and a full-actuated motion capability. Thanks to its compact size, 1.3 m long, 0.9 m wide, and 1.0 m tall, it is still considered a small-/medium-class ROV/AUV. Depending on the specific payload for each mission (dedicated sensor package, manipulation systems, etc.), the total weight can vary from 350 to 500 Kg in-air. The overall motion control is provided by a redundant and fault-tolerant thruster allocation with four vertical thrusters for vertical motion and four horizontal thrusters for the 2D horizontal positioning. In ROV mode, a fiber-optic based link provides real-time data transfer for both direct piloting/control/supervision of the robot, as well as on-line data gathering and analysis.

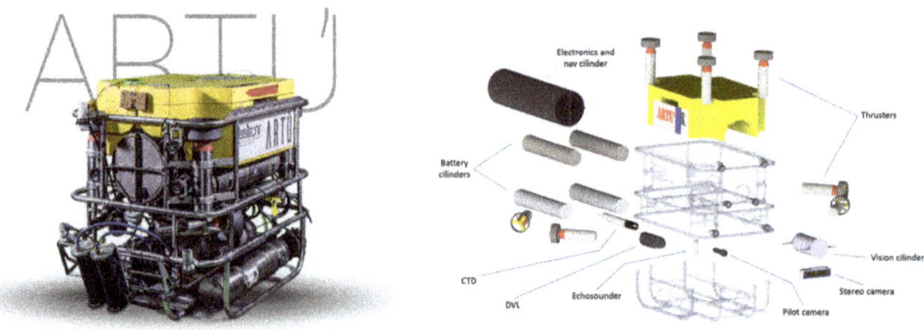

Figure 6. R2 (Artù) ROV. (**left**) a picture; (**right**) CAD design with sensors and actuators.

The R2 ROV/AUV during the trials was equipped with:

- IMU (Inertial Measurement Unit): Quadrans 3-axis Fiber Optic Gyro and 3DM-GX3-35 MicroStrain AHRS (Attitude Heading Reference System);
- GPS (Global Positioning System): 3DM-GX3-35 MicroStrain AHRS (Attitude Heading Reference System);
- Stereo camera: BumbleeBee XB3 13SC-38 3 sensor multi-baseline color camera;
- CTD (Conductivity, Temperature, and Depth sensors) : OceanSeven 304 Plus;

- Sonars: 2 Tritech PA500 echosounders (1 for seafloor detection and 1 for front obstacle detection);
- Lights: 6 front-mounted LED-based (Light-emitting diode) high intensity spot lights.

6. Trials at Sea and at Pool

Two experimental campaigns were carried out in 2015 [44,45], one in Biograd Na Moru (Croatia), at sea, in October and the second one in Genova (Italy), at pool, in November. Both campaigns were focused on the validation of the interaction capabilities of the robot, mostly related to the gesture recognition and compliant robot reactions to the desired commands. Five professional divers were involved, respectively four for the Biograd Na Moru campaign and one for the Genova one. All the divers were trained for about half an hour before the first dive, as the set of gestures used was minimal (i.e., 22 gestures) and it was mostly a subset of the "common gestures set" already used in the diving world (i.e., 15 out of 22). The Biograd Na Moru's trials were made in open sea at a 4 m depth over one week. Depending on the day, trials were carried out with the presence of currents or at calm sea. During the Genova campaign, trials were made at a 3-m-deep outdoor pool. Given that these trials were only focused on preliminary functional tests of the computer vision algorithms and aimed at gathering as much data as possible for their further refinement, the R2 ROV (see Figure 6), described in Section 5, was employed as the buddy AUV. According to the envisioned use-case scenarios of the CADDY project, trials were made up of four kinds of mission.

- **Movement Missions**—In this kind of mission, the diver issues a movement command with a number (i.e., "Go up 1 m").
- **"Take a photo" Missions**—In this kind of mission, the diver commands the AUV to take a picture from the point where it is stationing.
- **"Do a mosaic" Missions**—In this kind of mission, the diver commands the AUV to do a mosaic/tessellation of an area n x m of the seabed (see Section 3 under Works).
- **Complex Missions**—In this kind of mission, the diver commands the AUV to go to the boat and bring back a tool.

During the trials, a minimal subset of a modified version of *Caddian* has been used. Here, the syntax of the minimal language is followed:

$\langle S \rangle ::= \texttt{A} \langle \alpha \rangle \langle S \rangle \mid \forall$

$\langle \alpha \rangle ::= \langle direction \rangle \langle num \rangle \mid \langle place \rangle \mid \langle work \rangle$

$\langle direction \rangle ::= \texttt{forward} \mid \texttt{back} \mid \texttt{up} \mid \texttt{down}$

$\langle place \rangle ::= \texttt{B} \mid \texttt{H} \mid \Lambda$

$\langle work \rangle ::= \texttt{Te} \langle area \rangle \mid \texttt{Fo} \mid \texttt{carry} \mid \Lambda$

$\langle area \rangle ::= \langle num \rangle \langle num \rangle \mid \langle num \rangle$

$\langle num \rangle ::= \langle digit \rangle \langle num \rangle \mid \Psi$

$\langle digit \rangle ::= \texttt{1} \mid \texttt{2} \mid \texttt{3} \mid \texttt{4} \mid \texttt{5} \mid \texttt{6} \mid \texttt{7} \mid \texttt{8} \mid \texttt{9} \mid \texttt{0}$

Applying the syntax rules, the commands for each mission were as follows:

- **Movement missions:**

    - **"Go up 1 m":** A up 1 Ψ ∀
    - **"Go down 1 m":** A down 1 Ψ ∀
    - **"Go back 1 m":** A back 1 Ψ ∀
    - **"Go forward 1 m":** A forward 1 Ψ ∀

- **"Take a photo" mission:** A Fo ∀
- **"Do a mosaic" mission:** A Te 2 Ψ 4 Ψ ∀

- **Complex mission:** A B A carry A H ∀

The complex mission deserves a special mention because in its syntax we can observe three concatenated commands between the delimiters (i.e., "A" and "∀"). The first one is "go to the boat" ("B"), followed by "carry a tool" ("carry") and then "return here" ("H"). Consequently, the robot moves to the boat, opens the compartment to hold the tool requested, and returns to the point where the entire mission has been started.

7. Results

The first phase of the trials has been focused on simple gesture recognition, namely the recognition of single gestures by the robot and the execution of the associated command. Divers performing the gestures are shown in the following figures. Figure 3 shows a sample of the single-gestured commands used during the trials, while Figures 7–9 show gesture recognition in different kinds of environment to emphasize the different challenges that had to be solved to have correctly classified gestures. For example, Figure 7 shows the diver in different kinds of water at varying distance from the camera: the green and yellow circles shows that the recognition process has been successful for all four images.

Figure 7. Detected gestures, from the top left, clockwise: take a photo, carry equipment, start communication, and go to the boat.

Figure 8. "Go down 2 m" gesture sequence executed during trials.

Figure 9. CADDY simple gesture recognition in a harsh underwater environment.

Extensive tests focused on the complex gesture sequence recognition were carried out during the second phase of the trial. Complex sequences are enriched sets of gestures that represent dialogues or sentences containing more information that the robot has to recognize, separate, and interpret in order to achieve the desired goal requested by the diver. Examples of complex sequences are "go to boat and carry equipment here" or "execute a mosaic of N × M meters." Complex sequences have been tested 15 times each in order to evaluate the reliability and success rate of the system, obtaining good results in terms of recognition capabilities (i.e., all complex sequences had a success rate which varies from 87 to 99.9% except for the "do a mosaic" mission, which achieved 60%). An example of a complex sequence executed by a diver is the "Go down 2 m", depicted in Figure 8. From Figure 9, it is possible to appraise the harsh conditions that the CADDY system has to face while working in the underwater environment. The water visibility level can very badly affect the gesture interaction, as well as the illumination condition and gesture occlusions due to bubbles. The CADDY system was successful in the interaction through gestures even during days with a low visibility condition (2.5 m with suspended sediments) and with a strong current and strong waves; the robot lit up its lights to signal to the diver that his gestures were recognized and the command executed.

8. Cognition and Ease of Language Learning: Evaluation on Dry Land

The same gestures and relative missions performed during the trials were tested on dry land in order to evaluate the language learning curve and the language cognitive load in divers. The research team tested 22 volunteers: all of them completed the trials. Seven out of 22 volunteers were female and 4 out of 22 had previous diver experience. No replication of the experiment with the same volunteer has been made and missions have been provided sequentially with no order randomization. The evaluation of the language took place in the following stages: first, the language has been explained to the volunteers (the explanation had a duration of about five minutes); afterward, the volunteers were asked to repeat the six trials missions, which were as follows:

- **Mission 1:** "Go up 1 m";
- **Mission 2:** "Go down 1 m";
- **Mission 3:** "Go back 1 m";
- **Mission 4:** "Take a photo";
- **Mission 5:** "Do a mosaic";
- **Mission 6:** "Go to boat, bring me something (carry equipment), come back here."

Figure 10 describes the results of this evaluation. As can be observed, the error value decreases with the progress of the missions, proving that the volunteers learnt the language very quickly, given the fact that only five minutes were dedicated to the explanation of the whole language used in trials, made up of 22 gestures.

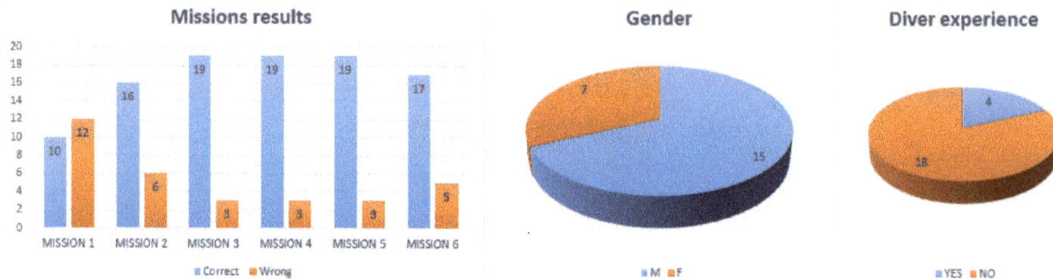

Figure 10. Evaluation of the six missions and description of volunteers data set.

The observed errors in the mission enunciation were as follows:

- **CLOSE_NUM_1:** forgetting "close number" when issuing the mission.
- **NUMBER_ONE:** confusing/swapping the "number one" with "close communication."
- **UP:** forgetting the "UP" gesture and issuing only the meters to go. The error of forgetfulness of a gesture appears only in the first and third mission. This error is supposed to be due to the initial nervousness of the candidate.
- **CLOSE_COMM:** confusing/swapping the "close communication' with "number one."
- **DOWN:** wrong orientation of the hand.
- **CLOSE_NUM_2:** confusing/swapping the "close number" with "close communication."
- **BACKWARDS:** forgetting the "BACKWARDS" gesture and issuing only the meters to go. The error of forgetfulness of a gesture appears only in the first and third missions. This error is supposed to be due to the nervousness of the candidate.
- **TAKE_PHOTO:** wrong orientation of the hand.
- **MOSAIC:** wrong orientation of the hands.
- **NUMBERS:** wrong orientation of the hand.
- **BOAT_CARRY:** the candidate did not remember the gestures for "boat" or "carry."
- **HERE_1:** the candidate did not remember the gestures for "come here."
- **START_MSG:** candidate uses, in the complex mission, the "start message" gesture to close the communication. This may also be an indication for a further improvement of the language where a single gesture is used to both open and close the communication. In this case, the interarrival time could be used as information to separate one communication from the following one.
- **HERE_2:** confusing/swapping the "here" gesture with "go backwards."

A more detailed view of the errors can be seen in Table 3.

Table 3. Summary of the observed errors.

Gesture	Semantic	Type of Error	Example	Occurrences
Ψ	Close number	CLOSE_NUM_1	A up 1 ∀	7
1	Number one	NUMBER_ONE	A up ∀ Ψ ∀	4
up	Go up	UP	A 1 Ψ ∀	2
∀	Close communication	CLOSE_COMM	A up 1 Ψ 1	5
down	down	DOWN	Wrong orientation	1
Ψ	Close number	CLOSE_NUM_2	A up 1 ∀ ∀	1
backwards	Go backwards	BACKWARDS	A 1 Ψ ∀	1
Fo	Take a photo	TAKE_PHOTO	Wrong orientation	3
Te	Do a mosaic	MOSAIC	Wrong orientation	2
2,4	Number two and four	NUMBERS	Wrong orientation	1
B	Go to the boat	BOAT_CARRY	A A A H ∀	1
H	Come back here	HERE_1	A B A carry A ∀	2
A	I'm starting a message	START_MSG	A B A carry A H A	1
H	Come back here	HERE_2	A B A carry A backwards ∀	1

9. Conclusions

In this paper, a novel gesture-based language for underwater human–robot interaction (UHRI) has been proposed: the description of the language, called *Caddian*, is provided with alphabet, syntax, semantics, and a communication protocol. The presented work has mostly focused on the definition of the language and on showing its potential and likely acceptance by the diving community. A description of the performed trials using a minimal modified subset of the language has been reported and is preliminary, but encouraging results are provided.

The classifier performed quite well both in normal and in stormy weather, thus testifying that gestures outside the standardized ones have been chosen correctly. Moreover, divers learnt the language very quickly, showing that the associated cognitive load on the divers is acceptable.

However, the trials also showed that the success rate of the classification can still be improved for some gestures (e.g., "do a mosaic"), and the syntax of natural numbers for humans is not as intuitive as preliminarily designed. Tests with the whole set of gestures have to be made because dynamic gestures, which were not involved during these trials, did not have the same rate of success as the static ones.

Regarding the characteristics of the current version of the language, it would be worth considering adding the ability of changing a mission with an updated parameter (for instance, a new "here" position), or setting macros and then using parameterized missions (i.e., functions).

A deeper study of the use of the whole language will be the next steps of the research, which might involve new scenarios with a consequent creation of new commands; more results about the performance of the classifier must be collected to prove the CADDY framework robust. On the other hand, acceptance by divers of the language will be taken into account. Through the study of their feedback, the language might be changed accordingly.

Author Contributions: Conceptualization, D.C., M.C., L.M. and P.C.; data curation, D.C., M.B. and A.R.; formal analysis, D.C.; funding acquisition, G.B., M.C., L.M. and P.C.; investigation, D.C., M.B., A.R. and E.Z.; methodology, D.C. and M.B.; project administration, M.C. and L.M.; resources, G.B. and M.C.; software, D.C., M.B., A.R. and E.Z.; supervision, G.B., M.C., L.M. and P.C.; validation, D.C., M.B., A.R. and E.Z.; visualization, D.C., M.B., A.R. and E.Z.; writing—original draft, D.C.; writing—review & editing, D.C., M.B., A.R. and E.Z.

Funding: The research leading to these results received funding from the European Union Seventh Framework Programme (FP7/2007-2013) under grant agreement no. 611373.

Acknowledgments: The authors would like to thank Giorgio Bruzzone and Edoardo Spirandelli for their invaluable assistance during trials and Mauro Giacopelli for the *Caddian* language pictures. We would also like to show our gratitude to M.Sc. Angelo Odetti and M.Sc. Roberta Ferretti for their support. We are also immensely grateful to M.Sc. Arturo Gomez Chavez who developed the vision system of CADDY and Dr Luca Caviglione who provided insight and expertise that greatly assisted the writing of this manuscript.

References

1. Cu, J.H.; Kong, J.; Gerla, M.; Zhou, S. The challenges of building mobile underwater wireless networks for aquatic applications. *IEEE Netw.* **2006**, *20*, 12–18.
2. Kilfoyle, D.; Baggeroer, A. The state of the art in underwater acoustic telemetry. *IEEE J. Ocean Eng.* **2000**, *25*, 4–27. [CrossRef]
3. Neasham, J.; Hinton, O. Underwater acoustic communications—How far have we progressed and what challenges remain. In Proceedings of the 7th European Conference on Underwater Acoustics, Delft, The Netherlands, 5–8 July 2004.
4. Confédération Mondiale des Activités Subaquatiques. Segni Convenzionali CMAS. Online pdf, 2003. Available online: https://www.cmas.ch/docs/it/downloads/codici-comunicazione-cmas/it-Codici-di-comunicazione-CMAS.pdf (accessed on 25 May 2018).

5. Recreational Scuba Training Council. Common Hand Signals for Recreational Scuba Diving. Online Pdf, 2005. Available online: http://www.neadc.org/CommonHandSignalsforScubaDiving.pdf (accessed on 25 May 2018).

6. Scuba Diving Fan Club. Most Common Diving Signals. HTML Page. 2016. Available online: http://www.scubadivingfanclub.com/Diving_Signals.html (accessed on 25 May 2018).

7. Jorge, M. Diving Signs You Need to Know. HTML Page. 2012. Available online: http://www.fordivers.com/en/blog/2013/09/12/senales-de-buceo-que-tienes-que-conocer/ (accessed on 25 May 2018).

8. Chiarella, D.; Bibuli, M.; Bruzzone, G.; Caccia, M.; Ranieri, A.; Zereik, E.; Marconi, L.; Cutugno, P. Gesture-based language for diver-robot underwater interaction. In Proceedings of the OCEANS 2015—Genova, Genoa, Italy, 18–21 May 2015; pp. 1–9. [CrossRef]

9. Tao, Y.; Wei, H.; Wang, T. A Speech Interaction System Based on Finite State Machine for Service Robot. In Proceedings of the 2008 International Conference on Computer Science and Software Engineering, Wuhan, China, 12–14 December 2008; Volume 1, pp. 1111–1114. [CrossRef]

10. Xu, Y.; Guillemot, M.; Nishida, T. An experiment study of gesture-based human-robot interface. In Proceedings of the 2007 IEEE/ICME International Conference on Complex Medical Engineering, Beijing, China, 23–27 May 2007; pp. 457–463. [CrossRef]

11. Waldherr, S.; Romero, R.; Thrun, S. A Gesture Based Interface for Human-Robot Interaction. *Auton. Rob.* **2000**, *9*, 151–173. [CrossRef]

12. Chomsky, N. Three models for the description of language. *IRE Trans. Inf. Theory* **1956**, *2*, 113–124. [CrossRef]

13. Backus, J.W. The Syntax and Semantics of the Proposed International Algebraic Language of the Zurich ACM-GAMM Conference. In Proceedings of the International Conference on Information Processing, UNESCO, Paris, France, 15–20 June 1959.

14. Dudek, G.; Sattar, J.; Xu, A. A Visual Language for Robot Control and Programming: A Human-Interface Study. In Proceedings of the 2007 IEEE International Conference on Robotics and Automation, Roma, Italy, 10–14 April 2007; pp. 2507–2513.

15. Xu, A.; Dudek, G.; Sattar, J. A natural gesture interface for operating robotic systems. In Proceedings of the 2008 IEEE International Conference on Robotics and Automation, ICRA, Pasadena, CA, USA, 19–23 May 2008; pp. 3557–3563.

16. Fiala, M. ARTag, a Fiducial Marker System Using Digital Techniques. In Proceedings of the 2005 IEEE Computer Society Conference on Computer Vision and Pattern Recognition (CVPR'05), San Diego, CA, USA, 20–25 June 2005; IEEE Computer Society: Washington, DC, USA, 2005; Volume 5, pp. 590–596. [CrossRef]

17. Sattar, J.; Bourque, E.; Giguere, P.; Dudek, G. Fourier tags: Smoothly degradable fiducial markers for use in human-robot interaction. In Proceedings of the Fourth Canadian Conference on Computer and Robot Vision (CRV '07), Montreal, QC, Canada, 28–30 May 2007; pp. 165–174. [CrossRef]

18. Islam, M.J.; Ho, M.; Sattar, J. Dynamic Reconfiguration of Mission Parameters in Underwater Human-Robot Collaboration. *arXiv* **2017**, arXiv:1709.08772.

19. Xu, P. Gesture-based Human-robot Interaction for Field Programmable Autonomous Underwater Robots. *arXiv* **2017**, arXiv:1709.08945.

20. Kim, B.; Jun, B.H.; Sim, H.W.; Lee, F.O.; Lee, P.M. The development of Tiny Mission Language for the ISiMI100 Autonomous Underwater Vehicle. In Proceedings of the OCEANS 2010 MTS/IEEE SEATTLE, Seattle, WA, USA, 20–23 September 2010; pp. 1–6. [CrossRef]

21. Garg, P.; Aggarwal, N.; Sofat, S. Vision based hand gesture recognition. *World Acad. Sci. Eng. Technol.* **2009**, *49*, 972–977.

22. Manresa, C.; Varona, J.; Mas, R.; Perales, F. Hand tracking and gesture recognition for human-computer interaction. *Electr. Lett. Comput. Vis. Image Anal.* **2005**, *5*, 96–104. [CrossRef]

23. Rautaray, S.S.; Agrawal, A. Vision Based Hand Gesture Recognition for Human Computer Interaction: A Survey. *Artif. Intell. Rev.* **2015**, *43*, 1–54. [CrossRef]

24. Biswas, K.; Basu, S.K. Gesture Recognition using Microsoft Kinect®. In Proceedings of the 5th International Conference on Automation, Robotics and Applications, Wellington, New Zealand, 6–8 December 2011; pp. 100–103.

25. Kawulok, M. Adaptive skin detector enhanced with blob analysis for gesture recognition. In Proceedings of the 2009 International Symposium ELMAR, Zadar, Croatia, 28–30 September 2009; pp. 37–40.

26. Elsayed, R.A.; Sayed, M.S.; Abdalla, M.I. Skin-based adaptive background subtraction for hand gesture segmentation. In Proceedings of the 2015 IEEE International Conference on Electronics, Circuits, and Systems (ICECS), Cairo, Egypt, 6–9 December 2015; pp. 33–36.

27. Chang, C.W.; Chang, C.H. A two-hand multi-point gesture recognition system based on adaptive skin color model. In Proceedings of the 2011 International Conference on Consumer Electronics, Communications and Networks (CECNet), XianNing, China, 16–18 April 2011; pp. 2901–2904. [CrossRef]

28. Ghaziasgar, M.; Connan, J.; Bagula, A.B. Enhanced adaptive skin detection with contextual tracking feedback. In Proceedings of the 2016 Pattern Recognition Association of South Africa and Robotics and Mechatronics International Conference (PRASA-RobMech), Stellenbosch, South Africa, 30 November–2 December 2016; pp. 1–6. [CrossRef]

29. Barneva, R.P.; Brimkov, V.E.; Hung, P.; Kanev, K. Motion tracking for gesture analysis in sports. In Proceedings of the 2016 IEEE Western New York Image and Signal Processing Workshop (WNYISPW), Rochester, NY, USA, 18 November 2016; pp. 1–5. [CrossRef]

30. Jiang, Y.; Hayashi, I.; Hara, M.; Wang, S. Three-dimensional motion analysis for gesture recognition using singular value decomposition. In Proceedings of the 2010 IEEE International Conference on Information and Automation, Harbin, China, 20–23 June 2010; pp. 805–810. [CrossRef]

31. Jost, C.; Loor, P.D.; Nédélec, L.; Bevacqua, E.; Stanković, I. Real-time gesture recognition based on motion quality analysis. In Proceedings of the 2015 7th International Conference on Intelligent Technologies for Interactive Entertainment (INTETAIN), Turin, Italy, 10–12 June 2015; pp. 47–56.

32. Czuszynski, K.; Ruminski, J.; Wtorek, J. Pose classification in the gesture recognition using the linear optical sensor. In Proceedings of the 2017 10th International Conference on Human System Interactions (HSI), Ulsan, Korea, 17–19 July 2017; pp. 18–24. [CrossRef]

33. Ng, C.W.; Ranganath, S. Gesture recognition via pose classification. In Proceedings of the 15th International Conference on Pattern Recognition, ICPR-2000, Barcelona, Spain, 3–7 September 2000; Volume 3, pp. 699–704. [CrossRef]

34. Yamashita, A.; Fujii, M.; Kaneko, T. Color registration of underwater images for underwater sensing with consideration of light attenuation. In Proceedings of the 2007 IEEE International Conference on Robotics and Automation, Roma, Italy, 10–14 April 2007; pp. 4570–4575.

35. Oliveira, M.; Sutherland, A.; Farouk, M. Two-stage PCA with interpolated data for hand shape recognition in sign language. In Proceedings of the 2016 IEEE Applied Imagery Pattern Recognition Workshop (AIPR), Washington, DC, USA, 18–20 October 2016; pp. 1–4. [CrossRef]

36. Birk, H.; Moeslund, T.B.; Madsen, C.B. Real-Time Recognition of Hand Alphabet Gestures Using Principal Component Analysis. In Proceedings of the 10th Scandinavian Conference on Image Analysis, Lappenranta, Finland, 9–11 June 1997.

37. Saxena, A.; Jain, D.K.; Singhal, A. Sign Language Recognition Using Principal Component Analysis. In Proceedings of the 2014 Fourth International Conference on Communication Systems and Network Technologies, Bhopal, India, 7–9 April 2014; pp. 810–813. [CrossRef]

38. Masurelle, A.; Essid, S.; Richard, G. Gesture recognition using a NMF-based representation of motion-traces extracted from depth silhouettes. In Proceedings of the 2014 IEEE International Conference on Acoustics, Speech and Signal Processing (ICASSP), Florence, Italy, 4–9 May 2014; pp. 1275–1279. [CrossRef]

39. Edirisinghe, E.M.P.S.; Shaminda, P.W.G.D.; Prabash, I.D.T.; Hettiarachchige, N.S.; Seneviratne, L.; Niroshika, U.A.A. Enhanced feature extraction method for hand gesture recognition using support vector machine. In Proceedings of the 2013 IEEE 8th International Conference on Industrial and Information Systems, Peradeniya, Sri Lanka, 17–20 December 2013; pp. 139–143. [CrossRef]

40. Chavez, A.G.; Pfingsthorn, M.; Birk, A.; Rendulić, I.; Miskovic, N. Visual diver detection using multi-descriptor nearest-class-mean random forests in the context of underwater Human Robot Interaction (HRI). In Proceedings of the OCEANS 2015—Genova, Genoa, Italy, 8–21 May 2015; pp. 1–7. [CrossRef]

41. Saha, H.N.; Tapadar, S.; Ray, S.; Chatterjee, S.K.; Saha, S. A Machine Learning Based Approach for Hand Gesture Recognition using Distinctive Feature Extraction. In Proceedings of the 2018 IEEE 8th Annual Computing and Communication Workshop and Conference (CCWC), Las Vegas, NV, USA, 8–10 January 2018; pp. 91–98. [CrossRef]

J. Mar. Sci. Eng. **2018**, *6*, 91

42. Hopcroft, J.E.; Motwani, R.; Ullman, J.D. chapter Context-Free Grammars and Languages. In *Introduction to Automata Theory, Languages, and Computation*, 2nd ed.; Addison-Wesley: Boston, MA, USA, 2001; pp. 169–217.
43. Jurafsky, D.; Martin, J.H. chapter Context-Free Grammars. In *Speech and Language Processing*; Prentice Hall: Upper Saddle River, NJ, USA, 2014; pp. 395–435.
44. Mišković, N.; Bibuli, M.; Birk, A.; Caccia, M.; Egi, M.; Grammer, K.; Marroni, A.; Neasham, J.; Pascoal, A.; Vasilijević, A.; Vukić, Z. CADDY—Cognitive Autonomous Diving Buddy: Two Years of Underwater Human-Robot Interaction. *Mar. Technol. Soc. J.* **2016**, *50*, 54–66. [CrossRef]
45. Mišković, N.; Pascoal, A.; Bibuli, M.; Caccia, M.; Neasham, J.A.; Birk, A.; Egi, M.; Grammer, K.; Marroni, A.; Vasilijević, A.; et al. CADDY Project, Year 2: The First Validation Trials. In Proceedings of the 10th IFAC Conference on Control Applications in Marine SystemsCAMS, Trondheim, Norway, 13–16 September 2016; Volume 49, pp. 420–425. [CrossRef]

Journal of
*Marine Science
and Engineering*

MDPI

Article

Fault-Tolerant Control for ROVs Using Control Reallocation and Power Isolation

Romano Capocci *, Edin Omerdic, Gerard Dooly and Daniel Toal

CRIS—Centre for Robotics & Intelligent Systems, University of Limerick, Castletroy, Limerick V94 T9PX, Ireland; edin.omerdic@ul.ie (E.O.); gerard.dooly@ul.ie (G.D.); daniel.toal@ul.ie (D.T.)
* Correspondence: romano.capocci@ul.ie; Tel.: +353-(0)85-759-3011

Received: 7 March 2018; Accepted: 10 April 2018; Published: 12 April 2018

Abstract: This paper describes a novel thruster fault-tolerant control system (FTC) for open-frame remotely operated vehicles (ROVs). The proposed FTC consists of two subsystems: a model-free thruster fault detection and isolation subsystem (FDI) and a fault accommodation subsystem (FA). The FDI subsystem employs fault detection units (FDUs), associated with each thruster, to monitor their state. The robust, reliable and adaptive FDUs use a model-free pattern recognition neural network (PRNN) to detect internal and external faulty states of the thrusters in real time. The FA subsystem combines information provided by the FDI subsystem with predefined, user-configurable actions to accommodate partial and total faults and to perform an appropriate control reallocation. Software-level actions include penalisation of faulty thrusters in solution of control allocation problem and reallocation of control energy among the operable thrusters. Hardware-level actions include power isolation of faulty thrusters (total faults only) such that the entire ROV power system is not compromised. The proposed FTC system is implemented as a LabVIEW virtual instrument (VI) and evaluated in virtual (simulated) and real-world environments. The proposed FTC module can be used for open frame ROVs with up to 12 thrusters: eight horizontal thrusters configured in two horizontal layers of four thrusters each, and four vertical thrusters configured in one vertical layer. Results from both environments show that the ROV control system, enhanced with the FDI and FA subsystems, is capable of maintaining full 6 DOF control of the ROV in the presence of up to 6 simultaneous total faults in the thrusters. With the FDI and FA subsystems in place the control energy distribution of the healthy thrusters is optimised so that the ROV can still operate in difficult conditions under fault scenarios.

Keywords: fault-tolerant control; thruster fault; fault detection and isolation; fault accommodation; ROV; remotely operated vehicle; underwater vehicle

---

1. Introduction

Over the past decades, the use of remotely operated vehicles (ROVs) has become more widespread. This is due to the reduction in costs driven by military and oil and gas research, making the technology available for other commercial and scientific purposes [1]. In more recent years ROVs have been employed for survey contracts and, with the push in the marine renewable energy (MRE) sector, ROVs will need to be capable of operating in more difficult environments to carry out close quarters inspections of the MRE devices and structures, thus reducing operational costs within the sector. The environment in which ROVs operate can be unpredictable, with the external parts of the system being subjected to seawater, changes in temperature, high pressures and interactions with solids drifting through the water column. These factors all contribute toward possibilities of thrusters becoming damaged or developing faults in their dynamic parts. In the past, it was common to abort a mission if a fault occurred in a thruster. Due to reduced weather windows at the sites of MRE converters [2], the expensive nature of ROV operations and the drive to reduce costs this method should be avoided, if possible. Fault-tolerant control system (FTC) within ROVs can be utilised to combat this. To accommodate faults and allow ROVs to manoeuvre in difficult environments they are usually designed so that they are over actuated. This increases the robustness of the system. Podder et al. utilised this configuration in a novel underwater vehicle approach for thruster force redistribution in the case of a fault [3]. Due to the particular advantages of FTC in ROV applications, many different techniques have been proposed:

Caccia et al. implemented thruster fault diagnosis by monitoring the motor current and revolutions per minute (RPM). If the monitored variables increased above a set threshold accommodation was performed by inhibiting the faulty thruster and by reconfiguring the distribution of the control actions cancelling the corresponding column in the thruster control matrix (TCM) [4].

Kim and Beale made use of Hotelling's T2 statistic to diagnosis a fault in an underwater vehicle. They compared measured variable basis data used for training, with actual variable data and carried out statistical analysis. If the results of the statistical analysis were above a certain threshold then a fault was present. Further analysis determined if the fault occurred in the stern plane (vertical) or rudder (horizontal). The system was designed so that the controller was reconfigurable, meaning that the type of fault in the system determined the type of controller to be utilised. Their tests were carried out in simulations and found that noise can increase fault detection times [5].

Montazeri et al. proposed fault diagnosis in the steering system of an autonomous underwater vehicle (AUV) through the use of two different neural network systems (multi-layer perceptron (MLP) and Adaline). This method, validated using simulations, was capable of detecting both partial and full faults. Results found that the speed of the MLP fault detection was lower than the Adaline method due to its larger complexity [6]. Zhu et al. utilised a neural network to increase the performance of on-line fault detection in thrusters on an open-frame ROV and provide appropriate control reallocation. This approach was tested in simulations but only total faults of the thruster were taken into consideration, which was different from the simulated fault situation of the thrusters [7].

A combination of tools was employed by Hai et al. for a fault-tolerable control scheme for an open-frame ROV. The methods combined were a petri network and a recurrent fuzzy neural network (PRFNN). This approach combines the advantages of low level learning, high level reasoning and reduced calculations. Simulation and experiments proved that the ROV FTC could accurately detect partial and full faults and accommodate this in the control [8].

Another previous approach was to integrate self-organizing maps and fuzzy logic clustering to achieve fault diagnosis. Upon diagnosing the fault a novel weighted pseudo-inverse scheme compensated in the control [9].

Liu and Zhu conducted thruster external fault diagnosis on an ROV by comparing expected heading values with actual heading values for a given control voltage and referencing it back to a fault code table [10]. Akmal et al. developed a fuzzy based thruster fault diagnosis and accommodation system, which monitored voltage and current, and compared it to pre-measured values. The values of the resulting residuals were then used to compare to a fuzzy fault code table, assigning PWM control constraints to the thrusters, depending on their state. This simple approach was successful but was unable to distinguish if there was an internal or external fault [11].

The FTC system for an ROV, proposed in [12], used a modified version of the Moore-Penrose pseudo inverse to redistribute the control effort to healthy thrusters if a fault occurs.

In the literature the majority of the fault-tolerant control methods proposed for underwater vehicles have been integrated and evaluated in a simulated environment. As a first step for evaluation this approach is beneficial but, when practical and affordable, real-world trials should be conducted to evaluate the system in the environment in which it is expected to operate. Real-world trials have been conducted in some cases to evaluate the FTC of the ROV [4,12], generating more accurate results.

This paper presents an active fault-tolerant control system for an ROV using a combination of thruster fault detection and isolation, faulty thruster power isolation and fault accommodation in the control of the ROV. The word "active" means that the method is based on active monitoring of relevant signals from the thrusters (currents, shaft speeds, applied voltages and temperatures of windings). For thrusters that cannot provide these signals, the authors are currently working on novel "passive" method for fault detection and isolation. It is expected to publish the main features of the "passive" method and comparison table of both methods in Spring 2018.

Most underwater vehicle fault detection schemes are model-based, and concern the dynamic relationship between actuators and vehicle behaviour or the specific input-output thruster dynamics [13]. The proposed thruster fault detection and isolation (FDI) approach is a model-free scheme based on a pattern recognition neural network, trained with simulated and real-world data. New features of the proposed FTC includes separation of the virtual control space into vertical and horizontal thruster subspaces (planes) and real-world implementation in the real-time ROV control system for various thruster configurations. The thruster FA subsystem receives thruster state data from the FDI and, in the case of a fault, reallocates thruster forces by reducing the saturation bounds of the faulty thruster (software-level action A). At the same time, in case of total faults, the thruster power is switched off (hardware-level action B), in order to prevent a threat to full power system and reduce the risk of damage to other ROV components.

The proposed FTC system has been successfully tested in a virtual environment (using a real time ROV simulator, created at CRIS, in the University of Limerick (UL)) and a real world environment (using VM5 thrusters from VideoRay in Pottstown, PA, USA) in a test tank at UL. The proposed FTC system is part of the OceanRINGS+ ROV smart control system, currently under development at CRIS, UL.

The paper outline is provided as follows. In Section 2 background information is provided, including links with other research projects and a short description of Inspection ROV (IROV). The architecture of the FTC, including description and implementation of the FDI & FA subsystems, is described in Section 3. Section 4 presents the testing and evaluation results of the proposed FTC in real-world and virtual (simulated) environments. Finally, concluding remarks and directions for future work are provided in Section 5.

2. Background

2.1. OceanRINGS⁺

OceanRINGS is an Internet/Ethernet-enabled ROV control system, based on robust control algorithms, deterministic network-oriented hardware and flexible, 3-layer software architecture [14]. ROV LATIS is a prototype platform developed at UL to test and validate OceanRINGS [15]. System validation and technology demonstration has been performed over the last eight years through a series of test trials with different support vessels. Operations include subsea cable inspection/survey, wave energy farm cable to shore routing, shipwreck survey, ROV-ship synchronisation and oil spill/HNS incident response.

Currently, researchers at UL are developing the next generation of the ROV control system (OceanRINGS⁺, the extended version of OceanRINGS). The highly adaptive 3-layer software architecture of OceanRINGS⁺ includes fault-tolerant control allocation algorithms in the bottom layer, transparent interface between an ROV and supporting platforms (surface platforms, surface/subsea garages and/or supporting vessels) in the middle layer and assistive tools for mission execution/monitoring/supervision in the top layer. Software modules have been developed for advanced control modes, such as auto compensation of ocean currents based on ROV absolute motion, robust speed/course controller with independent heading control, semi and full auto pilot capabilities, auto-tuning procedure for low-level controllers, ROV high precision dynamic position & motion control in absolute earth-fixed frame, or relative to target or support platform/vessel. This paper is focused on the description of the FTC, the module at the bottom layer of the full OceanRINGS⁺ control architecture.

2.2. Inspection-Class Remotely Operated Vehicle (I-ROV)

As part of the ongoing MaREI research project "Smart Inspection ROVs for Use in Challenging Conditions", researchers at the Centre for Robotics & Intelligent Systems (CRIS), University of Limerick have designed and developed a reconfigurable, inspection-class ROV (I-ROV) in the period of 2014–2018, aimed to perform periodic and post storm inspection of offshore MRE converters, moorings and foundations, reducing the need for commercial divers to be employed in this difficult and potentially dangerous environment. The I-ROV (Figure 1) is a reconfigurable system with the option to utilise two types of thrusters in different configurations.

Figure 1. Inspection-class remotely operated vehicle (I-ROV), developed at the Centre for Robotics & Intelligent Systems (CRIS) researchers, University of Limerick (UL).

The reconfigurable propulsion system of I-ROV includes two thruster configurations (Table 1):

- Configuration 1: Eight VideoRay M5 thrusters configured in two layers: Horizontal Layer with four thrusters and Vertical Layer with four thrusters. These thrusters provide active monitoring of relevant signals from thrusters (currents, shaft speeds, applied voltages and temperatures of windings).
- Configuration 2: Twelve Blue Robotics T200 thrusters configured in three layers: Horizontal Layers L0 and L1 with four thrusters each and Vertical Layer with four thrusters. These thrusters cannot provide monitoring of relevant signals and the passive FTC system for this class of thruster is under development.

It should be emphasized that the OceanRINGS$^+$ control architecture has been designed to be generic, i.e., not limited to exclusive use by IROV, but any ROV with standard physical layout of thrusters. Although testing and validation of proposed FTC is performed with IROV configured as Configuration 1, the active FTC proposed in this paper and implemented as part of OceanRINGS$^+$ is applicable to open-frame ROVs with a maximum of 12 thrusters subject to the constraint that each thruster can provide measurement of relevant signals (currents, shaft speeds, applied voltages and temperatures of windings). The FDI subsystem detects faults in thrusters regardless of their physical layout. However, for successful fault accommodation, the physical layout of thrusters plays an important role.

Table 1. Thruster configurations.

Configuration 1: 8 × M5 Thrusters	Configuration 2: 12 × T200 Thrusters
Horizontal Layer: HT1, HT2, HT3, HT4	Horizontal Layer (L0): HT1, HT2, HT3, HT4
	Horizontal Layer (L1): HT1, HT2, HT3, HT4
Vertical Layer: VT1, VT2, VT3, VT4	Vertical Layer: VT1, VT2, VT3, VT4

3. Fault-Tolerant Control (FTC) System

3.1. FTC Architecture

The overall functional architecture of the proposed FTC system is shown in Figure 2. The description of the architecture is provided in a hierarchical manner, such that the general description and the main idea are introduced first, while more details about individual components can be found in the following subsections. This architecture is an extension of the control architecture proposed in [16].

In contrast to the Fault Diagnosis and Accommodation System (FDAS), proposed in [17] and implemented in the original version of OceanRINGS, there are a number of new features in the FTC architecture shown in Figure 2 and implemented in OceanRINGS$^+$. Firstly, there is a clear separation between horizontal and vertical thrusters using decomposition of motion into horizontal and vertical subspaces (planes). Secondly, there is provision for two layers of horizontal thrusters and one layer of vertical thrusters. Thirdly, the Fault Detection Units utilize a pattern recognition neural network for real-time fault detection instead of self-organising maps.

The virtual control input τ for the control allocation is a normalised vector of forces and moments: $\tau = \begin{bmatrix} \tau_X & \tau_Y & \tau_Z & \tau_K & \tau_M & \tau_N \end{bmatrix}^T$, where τ_X is surge force, τ_Y is sway force, τ_Z is heave force, τ_K is roll moment, τ_M is pitch moment, and τ_N is yaw moment. Decomposition of this vector into the horizontal and vertical planes is presented in Table 2.

The Control Clusters (Figure 2 and Table 2) are links between the FTC module in the bottom layer and upper layers in control architecture. The HT Control Cluster consists of two subclusters: Virtual Joystick (to mimic direct surge & sway forces and yaw moment generated by human or virtual pilot) and a set of settings for Low-Level Controllers (set points for surge speed u_d (m/s), sway speed v_d (m/s) and heading Y_d (°), feed-forward inputs and on/off switches to enable/disable individual controllers). The VT Control Cluster has two subclusters: Virtual Joystick (to mimic direct heave force and roll & pitch moments generated by human or virtual pilot) and a set of settings for Low-Level Controllers (set points for depth/altitude z_d (m), roll R_d (°) and pitch P_d (°), feed-forward inputs and on/off switches to enable/disable individual controllers). Inside the Synthesis module the LLC

Settings subclusters are used as one of the inputs to the LLC loops (the other inputs are navigation data and parameters of controllers). There is a single controller for each degree of freedom (DOF). Surge and Sway controllers are velocity controllers, while Heave, Roll, Pitch & Yaw are position controllers. Further information about the internal structure of LLC loops can be found in [9]. Individual outputs of LLC loops are bundled into vectors τ_{LLC}. The final outputs of the Synthesis module are vectors τ_{HT} and τ_{VT}, obtained by summation and normalisation of corresponding vectors τ_{VJ} and τ_{LLC} for each subspace, respectively.

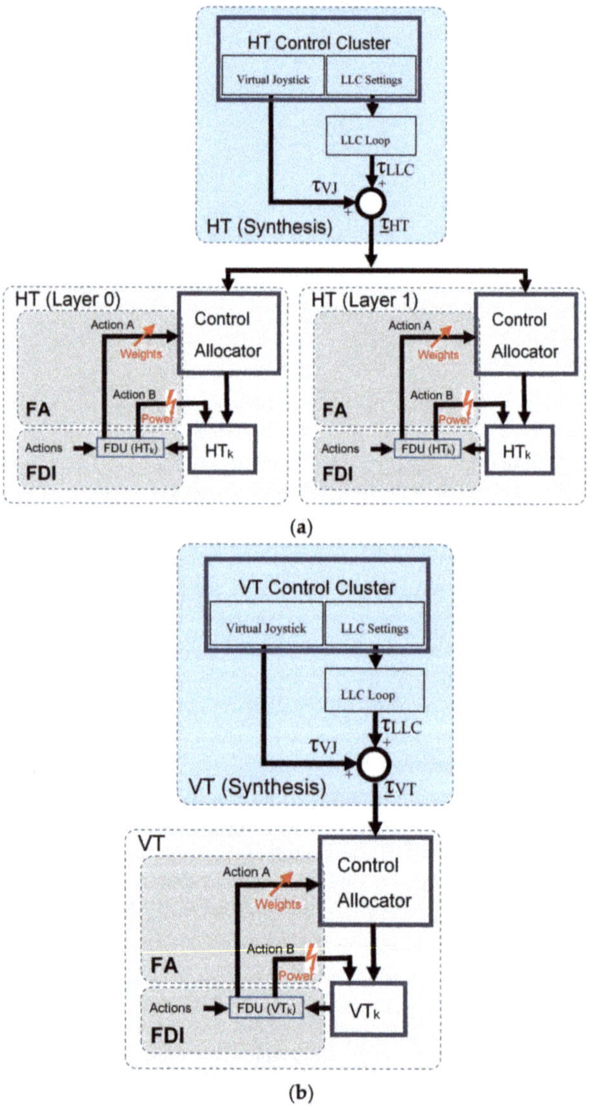

Figure 2. Architecture of the Fault Tolerant Control (FTC) system: (**a**) Horizontal Subspace (Plane); (**b**) Vertical Subspace (Plane).

Table 2. Decomposition into Horizontal and Vertical Subspaces (Planes).

Virtual Control Input	Horizontal Plane τ_{HT}		Vertical Plane τ_{VT}	
$\tau = \begin{bmatrix} \tau_X \\ \tau_Y \\ \tau_Z \\ \tau_K \\ \tau_M \\ \tau_N \end{bmatrix}$	$\tau_{HT} = \begin{bmatrix} \tau_X \\ \tau_Y \\ \tau_N \end{bmatrix}$	τ_X—Surge Force τ_Y—Sway Force τ_N—Yaw Moment	$\tau_{VT} = \begin{bmatrix} \tau_Z \\ \tau_K \\ \tau_M \end{bmatrix}$	τ_Z—Heave Force τ_K—Roll Moment τ_M—Pitch Moment

Each layer of horizontal thrusters has its own FDI, FA and Control Allocator module. It should be emphasized that both layers of horizontal thrusters are independent from each other, i.e., they can have a different physical layout and number of thrusters. In the Horizontal Plane, both layers have the same input (vector τ_{HT}), which is the output of HT Synthesis module.

3.2. Fault Detection and Isolation (FDI) Subsystem

3.2.1. Fault Classification

Thrusters are liable to different fault types during the underwater mission e.g., propellers can be jammed, broken or lost, water can penetrate inside the thruster enclosure, communication between the thruster and the master node can be lost, applied voltage or temperature of the winding can exceed the threshold, etc. Some of these faults (partial faults) are not critical and the thruster is able to continue operation in the presence of a fault with the restricted usage, i.e., reduced maximum velocity. In other cases (total faults—failures) the thruster must be switched off and the mission has to be continued with remaining operable thrusters. Thruster faults are classified into two main classes:

- *Internal faults* (e.g., temperature of the windings is out of range, drop in bus voltage etc.),
- *External faults* (e.g., lightly jammed, jammed, heavily jammed, lost or broken propeller).

3.2.2. Fault Code Table

Relationships between thruster states, fault types and remedial actions are stored in the thruster fault code table (Table 3). It must be emphasized, at this point, that this fault code table is just a suggestion, intended to reveal the main ideas of the proposed FTC system. New states (rows) can be added, and the existing relationships can be changed, in order to accommodate specific requirements and available thruster data. For example, other faults like thruster shaft misalignment or damaged bearings can cause excessive vibration, increased temperature and, in worst case scenarios, cause flooding through broken shaft or damaged enclosure [18].

Table 3. Thruster Fault Code Table.

Thruster State	Class	Type	Action A: Saturation Bounds	Action B: Thruster Power
Invalid	-	-	1.00	ON
Healthy	-	-	1.00	ON
Lightly Jammed	External	Partial	0.75	ON
Jammed	External	Partial	0.25	ON
Heavily Jammed	External	Total	0.00	OFF
Broken Propeller	External	Total	0.00	OFF
Unknown	External	Total	0.00	OFF
Voltage outside threshold	Internal	Total	0.00	OFF
Temp. outside threshold	Internal	Total	0.00	OFF

3.2.3. Fault Detection Unit (FDU)

It is assumed that each thruster is driven by a Thruster Control Unit (TCU), with integrated power amplifiers and a microcontroller (Figure 3). The input to the TCU is desired shaft speed n_d (%). The outputs are current I (A), shaft speed n (rpm), bus voltage V (V) and temperature T (°C) of windings. The outputs of each thruster are sent to the associated FDI module in real-time. The FDI module utilises the Fault Detection Units (FDUs) to monitor the state of the thrusters. The FDU is a software module associated with the thruster, able to detect internal faults and external faults. The output of the FDU is a fault state vector f_i. Connections between the FDU and the TCU for an arbitrary thruster T_i are indicated in Figure 3.

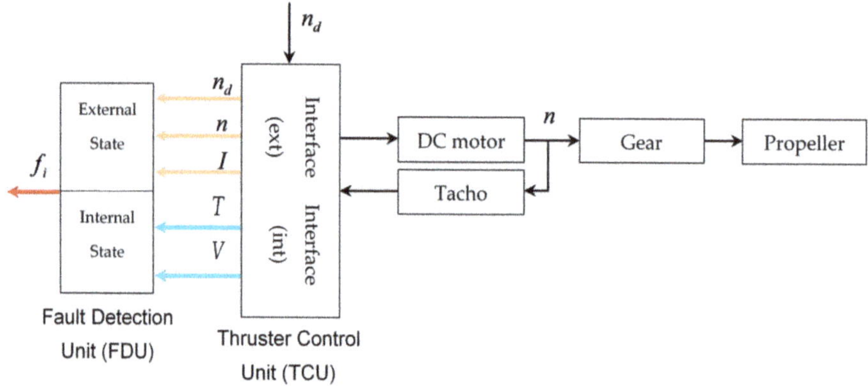

Figure 3. Block diagram showing connections between the fault detection unit (FDU) and the thruster control unit (TCU) for thruster T_i.

Signals for detection of internal faults are already available in existing TCUs for VideoRay thrusters M5. In particular, the communication protocol for the M5 provides monitoring of the winding temperature T (°C) and bus voltage V (V) of each thruster. In order to build a universal FDU, able to detect both internal and external faults, it is necessary to augment the existing internal protection with a software module for fast and reliable detection of external faults.

For detection of external faults available signals are desired: shaft speed n_d (%), actual shaft speed n (rpm) and current consumption I (A) of the thruster. By monitoring n and I, together with desired shaft speed n_d, obtained as the output of the Control Allocation, the FDU is designed to detect, isolate and categorise external thruster faults using Pattern Recognition Neural Network (PRNN).

Finally, the universal FDU integrates both parts (internal and external) into one unit, which is able to detect both internal and external faults (Figure 3). Integration includes a priority scheme, where total faults have higher priority than partial faults. The fault state vector f_i, the output of the FDU, is the code of the fault.

3.2.4. Implementation

Implementation involves two phases: *off-line training* and *on-line fault detection*.

Off-Line Training Phase

The first stage in the training phase is acquisition of training data. In the virtual environment, thruster faults are simulated by varying properties of the thruster dynamic model (load, friction, etc.) inside the propulsion subsystem of the ROV dynamics simulator. In the real-world environment, various jammed propeller faults are simulated such that the objects of different sizes, shapes and weights were attached to the blades, while a broken propeller was simulated with all blades removed from the shaft. Further details about acquisition of training data in real-world environment is given in the following. A normal state and four different fault cases were considered (lightly jammed, jammed, heavily jammed and broken (lost) propeller). A test rig has been set up and thruster mounted in a test tank in the University of Limerick. To mimic the jammed thruster states various objects ("blockages") have been attached to the thruster propeller during tests. The thruster test setup and "blockages" are shown in Figure 4. Each "blockage" reduces efficiency of the thruster due to increased load on the shaft and reduced flow of the water through the duct.

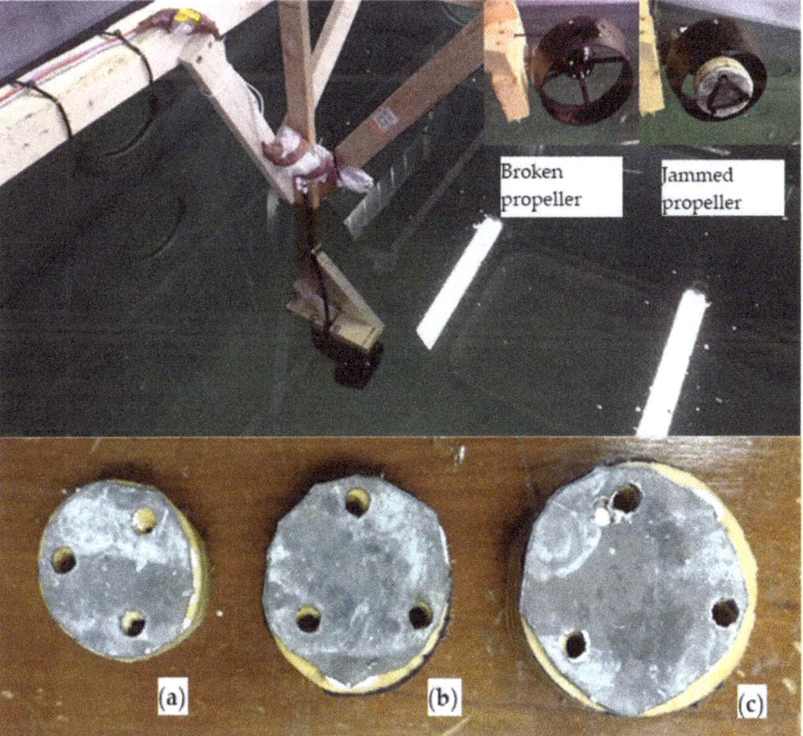

Figure 4. Thruster setup for acquisition of real-world data: (**a**) 30% area "blockage"; (**b**) 60% area "blockage"; (**c**) 90% area "blockage".

The relationship between the thruster states and setup for real-world data acquisition is given in Table 4. For each state in Table 4, the thruster was actuated with a saw-like command signal n_d (%) with the following pattern: 0% » $MAX\%$ » 0% » $-MAX\%$ » 0%, with a step size 1%. The total duration of signal was 20 s, sampling period 50 ms and max. value $MAX = 100\%$. In each iteration the desired shaft speed n_d (%), actual shaft speed n (rpm) and current consumption I (A) have been logged. The real-world raw data acquired for each thruster state are presented in Figure 5.

Table 4. Thruster states & setup for real-world data acquisition.

Thruster State	Setup
Healthy	No blockage
Lightly Jammed	Blockage (30% area)
Jammed	Blockage (60% area)
Heavily Jammed	Blockage (90% area)
Broken Propeller	Propeller detached

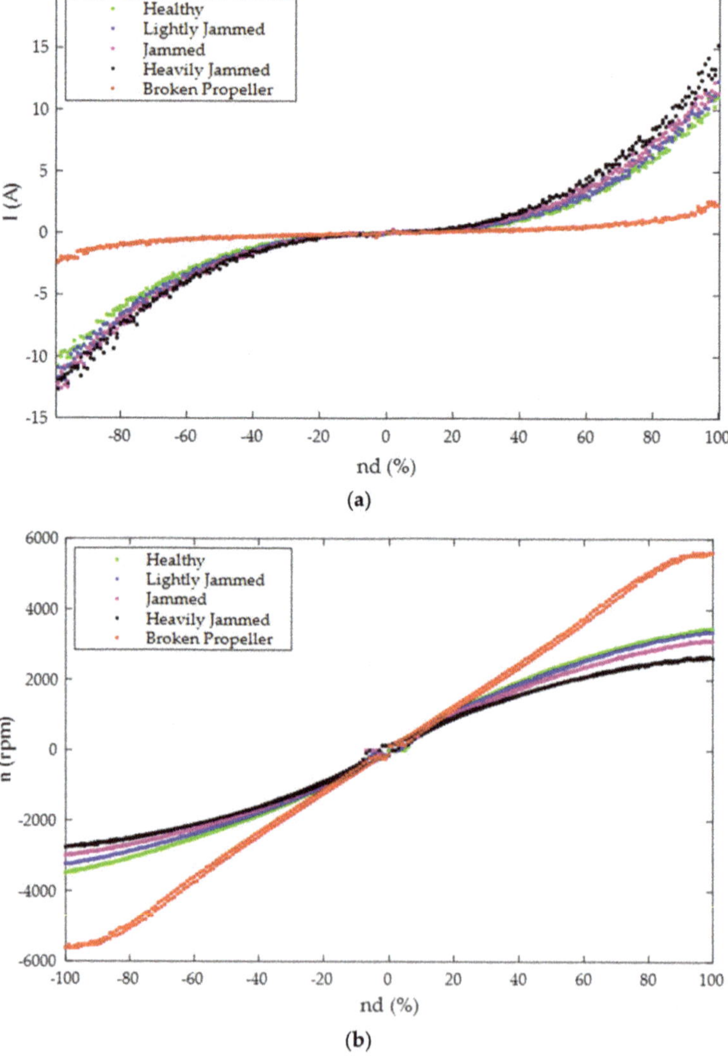

Figure 5. Diagrams of raw training data: (**a**): I versus n_d plot; (**b**) n versus n_d plot.

Analysing the distribution of the training data in Figure 5, the first feature that can be noticed is that each fault type creates a certain pattern. The presence of measurement noise is noticeable in acquired data for currents, resulting in patterns which exhibit a fuzzy ("cloudy") look. The second feature is that it is very difficult to distinguish individual patterns in the zone around $n_d \approx 0$ (called the *critical zone*). This makes successful fault detection and isolation in the critical zone difficult to achieve. In particular, for the near-zero velocity case $n_d \approx 0$ the thruster does not rotate or rotates very slowly, reliable fault detection is impossible. The solution to this issue is the exclusion of the critical zone from FDI during the on-line fault detection phase. For this reason, the critical zone is called the *forbidden zone* with associated "Invalid" thruster state. When desired shaft speed is in this zone, the FDI algorithm goes to sleep mode and outputs the "Invalid" state without any action.

Each fault type in Figure 5 is characterised by specific features, which makes them different from the other types. These features are discussed in the following. In general, all the variables (n_d, I and n) are correlated, i.e., they tend to rise and fall together in a non-linear way. For lightly jammed, jammed and heavily jammed propeller states, objects ("blockages") attached to the blades generate an additional load for the motor, leading to higher current I and lower n than in the fault-free case for the same value of n_d. In the case of a broken propeller, the absence of the blades means that the load for the motor is much smaller than in other cases, yielding a reduction in current consumption and a significant increase in shaft speed. However, the thruster does not generate any propulsion force in this case.

The main idea of the *second stage* in the training phase is to use the acquired data from the first stage to train a Pattern Recognition Neural Network (PRNN). In order to improve the PRNN classification accuracy, the raw data shown in Figure 5 have been first replaced with best fit curves, as shown in Figures 6 and 7, before creating input-output data sets for NN training. The MATLAB App "Curve Fitting" has been employed to generate best fit curves. Details of the Curve Fitting functions and their parameters can be found in Table 5.

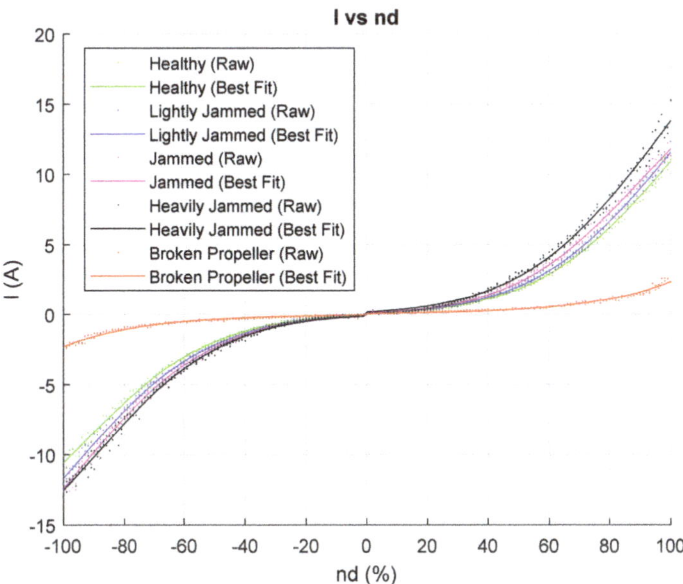

Figure 6. Diagrams of raw & best fit training data: *I* versus n_d plot.

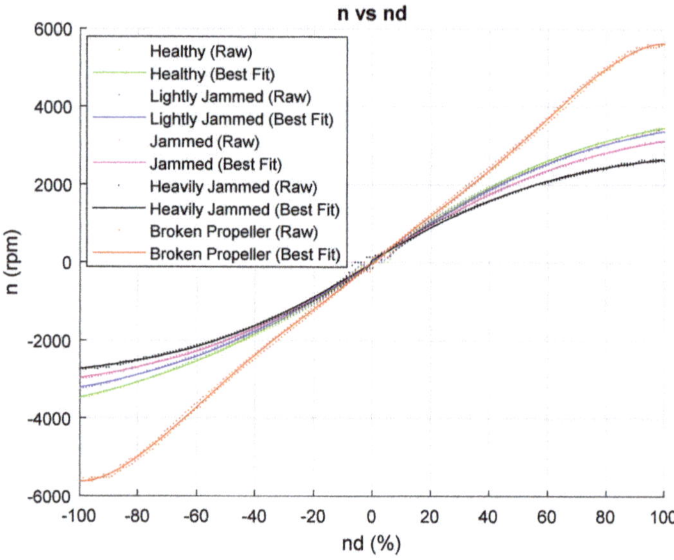

Figure 7. Diagrams of raw & best fit training data: n versus n_d plot.

Table 5. Curve Fitting functions for thruster states.

Thruster State Dataset Plot	Curve Fitting Function Type	Curve Fitting Function	Coefficients
Healthy I v n_d positive	Gaussian	$f(x) = a * \exp(-((x - b)/c)^2)$	a = 19.15 b = 147.8 c = 63.78
Healthy I v n_d negative	Gaussian	$f(x) = a * \exp(-((x - b)/c)^2)$	a = −16.07 b = −140.2 c = 62.24
Healthy n v n_d positive	Hyperbolic tangent	$f(x) = a * \tanh(b * x) + c$	a = 4008 b = 0.01298 c = 7.086
Healthy n v n_d negative	Hyperbolic tangent	$f(x) = a * \tanh(b * x) + c$	a = 4212 b = 0.01134 c = −42.63
Lightly Jammed I v n_d positive	Gaussian	$f(x) = a * \exp(-((x - b)/c)^2)$	a = 20.04 b = 147.8 c = 64.41
Lightly Jammed I v n_d negative	Gaussian	$f(x) = a * \exp(-((x - b)/c)^2)$	a = −19.26 b = −145.9 c = 65.05
Lightly Jammed n v n_d positive	Hyperbolic tangent	$f(x) = a * \tanh(b * x) + c$	a = 3908 b = 0.01274 c = 30.12
Lightly Jammed n v n_d negative	Hyperbolic tangent	$f(x) = a * \tanh(b * x) + c$	a = 3764 b = 0.01237 c = −37.82
Jammed I v n_d positive	Gaussian	$f(x) = a * \exp(-((x - b)/c)^2)$	a = 16.27 b = 133.9 c = 59.91
Jammed I v n_d negative	Gaussian	$f(x) = a * \exp(-((x - b)/c)^2)$	a = −20.4 b = −146.1 c = −65.54
Jammed n v n_d positive	Hyperbolic tangent	$f(x) = a * \tanh(b * x) + c$	a = 3565 b = 0.01289 c = 60.38

Table 5. *Cont.*

Thruster State Dataset Plot	Curve Fitting Function Type	Curve Fitting Function	Coefficients
Jammed n v n_d *negative*	Hyperbolic tangent	$f(x) = a * \tanh(b * x) + c$	$a = 3352$ $b = 0.01355$ $c = -29.18$
Heavily Jammed I v n_d *positive*	Gaussian	$f(x) = a * \exp(-((x - b)/c)^2)$	$a = 22.92$ $b = 147.3$ $c = 66.57$
Heavily Jammed I v n_d *negative*	Gaussian	$f(x) = a * \exp(-((x - b)/c)^2)$	$a = -17.17$ $b = -134$ $c = 60.65$
Heavily Jammed n v *positive*	Hyperbolic tangent	$f(x) = a * \tanh(b * x) + c$	$a = 2875$ $b = 0.01468$ $c = 68.49$
Heavily Jammed n v *negative*	Hyperbolic tangent	$f(x) = a * \tanh(b * x) + c$	$a = 3007$ $b = 0.01488$ $c = -14.55$
Broken Propeller I v n_d *positive*	Exponential	$f(x) = a * \exp(b * x)$	$a = 0.05542$ $b = 0.03747$
Broken Propeller I v n_d *negative*	Exponential	$f(x) = a * \exp(b * x)$	$a = -0.05494$ $b = -0.03735$
Broken Propeller n v n_d *positive*	Polynomial	$f(x) = p1 * x^4 + p2 * x^3 + p3 * x^2 + p4 * x + p5$	$p1 = -0.0001074$ $p2 = 0.01742$ $p3 = -0.8264$ $p4 = 72.56$ $p5 = -46.07$
Broken Propeller n v n_d *negative*	Polynomial	$f(x) = p1 * x^4 + p2 * x^3 + p3 * x^2 + p4 * x + p5$	$p1 = 0.0001022$ $p2 = 0.0163$ $p3 = 0.7478$ $p4 = 70.58$ $p5 = 34.14$

A two-layer feed-forward network, with 16 sigmoid hidden and 5 softmax output neurons has been trained to classify input vectors. The architecture of the Pattern Recognition Neural Network is shown in Figure 8.

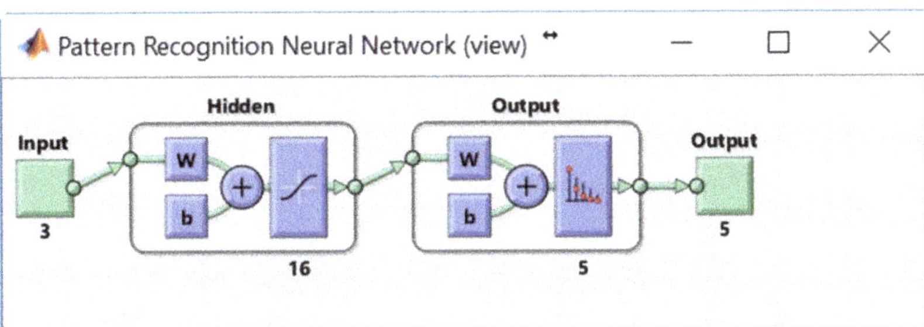

Figure 8. The architecture of Pattern Recognition Neural Network.

As indicated in Table 4, there are five classes (Healthy, Lightly Jammed, Jammed, Heavily Jammed and Broken Propeller). The input data set (matrix thrusterInputs_RWE) has dimension 3×1005 and consists of five Input Blocks: thrusterInputs0_RWE, thrusterInputs1_RWE, thrusterInputs4_RWE (one block for each class, see Table 6). Network inputs are stored in columns of the matrix thrusterInputs_RWE. For each class, values of n_d are -100, -99, $+99$, $+100$.

Table 6. Structure of the input training data set thrusterInputs_RWE.

Class	Input Block	Size	Column(j)
Healthy	thrusterInputs0_RWE	3×201	$n_d(j); I(j); n(j)$
Lightly Jammed	thrusterInputs1_RWE	3×201	$n_d(j); I(j); n(j)$
Jammed	thrusterInputs2_RWE	3×201	$n_d(j); I(j); n(j)$
Heavily Jammed	thrusterInputs3_RWE	3×201	$n_d(j); I(j); n(j)$
Broken Propeller	thrusterInputs4_RWE	3×201	$n_d(j); I(j); n(j)$

The target data set (matrix thrusterTargets_RWE) has dimension 5×1005 and consists of five Target Blocks: thrusterTargets0_RWE, thrusterTargets1_RWE, thrusterTargets4_RWE (one block for each class, see Table 7). For class k columns of corresponding Target Block have *1* at position k, while all other column elements have value 0.

Table 7. Structure of the output training data set thrusterTargets_RWE.

Class	Target Block	Size	Column
Healthy	thrusterTargets0_RWE	5×201	1; 0; 0; 0; 0
Lightly Jammed	thrusterTargets1_RWE	5×201	0; 1; 0; 0; 0
Jammed	thrusterTargets2_RWE	5×201	0; 0; 1; 0; 0
Heavily Jammed	thrusterTargets3_RWE	5×201	0; 0; 0; 1; 0
Broken Propeller	thrusterTargets4_RWE	5×201	0; 0; 0; 0; 1

Algorithms used in PRNN training are provided in Table 8.

Table 8. Pattern Recognition Neural Network training algorithms.

	Algorithms
Data Division	Random (dividerand)
Training	Scaled Conjugate Gradient (trainscg)
Performance	Cross-Entropy (crossentropy)
Calculations	MEX

The input matrix thrusterInputs_RWE has been randomly divided up into training samples (70%), Validation samples (15%) and Testing samples (15%). These samples were presented to the network during training, and the network was adjusted according to its error. Validation samples were used to measure network generalization, and to halt training when the generalization stops improving. The testing samples have no effect on training and provide an independent measure of network performance during and after training. The training, Validation and Confusion Matrices are shown in Figure 9. On the confusion matrix plot, the rows correspond to the predicted class (Output Class), and the columns show the true class (Target Class). The diagonal cells show for how many (and what percentage) of the examples the trained network correctly estimates the classes of observations. That is, it shows what percentage of the true and predicted classes match. The off diagonal cells show where the classifier has made mistakes. The column on the far right of the plot presents the accuracy for each predicted class, while the row at the bottom of the plot shows the accuracy for each true class. The cell in the bottom right of the plot shows the overall accuracy.

Figure 10 displays the Neural Network Cross-Entropy and Performance plots. Minimizing Cross-Entropy within the neural network results in enhanced classification.

The Receiver Operating Characteristic (ROC) is a metric employed to check the quality of classifiers. For each class of a classifier, the ROC applies threshold values across the interval [0, 1] to outputs. For each threshold, two values are calculated, the True Positive Ratio (TPR) and the False Positive Ratio (FPR). For a particular class i, the TPR is the number of outputs whose actual and predicted class is class *i*, divided by the number of outputs whose predicted class is class *i*. The FPR is the number of outputs whose actual class is not class *i*, but predicted class is class *i*, divided by the number of outputs whose predicted class is not class *i*. Figure 11 displays the ROC for each output class of PRNN. The more each curve hugs the left and top edges of the plot, the better the classification.

The classification performance of the PRNN in the real-world environment is verified in Section 4.

In the third and last stage of the training phase, the structure of the trained PRNN is saved as a MATLAB function nn_pr_16_RWE on the hard disk for future use. In this way, time consuming training calculations are performed off-line, during the training phase, which enables fast and efficient detection during the on-line phase.

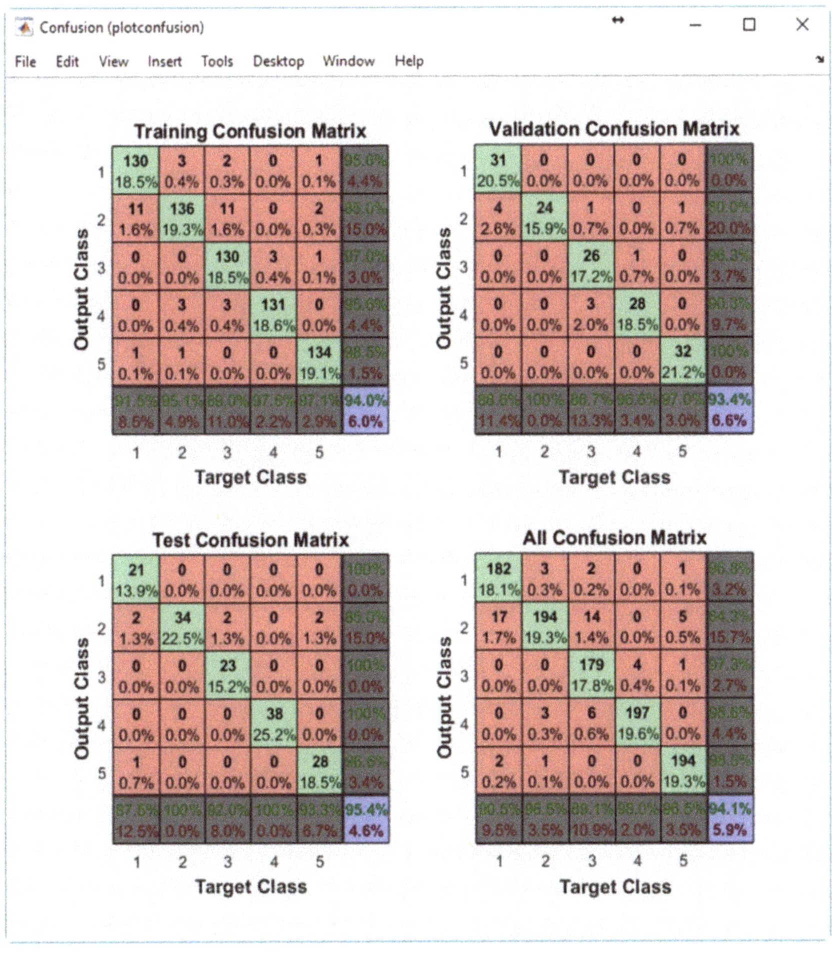

Figure 9. Plots of Training, Validation, Test and All Confusion Matrices.

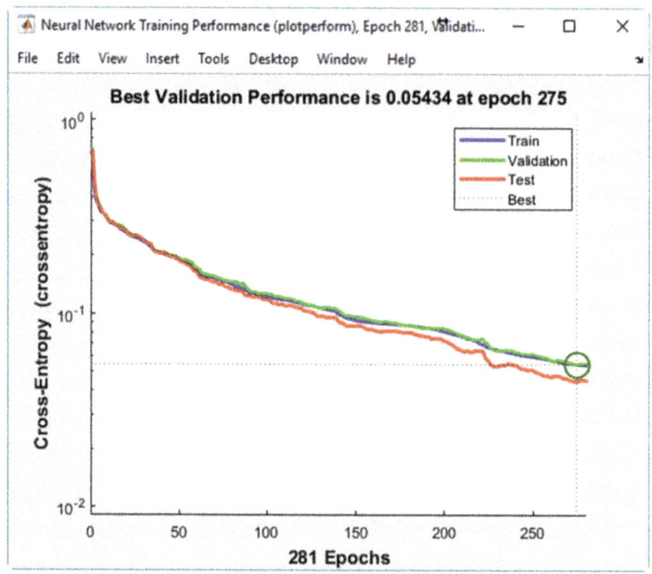

Figure 10. Neural Network Cross-Entropy and Performance plots.

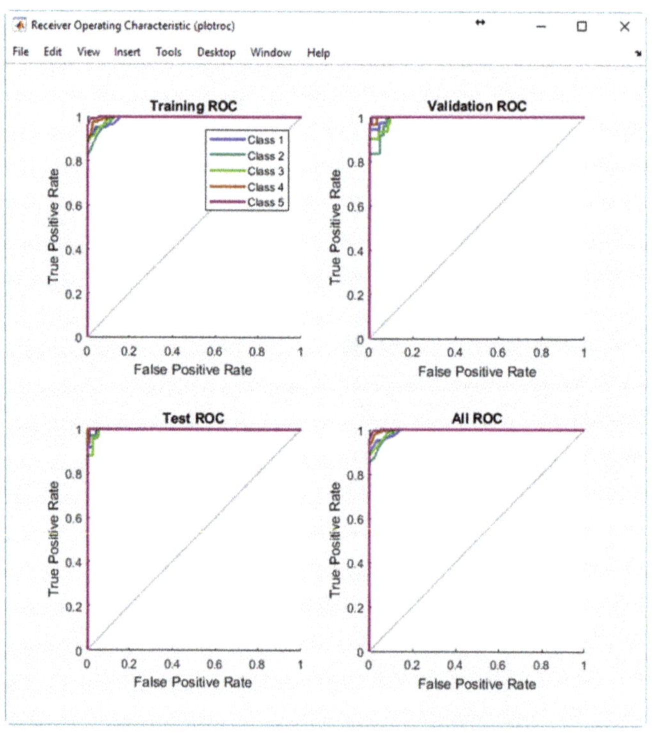

Figure 11. Receiver Operating Characteristic plots.

On-Line Fault Detection Phase

From the preceding discussion, the problem of thruster fault detection is considered as a pattern recognition problem. An original method for on-line fault detection, adapted to the specific features of the underlying pattern recognition problem, will now be described.

During the initialisation stage of the on-line fault detection phase, training data for each class (acquired in the off-line training phase) are loaded into memory and displayed as separate static background plots I versus n_d and n versus n_d for each layer (HT Layer 0, HT Layer 1 and VT Layer). These plots are utilised to represent the relationship between variables for different thruster states and for visualisation of actual thruster measurements in real-time. Additional activities during initialisation phase include memory allocation for buffers, reading fault code table settings from file and the creation of action lists.

After the initialisation is finished, the fault detection is performed by repeating the steps from the FDU Algorithm (Table 9) for each thruster at each programme cycle.

Table 9. FDU Algorithm—On-line fault detection.

	FDU Algorithm
1	Read values for external faults (n_d, I and n) and internal faults (V and T) from TCU.
2	Create vector $x = [\ n_d\ \ \ I\ \ \ n\ ]^T$ and execute $y = \text{nn\_pr\_16\_RWE}(x)$; $y = round(y)$; [1]
3	Combine fault code table (Table 3) and thruster target table (Table 6) to determine thruster external fault state and corresponding actions A & B.
4	Determine internal fault state and actions A & B from thruster fault code table by examining if values of V and T exceed the limits.
5	(Optional) Use prioritisation scheme to resolve simultaneous appearance of external and internal faults: total faults have higher priority than partial faults.
6	Deliver the final thruster state and actions A & B as output.

[1] Vector y will have one of the following six values: [1; 0; 0; 0; 0], [0; 1; 0; 0; 0], [0; 0; 1; 0; 0], [0; 0; 0; 1; 0], [0; 0; 0; 0; 1], [0; 0; 0; 0; 0]. The last value is obtained in cases when a thruster operates in an unknown regime i.e., out-of-normal regime, different from faulty cases shown in Table 5. Typical examples for this case from the real-world environment include propeller jammed with rope or seaweed. The thruster state associated with value y = [0; 0; 0; 0; 0] is "Unknown" (see Thruster Fault Code Table in Table 3).

It should be mentioned that, in order to avoid false detection due to outliers and measurement noise, the outputs of FDU Algorithm are buffered i.e., the final decision about thruster fault is not derived from a single measurement, but is accomplished using present and past thruster states (FDU Algorithm outputs), which are stored in the buffer. This buffer operates similar to the shift register: when the new state is pushed into the buffer, the other states are pushed (shifted) down and the "oldest" state is pushed (shifted) out. Elements of the buffer are compared to each other, and if all buffer elements have the same value (state), then the aggregate thruster state is set to this value. Otherwise, the previous value is kept as the aggregate state.

3.3. Fault Accommodation (FA) Subsystem

The proposed FA subsystem is an extension of the hybrid approach for control allocation based on integration of the pseudoinverse and the fixed-point iteration method which compensates the thruster fault effect [15,16]. It is implemented as a two-step process. The pseudoinverse solution is found in the first step. Then the feasibility of the solution is examined, analysing its individual components. If violation of actuator constraint(s) is detected, the fixed-point iteration method is activated in the second step. In this way, the hybrid approach is able to allocate the exact solution, optimal in the l_2 sense, inside the entire attainable command set. This solution minimises a control energy cost function, the most suitable criteria for underwater applications.

As stated in Section 3.1, the HT Synthesis and VT Synthesis modules create virtual control vectors τ_{HT} and τ_{VT}. These vectors represent the total control effort (normalised forces and moments) to be produced by the actuators (thrusters). Control Allocators in Figure 2 find individual actuator settings (true control vectors u_{HT} and u_{VT} for horizontal and vertical thrusters, respectively) to be applied in order to produce desired control effort. The Fault Accommodation (FA) subsystem uses FDI outputs (aggregate thruster states) and associated software-level actions (action A) and hardware-level actions (action B) to solve the control allocation problem for each thruster layer in presence of partial/total thruster faults. Action A (see Table 3) includes penalisation of faulty thruster in the solution of control allocation problem by restricting saturation bounds i.e., by increasing the corresponding weight in the weighting matrix [16]. Action B includes power isolation of faulty thrusters (total faults only in Table 3) such that the entire ROV power system is not compromised. Hence, if the aggregated thruster state is a partial fault, the thruster is penalised, but will continue to operate. However, in the case of a total fault, the thruster is removed from the control allocation process, and its power is switched off in parallel.

3.4. Software Implementation

The proposed FTC has been implemented as a LabVIEW VI named "Thruster Active FDI", which is a software module integrated with other OceanRINGS⁺ modules. The user interface, which shows FDI results, is presented in Figure 12. This image is just an example; it has been artificially generated by injecting faults in the simulated thruster models, and it is intended to illustrate various options available in the software to detect, isolate and accommodate faults. The display is divided into 3 groups: horizontal thrusters (Layer 0 & Layer 1) and vertical thrusters. Each state has an associated colour code box. The external faults are represented with a column named "States", while internal faults are displayed as columns named "Voltages" and "Temperatures". The fault code table for the individual layers can be edited on the configuration tab of the main application. The fault code table for the horizontal thrusters (Layer 0) can be viewed in Figure 13.

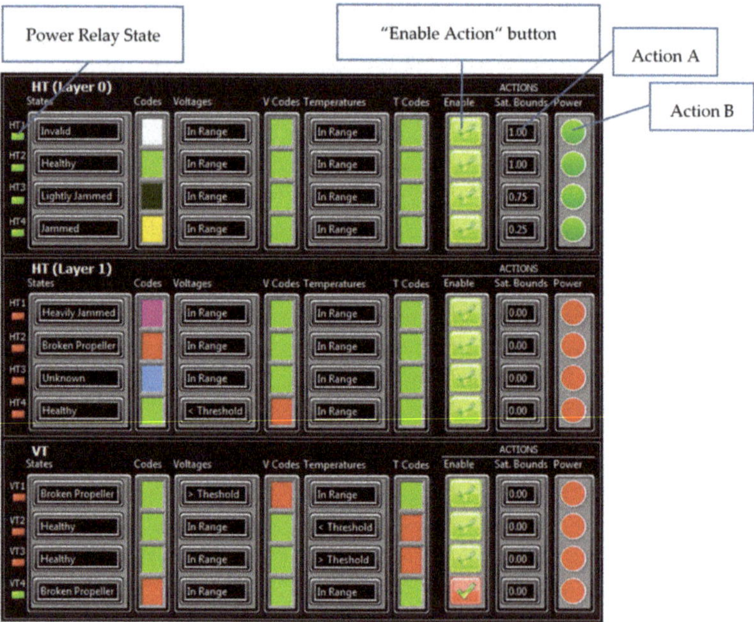

Figure 12. User Interface showing thruster FDI states and actions.

In order to increase the flexibility of the proposed FTC scheme, each thruster has associated individual "Enable Action" buttons. If the "Enable Action" is true, actions A & B will be applied. If the "Enable Action" is false, no actions will be applied i.e., the FDI subsystem will detect fault, but no action will be executed by the FA subsystem.

Thrusters HT1, HT2, HT3 & HT4 in Layer 0 in Figure 12 have "Enable Action" buttons set to true. The FDI subsystem has detected external faults (partial faults) in HT3 (Lightly Jammed) and HT4 (Jammed) and no presence of internal faults. All four thrusters remain powered on, while the software-level actions resulted in Saturation Bounds set to 0.75 and 0.25 for HT3 and HT4, respectively.

In a similar way, thrusters HT1, HT2, HT3 & HT4 in Layer 1 in Figure 12 have "Enable Action" buttons set to true. In this case, the FDI has detected external faults (total faults) in HT1 (Heavily Jammed), HT2 (Broken) and HT3 (Unknown), and presence of an internal fault in HT4 (Voltage < Threshold). Hardware-level actions resulted that the power to all four thrusters is switched off (as was confirmed by the power relay states for HT1–HT4, Layer 1), while software-level actions resulted in Saturation Bounds set to 0 for all four thrusters.

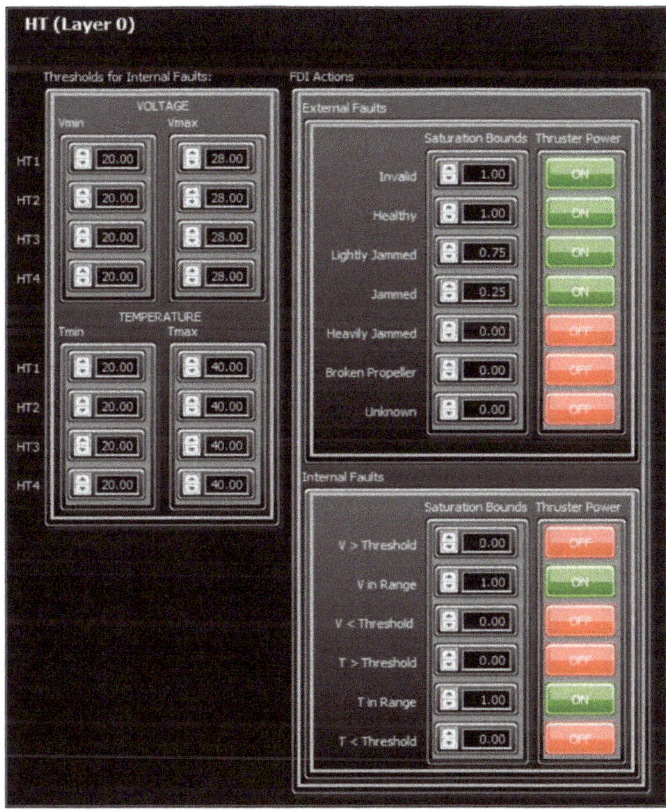

Figure 13. Fault Code Table for horizontal thrusters (Layer 0).

Finally, thrusters VT1, VT2 & VT3 have "Enable Action" buttons set to true. The FDI has detected external fault (total fault) in VT1 (Broken Propeller), and internal faults (total faults) in VT2 (Temperature < Threshold) and VT3 (Temperature > Threshold). Hardware-level actions resulted that the power to all three thrusters is switched off (as confirmed by Power Relay States), while software-level actions resulted in Saturation Bounds set to 0 for all three thrusters. Thruster VT4 has "Enable Action" button set to false. The FDI has detected an external fault (total fault) in VT4 (Broken Propeller) and no presence of internal faults. However, the FA subsystem did not execute action A (set Saturation Bounds to 0) and action B (switch off power to the thruster, as confirmed with power relay state for VT4).

An event log table is used to log Fault Detection and Isolation (FDI) events i.e., details of thruster external and internal faults as they occur. The FDI Event Log Table data includes event I.D., date and timestamp of fault occurrence, thruster in which fault occurred, type of external and internal fault and, if the "Enable Action" button is set to true, the software-level and hardware-level actions that were applied. An example of the FDI Event Log Table is shown in Figure 14.

FDI Event Log

ID	Date	Time	Thruster	State	Voltage	Temperature	Action (S)	Action (H)
0014	22/01/2018	12:06:45	VT4	Lightly Jammed	In Range	In Range	NONE	NONE
0013	22/01/2018	12:06:34	VT4	Lightly Jammed	In Range	In Range	NONE	NONE

Figure 14. Thruster Fault Detection and Isolation (FDI) Event Log table.

4. Testing and Evaluation of the FTC

The performance of the FTC, proposed and described in Section 3, is evaluated and tested in a real-world and virtual (simulated) environment. In the real-world environment, a single M5 VideoRay thruster has been utilised to evaluate the FTC performance for various fault conditions. In the virtual (simulated) environment, various thruster fault conditions were simulated in order to examine the behaviour of the FTC in different situations (open-loop/closed-loop control performance with enabled/disabled FA actions).

4.1. Evaluation of the FTC in Real-World Environment

A single M5 VideoRay thruster, configured as VT4, has been used for post-training validation of the trained PRNN and to evaluate the performance of the FTC in fault-free mode (no object attached to the propeller) and faulty modes (with "blockages" of various sizes and shapes attached to the blades, see Figure 4). The main objective was to evaluate the capability of the FDI subsystem to detect thruster states for external faults only, with a disabled FA subsystem (the "Enable Action" button set to false for VT4). The thruster was actuated with the same signal that has been used to acquire training data (Figure 15) i.e., with saw-like command signal n_d (%) with the following pattern: 0% » $MAX\%$ » 0% » $-MAX\%$ » 0%, step size 1%. The total duration of signal was $T = 20$ s, sampling period 50 ms and max. value $MAX = 100\%$. The critical (forbidden) zone has been set to $n_d \in [-20\%, 20\%]$. Screenshots of the FDI results for various thruster states are shown in Figures 16–20. Note that the results are for VT4 only. All other thruster states are shown as Invalid.

The FDI subsystem registers events only when the command signal n_d leaves the critical (forbidden) zone. Two events have been logged for each fault: the Event 1 with timestamp t_1 (time instance when the command signal n_d left the critical zone for the first time, see Figure 15), and the Event 2 with timestamp t_2 (time instance when the command signal n_d left the forbidden zone for the second time). While passing through the forbidden zone, the FDI subsystem went to "sleep" mode indicating a thruster state = "Invalid". The PRNN correctly recognised a thruster state outside the forbidden zone in all five cases. Test results confirmed that the PRNN has 100% classification accuracy outside the forbidden zone, and that the miss-classification appears exclusively inside the critical (forbidden) zone, as predicted in Section 3.1.

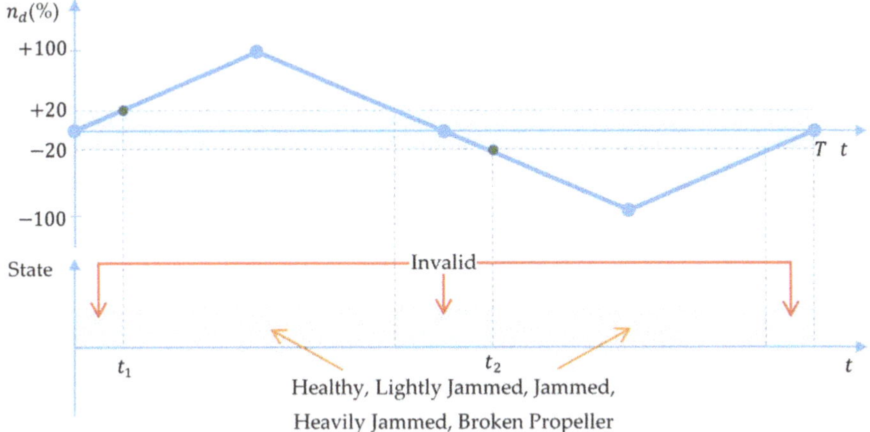

Figure 15. Time responses of command signal and thruster states.

Figure 16. Thruster FDI result for thruster state = "Healthy".

Figure 17. Thruster FDI result for thruster state = "Lightly Jammed".

Figure 18. Thruster FDI result for thruster state = "Jammed".

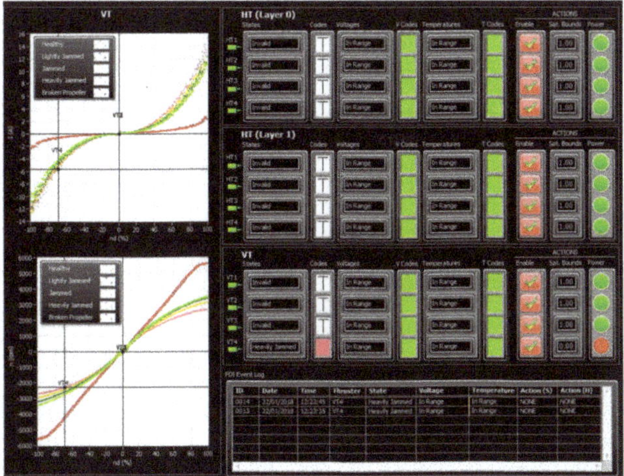

Figure 19. Thruster FDI result for thruster state = "Heavily Jammed".

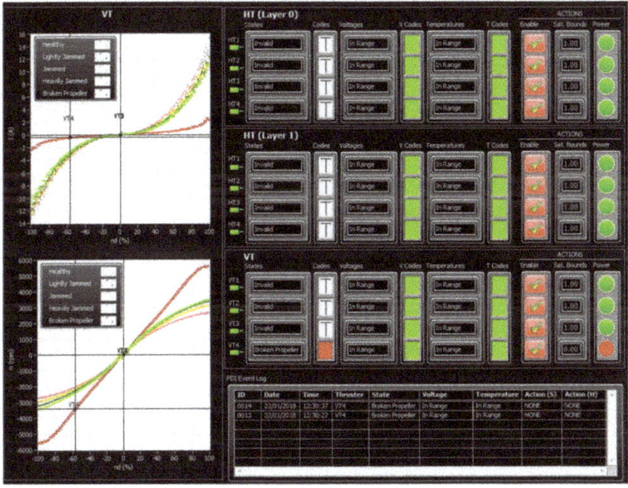

Figure 20. Thruster FDI result for thruster state = "Broken Propeller".

4.2. Evaluation of the FTC in Virtual (Simulated) Environment

As mentioned in Section 3.2, in the virtual (simulated) environment thruster faults are simulated by varying properties of the thruster dynamic model (load, friction, etc.) inside the propulsion subsystem of the ROV dynamics simulator. Off-line training and on-line fault detection phases are performed in the same way as described in Section 3.2 for the real-time environment but, in this case, faults are simulated by varying efficiency of thruster or by pushing the "Broken Propeller" button on the Thruster Configuration tab, shown in Figure 21.

Figure 21. Thruster Configuration tab for horizontal thrusters (Layer 0) in the ROV dynamics simulator.

A description of the simulation test cases, presented in this section, is given in Table 10.

Table 10. Description of simulation tests.

ID	Type	LLC Settings	Thruster	Fault	"Enable Action"
	Open-Loop				
1	$\tau_X(VJ) = 0.3$ $\tau_Y(VJ) = 0.0$ $\tau_N(VJ) = 0.0$	Heading: OFF Surge Speed: OFF Sway Speed: OFF	HT2 (L0)	Broken Propeller	False
2	*Closed-Loop* $SOG_d = 0.54$ m/s $COG_d = 60°$	$v_d = \text{proj}(SOG_d)_y$ $u_d = \text{proj}(SOG_d)_x$ $Y_d = COG_d$	HT2 (L0)	Broken Propeller	False
3	*Closed-Loop* $SOG_d = 0.54$ m/s $COG_d = 60°$	$Y_d = COG_d$ $u_d = \text{proj}(SOG_d)_x$ $v_d = \text{proj}(SOG_d)_y$	HT2 (L0)	Broken Propeller	True

The test cases described in Table 10 enable investigation of the influence of thruster faults on the overall control performance of the ROV and demonstrate the ability of the FTC to accommodate the faults and minimise the negative impact on control performance. It should be emphasized that similar test cases have been conducted with total and partial faults in vertical thrusters, and the main conclusion is that the ROV control system with active FDI and FA subsystems is capable of maintaining full 3 DOF in the vertical plane in the presence of a partial or total fault in a single vertical thruster.

4.2.1. Test 1: Impact of "Broken Propeller" in Open-Loop Mode with Disabled Fault Accommodation

In this case, the Surge, the Sway and the Yaw LLCs are disabled and the virtual control vector $\tau_{HT} = [\ 0.3 \quad 0.0 \quad 0.0\ ]^T$ is constant during the duration of the test. At time instance t_0 the "Broken Propeller" fault occurs in HT2, Layer 0. The FDI subsystem detects the fault and logs the FDI event, but the FA subsystem does not perform any action, since the "Enable Action" button is set to false (Figure 22). For this reason, the Control Allocator still assumes that the thruster HT2 is healthy and does not modify the solution of the control allocation problem. However, since HT2 does not produce any thrust/moment, the ROV drifts to the right side, with continuous change in heading (Figure 23).

4.2.2. Test 2: Impact of "Broken Propeller" in Closed-Loop Mode with Disabled Fault Accommodation

In this case, the Surge, the Sway and the Yaw LLCs are enabled, while all components of the Virtual Joystick vector are set to zero. The Speed Mode is set to SM2 (Follow Speed and Course) and the Heading Mode is set to HM2 (Follow Course). The Desired Speed Over Ground (SOG_d) is set to 0.54 m/s, and Desired Course Over Ground (COG_d) is set to 60°. The set points u_d for Surge Speed LLC and v_d for Sway Speed LLC are found as projections of the velocity vector in the horizontal plane on the x and y axis of the body-fixed {b} frame. At time instance t_0 the "Broken Propeller" fault occurs in HT2, Layer 0. The FDI subsystem detects the fault and logs the FDI event, but the FA subsystem does not perform any action, since the "Enable Action" button is set to false (Figure 24). Similar to Test 1, the Control Allocator still assumes that the thruster HT2 is healthy. However, robustness of closed-loop low-level controllers limit the ROV drift to the starboard side with a heading offset of 4.833° (Figure 25).

4.2.3. Test 3: Impact of "Broken Propeller" in Closed-Loop Mode with Enabled Fault Accommodation

Similar to Test 2, the Surge, the Sway and the Yaw LLCs are enabled, all components of Virtual Joystick vector are set to zero, the Speed Mode is set to SM2 (Follow Speed and Course) and the Heading Mode is set to HM2 (Follow Course). At time instance t_0 the "Broken Propeller" fault occurs in HT2, Layer 0. Since the "Enable Action" button is set to true (Figure 26), the FDI subsystem detects the fault and logs the FDI event, and the FA subsystem sets Saturation Bounds to zero for HT2 (action A) and switches off the thruster power (action B). The Control Allocator penalises (excludes) the faulty thruster in the solution of the control allocation problem and reallocates control energy among the operable thrusters (Figure 27). Small changes in the surge & sway speeds and heading are visible during the transient stage. The appearance of these transients is unavoidable, since operable thrusters need time to spin to new setpoints. However, after a short time period, all these changes diminish and the ROV continues in a straight-line motion, following the desired speed and course without error (Figure 27).

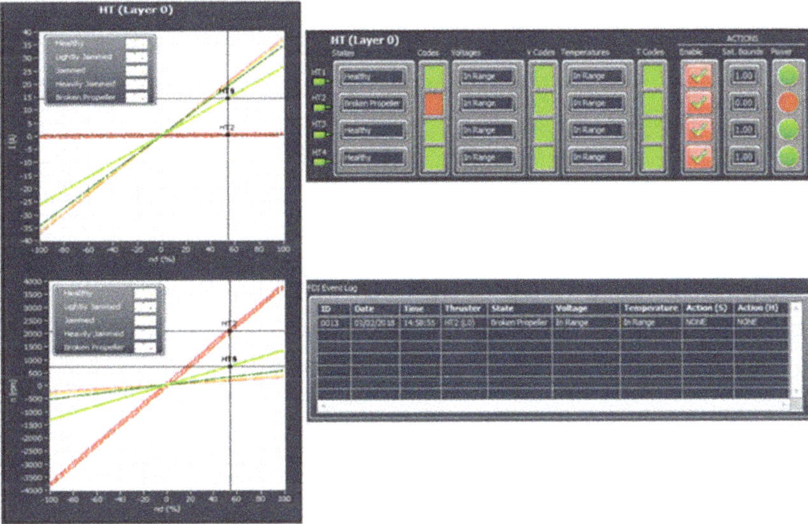

Figure 22. Test 1 (Open-Loop): Total fault in HT2 ("Broken Propeller") with disabled fault accommodation.

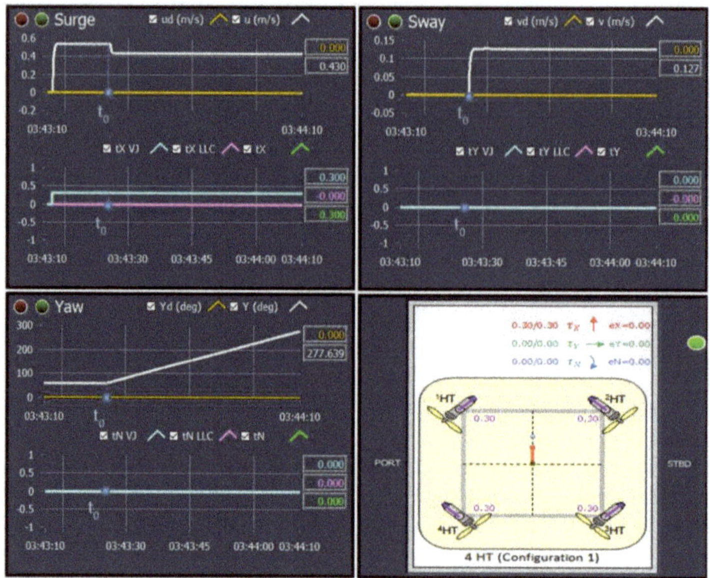

Figure 23. Test 1 (Open-Loop): Total fault in HT2 ("Broken Propeller") occurred at t_0; the lack of fault accommodation causes the ROV to drift with continuous change in heading.

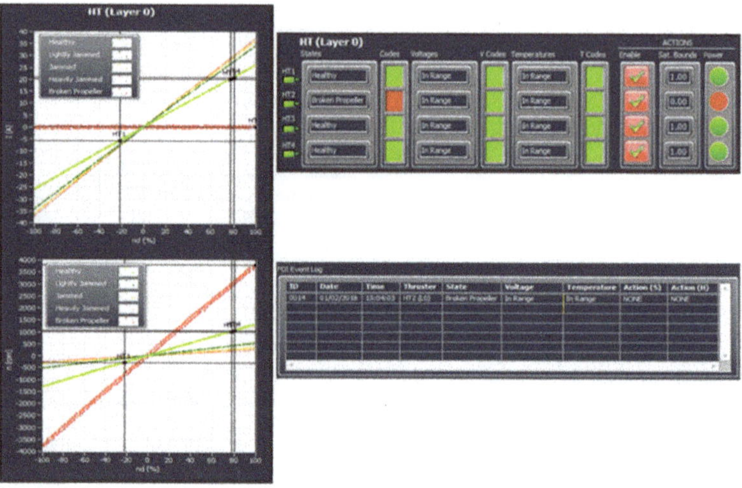

Figure 24. Test 2 (Closed-Loop): Total fault in HT2 ("Broken Propeller") with disabled fault accommodation.

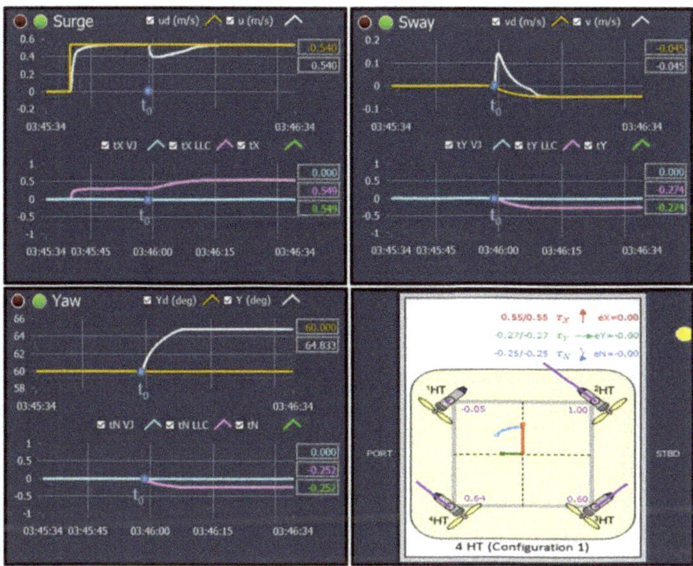

Figure 25. Test 2 (Closed-Loop): Total fault in HT2 ("Broken Propeller") occurred at t_0; the lack of fault accommodation and robustness of the closed-loop controllers causes limited drift to the starboard side with a heading offset of 4.833°.

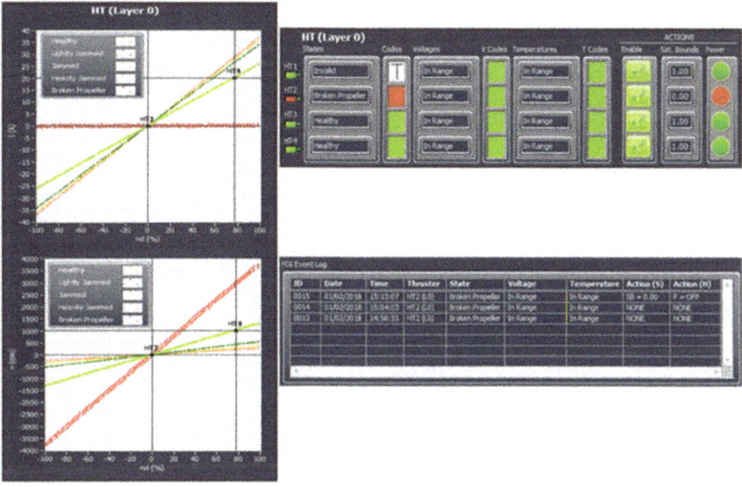

Figure 26. Test 3 (Closed-Loop): Total fault in HT2 ("Broken Propeller") with enabled fault accommodation.

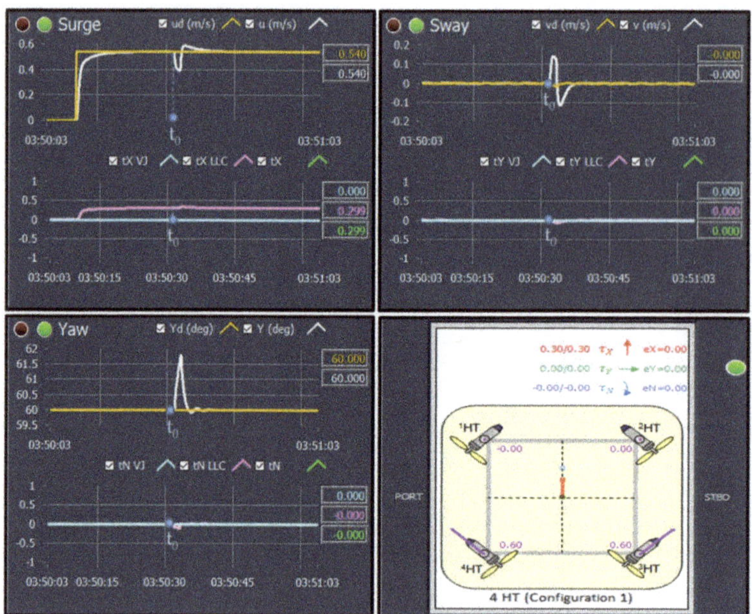

Figure 27. Test 3 (Closed-Loop): Total fault in HT2 ("Broken Propeller") occurred at t_0; excellent control performance is achieved due to fault accommodation and robustness of closed-loop controllers.

5. Conclusions and Future Work

This paper described a novel thruster fault-tolerant control system (FTC) for open-frame remotely operated vehicles (ROVs), developed by researchers at the Centre for Robotics & Intelligent Systems (CRIS), University of Limerick. The proposed FTC system is part of the OceanRINGS+ ROV smart control system, currently under development.

The proposed FTC consists of two subsystems: a model-free thruster fault detection and isolation subsystem (FDI) and a fault accommodation subsystem (FA). The FDI subsystem employs fault detection units (FDUs), associated with each thruster, to monitor their state. The robust, reliable and adaptive FDUs use a model-free pattern recognition neural network (PRNN) to detect internal and external faulty states of thrusters in real time. The FA subsystem combines information provided by the FDI subsystem with predefined, user-configurable actions to accommodate partial and total faults and to perform an appropriate control reallocation. Software-level actions include penalisation of faulty thrusters in solution of the control allocation problem and reallocation of control energy among the operable thrusters. Hardware-level actions include power isolation of faulty thrusters (total faults only) such that the entire ROV power system is not compromised.

Using simulations and real-world tests, the performance of the FTC was evaluated through a series of representative test cases, in order to examine the behaviour of the FTC in fault-free and faulty conditions. In the real-world environment, a single M5 VideoRay thruster has been used to evaluate the FTC performance for various fault conditions. Test results confirmed that the PRNN has 100% classification accuracy outside the forbidden zone, and that the miss-classification appears exclusively inside the critical (forbidden) zone. In the virtual (simulated) environment, various thruster fault conditions were simulated in order to examine the behaviour of the FTC in different situations (open-loop/closed-loop control performance with enabled/disabled FA actions). Simulation results show that the FTC provides automatic reallocation in faulty conditions, keeping all the DOF in the horizontal plane fully controllable and providing the opportunity to continue the mission with a minimal loss of control performance.

In the worst case scenario, the proposed FTC is able to maintain full 6 DOF control of the ROV in presence of up to 6 simultaneous total faults in thrusters (one layer with four horizontal thrusters fully disabled plus one disabled thruster in the other layer of horizontal thrusters plus one disabled vertical thruster).

Future work includes extension of the proposed FTC to interface with other ROVs and development of a "passive" method for fault detection and isolation for ROVs with thrusters that cannot provide active real-time monitoring of relevant signals from thrusters (currents, shaft speeds, applied voltages and temperatures of windings).

Acknowledgments: This paper is based upon works supported by the Science Foundation Ireland under Grant No. 12/RC/2302 for the Marine Renewable and Energy Ireland (MaREI) centre. SFI funding has been in collaboration with industry partners IDS Monitoring, Shannon Foynes Port Company, Commissioners of Irish Lights and Teledyne BlueView.

Author Contributions: R.C. and E.O. wrote the paper. R.C. designed the hardware and software for the FDI and constructed the test rig for the real-world tests, under E.O. supervision. G.D. provided guidance in the selection of hardware for the system. G.D. and D.T. proof read and provided guidance on the correct structure of the paper.

Conflicts of Interest: The authors declare no conflict of interest. The founding sponsors had no role in the design of the study; in the collection, analyses, or interpretation of data; in the writing of the manuscript, and in the decision to publish the results.

References

1. Capocci, R.; Dooly, G.; Omerdić, E.; Coleman, J.; Newe, T.; Toal, D. Inspection-class remotely operated vehicles—A review. *J. Mar. Sci. Eng.* **2017**, *5*, 13. [CrossRef]
2. O'Connor, M.; Lewis, T.; Dalton, G. Weather window analysis of Irish west coast wave data with relevance to operations & maintenance of marine renewables. *Renew. Energy* **2013**, *52*, 57–66.
3. Podder, T.K.; Antonelli, G.; Sarkar, N. Fault tolerant control of an autonomous underwater vehicle under thruster redundancy: Simulations and experiments. In Proceedings of the 2000 ICRA, IEEE International Conference on Robotics and Automation, Millennium Conference, Symposia Proceedings (Cat. No. 00CH37065), San Francisco, CA, USA, 24–28 April 2000; Volume 2, pp. 1251–1256.
4. Caccia, M.; Bono, R.; Bruzzone, G.; Bruzzone, G.; Spirandelli, E.; Veruggio, G. Experiences on Actuator Fault Detection, Diagnosis and Accommodation for ROVs. In *International Symposium Unmanned Untethered Submersible Technology*; University of New Hampshire, Marine Systems Engineering Laboratory: Durham, NH, USA, 2001.
5. Kim, J.H.; Beale, G.O. Fault Detection and Classification in Underwater Vehicles Using the T2 Statistic. *Autom. J. Control Meas. Electron. Comput. Commun.* **2002**, *43*, 29–37.
6. Montazeri, M.; Kamali, R.; Askari, J. Fault diagnosis of autonomous underwater vehicle using neural network. In Proceedings of the 2014 22nd Iranian Conference on Electrical Engineering (ICEE), Tehran, Iran, 20–22 May 2014; pp. 1273–1277.
7. Zhu, D.; Liu, Q.; Yang, Y. An Active Fault-Tolerant Control Method of Unmanned Underwater Vehicles with Continuous and Uncertain Faults. *Int. J. Adv. Robot. Syst.* **2008**, *5*, 1729–8806. [CrossRef]
8. Huang, H.; Wan, L.; Chang, W.-T.; Pang, Y.-J.; Jiang, S.-Q. A Fault-tolerable Control Scheme for an Open-frame Underwater Vehicle Regular Paper. *Int. J. Adv. Robot. Syst.* **2014**, *11*, 77.
9. Omerdić, E.; Roberts, G. Thruster fault diagnosis and accommodation for open-frame underwater vehicles. *Control Eng. Pract.* **2004**, *12*, 1575–1598. [CrossRef]
10. Liu, Q.; Zhu, D. Fault-tolerant Control of Unmanned Underwater Vehicles with Continuous Faults: Simulations and Experiments. *Int. J. Adv. Robot. Syst.* **2009**, *6*, 1729–8806. [CrossRef]
11. Yusoff, M.A.M.; Arshad, M.R. Active Fault Tolerant Control of a Remotely Operated Vehicle Propulsion System. *Procedia Eng.* **2012**, *41*, 622–628.
12. Baldini, A.; Ciabattoni, L.; Felicetti, R.; Ferracuti, F.; Monteriù, A.; Fasano, A.; Freddi, A. Active fault tolerant control of remotely operated vehicles via control effort redistribution. In Proceedings of the ASME Design Engineering Technical Conference, Cleveland, OH, USA, 6–9 August 2017; Volume 9, pp. 1–7.

13. Antonelli, G. A Survey of Fault Detection/Tolerance Strategies for AUVs and ROVs. In *Fault Diagnosis and Fault Tolerance for Mechatronic Systems: Recent Advances*; Caccavale, F., Villani, L., Eds.; Springer: Berlin/Heidelberg, Germany, 2003; pp. 109–127.

14. Omerdic, E.; Toal, D.; Dooly, G. OceanRINGS: Smart Technologies for Subsea Operations. In *Advanced in Marine Robotics*; Lambert Academic Publishing: Salbruggen, Germany, 2013.

15. Omerdic, E.; Toal, D.; Nolan, S.; Ahmad, H. ROV LATIS: Next Generation Smart Underwater Vehicle. In *Further Advances in Unmanned Marine Vehicles*; Institution of Engineering & Technology: Stevenage, UK, 2012; Chapter 2.

16. Omerdic, E.; Roberts, G. Extension of Feasible Region of Control Allocation for Open-frame Underwater Vehicles. In Proceedings of the IFAC Conference on Control Applications in Marine Systems (CAMS 2004), Ancona, Italy, 7–9 July 2014.

17. Omerdic, E.; Toal, D.; Nolan, S.; Ahmad, H. Smart ROV LATIS: Control Architecture. In Proceedings of the UKACC International Conference on CONTROL 2010, Coventry, UK, 7–10 September 2010.

18. U.S. Department of Energy. *The Importance of Motor Shaft Alignment*; U.S. Department of Energy: Washington, DC, USA, 2012.

MDPI

St. Alban-Anlage 66

4052 Basel

Switzerland

Tel. +41 61 683 77 34

Fax +41 61 302 89 18

www.mdpi.com

Journal of Marine Science and Engineering Editorial Office

E-mail: jmse@mdpi.com

www.mdpi.com/journal/jmse